AF892522

Geographies of Tourism and Global Change

Series Editors
Dieter K. Müller, Department of Geography, Umeå University, Umeå, Sweden
Jarkko Saarinen, Geography Research Unit, University of Oulu, Oulu, Finland
Carolin Funck, Hiroshima University Graduate School of Humanities and Social Sciences, Higashi-Hiroshima City, Hiroshima, Japan

Now accepted by Scopus! Content available on the Scopus site in the second half of 2023.

In a geographical tradition and using an integrated approach, this book series acknowledges the interrelationship of tourism to wider processes within society and environment. This is done at local, regional, national, and global scales demonstrating links between these scales as well as outcomes of global change for individuals, communities, and societies. Local and regional factors will also be considered as mediators of global change in tourism geographies affecting communities and environments. Thus *Geographies of Tourism and Global Change* applies a truly global perspective highlighting development in different parts of the world and acknowledges tourism as a formative cause for societal and environmental change in an increasingly interconnected world.

The scope of the series is broad and preference will be given to crisp and highly impactful work. Authors and Editors of monographs and edited volumes from across the globe are welcome to submit proposals. The series insists on a thorough and scholarly perspective, in addition authors are encouraged to consider practical relevance and matters of subject specific importance. All titles are thoroughly reviewed prior to acceptance and publication, ensuring a respectable and high quality collection of publications.

Lesego S. Stone • Moren T. Stone
Editors

Wildlife Tourism Dynamics in Southern Africa

Contemporary Issues, Challenges and Prospects for Sustainable Development

Springer

Editors
Lesego S. Stone
Faculty of Business, Department Tourism and Hospitality Management
University of Botswana
Gaborone, Botswana

Moren T. Stone
Department of Environmental Science
University of Botswana
Gaborone, Botswana

ISSN 2366-5610 ISSN 2366-5629 (electronic)
Geographies of Tourism and Global Change
ISBN 978-3-031-57251-7 ISBN 978-3-031-57252-4 (eBook)
https://doi.org/10.1007/978-3-031-57252-4

© The Editor(s) (if applicable) and The Author(s), under exclusive license to Springer Nature Switzerland AG 2024
This work is subject to copyright. All rights are solely and exclusively licensed by the Publisher, whether the whole or part of the material is concerned, specifically the rights of translation, reprinting, reuse of illustrations, recitation, broadcasting, reproduction on microfilms or in any other physical way, and transmission or information storage and retrieval, electronic adaptation, computer software, or by similar or dissimilar methodology now known or hereafter developed.
The use of general descriptive names, registered names, trademarks, service marks, etc. in this publication does not imply, even in the absence of a specific statement, that such names are exempt from the relevant protective laws and regulations and therefore free for general use.
The publisher, the authors, and the editors are safe to assume that the advice and information in this book are believed to be true and accurate at the date of publication. Neither the publisher nor the authors or the editors give a warranty, expressed or implied, with respect to the material contained herein or for any errors or omissions that may have been made. The publisher remains neutral with regard to jurisdictional claims in published maps and institutional affiliations.

This Springer imprint is published by the registered company Springer Nature Switzerland AG
The registered company address is: Gewerbestrasse 11, 6330 Cham, Switzerland

Paper in this product is recyclable.

A Synopsis About the Book

This book discusses wildlife tourism in the southern African region. For many countries in the region, wildlife tourism is promoted for community development and conservation purposes. The book provides various case studies from across the region to showcase the contribution and challenges the region faces in its pursuit of sustainability using wildlife resources. The book addresses contemporary issues related to wildlife tourism including the benefits, challenges, governance issues, forms of wildlife tourism, and drivers and deterrents of wildlife tourism in the region. A broad range of multidisciplinary scholars contributed to the volume.

Acknowledgments

We would like to thank several people who were instrumental in ensuring that this book is successfully completed. First, the contributors, who dedicated their time to writing the chapters and ensuring that they are delivered on time. Second, our children who were patient with us and gave us the time to work on the project from the beginning to the end. Lastly, we thank the publisher for giving us this opportunity and giving us enough time to accomplish the task.

Contents

Editors and Contributors

About the Editors

Lesego S. Stone is an Associate Professor of Tourism Management in the Department of Tourism and Hospitality Management at the University of Botswana. Her research interests are in sustainable tourism development with specific reference to nature-based tourism, community-based tourism, and community participation in tourism.

Moren T. Stone is an Associate Professor of Environmental Science and Tourism Studies in the Department of Environmental Science at the University of Botswana. His research interests include sustainable tourism development and management, ecotourism, community-based tourism, protected areas conservation, and community livelihood dynamics.

Contributors

Margaret Ndapewa Angula is a Senior Lecturer in the Department of Environmental Sciences at the University of Namibia. She has over 20 years' research experience in environmental management and sustainable livelihoods. She specializes in gender and climate change, climate resilient development, and capacity building for climate action. She has been a project coordinator for international and multi-institutional projects on climate change adaptation since 2014 on behalf of Namibia.

Felix Bello is a Senior Lecturer in the Division of Tourism Management, Department of Marketing Management at the University of Pretoria, South Africa. His research is on sustainable tourism development and planning, mainly focusing on protected area-based tourism, community participation, community-based tourism,

Sustainable Development Goals (SDGs), and corporate social responsibility (CSR). He has published in leading international tourism journals and is a National Research Foundation (South Africa) rated researcher.

Michael Chambwe holds a PhD in Economic and Management Sciences with Tourism Management from North-West University in South Africa, where he is currently a Post-Doctoral Researcher with Tourism Research in Economics, Environs and Society (TREES). His research interests lie in nature areas and their sustainability, the equitable development of communities surrounding protected areas, rural tourism livelihoods and entrepreneurship, and community-based tourism.

Gaofenngwe Dibotlhale is an MPhil/PhD student in the Department of Environmental Science at the University of Botswana. Her research interests include cultural and heritage tourism, community-based natural resources management, and sustainable tourism development.

Edson Gandiwa is a Director of Scientific Services at the Zimbabwe Parks and Wildlife Management Authority. He has a PhD in Wildlife Conservation and Management from Wageningen University, the Netherlands. He has widely published in the areas of wildlife conservation and management.

Hesekia Garekae is a Lecturer in the Department of Environmental Science, University of Botswana. He holds a PhD in Environmental Science from Rhodes University, South Africa. His research interests revolve around forests, livelihoods, and conservation.

Owen Gohori is a Post-Doctoral Research Fellow at North-West University's Tourism Research in Economics, Environs and Society (TREES) research entity, Potchefstroom campus, South Africa. He worked in the tourism industry for 7 years and for five of those years he was an entrepreneur running a destination management company in Zimbabwe. He later advanced his studies and graduated with a PhD in Tourism Management in 2020. His research interests include pro-poor tourism, community-based tourism, sustainable tourism, natural area tourism, and the incorporation of indigenous knowledge systems in tourism development. Owen has extensive knowledge and experience working with inbound tourists and promotes the development of community-based tourism in rural communities as a strategy to benefit disadvantaged local people.

Bongani G. Gumbo holds a PhD in Historical Studies from the University of Cape Town (2010). He is a Senior Lecturer of Economic History in the Department of History at the University of Botswana. Dr. Gumbo's research interests focus largely on the dynamics in livelihood strategies of riparian communities in the tourism regions of Chobe and Ngamiland in northern Botswana. He has published in refereed journals and book chapters on the political economy of development in these wetland regions. He has also externalized graduate and postgraduate theses and dissertations for various universities in the Southern African region.

Raili Hasheela currently works in the GEF-funded project on human-wildlife conflict and wildlife crime in the Ministry of Environment, Forestry and Tourism of Namibia. She has a valuable background on environmental management and development. Her areas of expertise are research and project management. She has conducted extensive research in the areas of knowledge management, protected area management, community-based natural resource management (CBNRM), biodiversity conservation, climate change, waste management, water resource management, and coastal management. She has also coordinated several environmental management–related projects and programs. At the moment she is managing a donor-funded project that focuses on addressing the twin-challenge of human-wildlife conflict and wildlife crime.

Jona Heita is a Lecturer of Tourism Studies in the Department of Wildlife Management and Tourism Studies at the University of Namibia. He obtained his MA in Culture and Environment in Africa (CEA) from the University of Cologne, Germany. Apart from teaching tourism studies courses, he has been involved in tourism-related research activities in communal conservancies in Namibia. He is currently a PhD in Geography candidate. His PhD research focuses on factors influencing human-predator conflicts in Sheya Uushona communal conservancy in Namibia.

Lyarukwa G. Jerome is a Lecturer at the College of African Wildlife Management, Mweka, Tanzania, and a PhD fellow in the Department of Recreations Parks and Tourism Management of Pennsylvania State University. Gasto has an undergraduate degree in Wildlife Management from Sokoine University of Agriculture in Tanzania and an MSc in Biological Sciences with a specialization in Zoology from the University of Western Australia. His research focuses on understanding human-wildlife conflicts, coexistence, and how nature-based tourism benefits communities adjacent to national parks and promotes sustainable biodiversity conservation. He is also interested in human-wildlife interactions and the relationships between the interactions of community livelihoods and sustainable wildlife utilization.

Gasto has 14 years of experience working in the wildlife sector in Tanzania, where he worked as a park warden in charge of law enforcement and ecotourism in Serengeti National Park. He teaches wildlife management and biology courses at the College of African Wildlife Management, Mweka. He also has taught park rangers in Tanzania National Parks on using Spatial Monitoring and Reporting Tool (SMART) to improve wildlife monitoring and patrol performance. He has also been involved in wildlife reintroduction programs at Sharaan Nature Resources in Saudi Arabia, training rangers and monitoring re-introduced wildlife.

Tariro Kamuti is a Research Fellow at the Centre for Gender and Africa Studies of the University of the Free State, South Africa. He holds a joint PhD in Geography from Vrije University Amsterdam, The Netherlands, and the University of the Free State, South Africa. His research interests include land and environmental policy, governance issues surrounding sustainable utilization of natural resources, the social dimensions of biodiversity conservation, climate change, and sustainable development. He is a member of the Society of South African Geographers, the Regional Studies Association, United Kingdom, and the International Society of Sustainability Professionals, USA. He is a scholar of the Climate Social Science Network (CSSN), an international network of scholars headquartered at the Institute at Brown for Environment and Society, USA. Tariro is also a Research Fellow of Earth System Governance (ESG), a global research alliance, which is a social science research network in the area of governance and global environmental change, based at Utrecht University, The Netherlands.

Mathews Lambulira is Lecturer in the Department of Tourism at Mzuzu University, Malawi. He is recognized for his expertise in tourism planning and development. His research focus is on tourism planning and development, ecotourism, nature-based tourism, and tourism education. His scholarly works illuminate the intricate interplay between tourism development, socio-economic growth, and environmental conservation.

Refiloe Lekgau holds a PhD in Tourism and Hospitality from the University of Johannesburg. She is currently working as a Lecturer in the School of Tourism and Hospitality at the University of Johannesburg. Her teaching and research interest lies in events management and tourism development. Relatedly, she conducts research on the following themes: MICE tourism development, wildlife tourism, and tourism and resilience.

Selma Lendelvo is an Associate Research Professor in Life Sciences and currently the Director for the Centre for Grants Management and Resource Mobilization at the University of Namibia (UNAM) with research and project management experience spanning over 20 years. Prof. Lendelvo has been extensively involved in local and international research since 2001 and published extensively in cross-cutting fields of community-based tourism, conservation and environmental management, rural development, gender, and climate change. Prof. Lendelvo serves on the Namibia National Committee for the Rio Conventions, the Namibian Nature Conservation Board and the Namibia Association of CBNRM Support Organisations (NACSO).

Mfundo Mlilo holds the position of Research Associate at the Institute of Geography, University of Cologne. He is currently pursuing his PhD in Economic

Geography under the auspices of the DFG-funded CRC 228, Future Rural Africa project in which he investigates Agency, Entrepreneurship and Exclusion in the evolution of tourism value chains in the Southern African region, with a specific focus on Botswana, Namibia, and Zimbabwe. In addition, he has also been associated with the Global South Studies Center at the same institution.

Chiedza N. Mutanga is a Senior Research Fellow in Sustainable Tourism at the Okavango Research Institute (ORI), University of Botswana. She holds a PhD in Tourism and Wildlife Conservation from Chinhoyi University of Technology, Zimbabwe. Her research interests are in sustainable tourism development and wildlife resource conservation, with a focus on nature-based tourism, community-based tourism and community livelihoods, ecotourism, protected area tourism, community-based natural resource management, and biodiversity conservation.

Emmanuel Mwakiwa is a Senior Lecturer and Researcher in the Department of Agricultural Business Development and Economics in the Faculty of Agriculture, Environment and Food Systems at the University of Zimbabwe. He is a Research Director for Zimbabwe in the Africa Network of Agricultural Policy Research Institutes. Emmanuel has a PhD in Environment and Resource Economics from Wageningen University, Netherlands. Emmanuel's PhD research work is titled "Cooperation or competition: Dilemma for resource managers in sustainable wildlife utilization." In this work he analyzed the wildlife management interactions between public and private landowners for the Kruger National Park and surrounding landowners in South Africa. He was involved in the review of the Communal Areas Management Programme for Indigenous Resources (CAMPFIRE). Emmanuel has over 25 years' comprehensive experience in development work in the following fields: wildlife management, natural resource management, communal resource management, climate change, agricultural value chain analyses, rural livelihoods, community resilience, food security, and rural development. He has researched and written extensively in these areas in Africa and has published over 30 publications in international journals. He has undertaken assignments in several African countries including Zimbabwe, South Africa, Mozambique, Zambia, and the southern, central, and western Africa regions.

Edwin Sabuhoro is an Assistant Professor of Recreation, Park, and Tourism Management and African Studies at Penn State University. He earned an undergraduate degree in Law from the National University of Rwanda in 2002 and a Master of Science in Conservation and Tourism from the University of Kent at Canterbury in 2006. In 2018, Sabuhoro graduated from Clemson University with a PhD in parks, recreation, and tourism management.

His work focuses on integrated biodiversity conservation and community development, human-wildlife coexistence, and tourism's impact on local livelihoods. As part of his community outreach program, he engages wildlife poachers in conservation and community development initiatives in East Africa. In addition to sustainable management and use of parks and protected areas, he also has an interest in community development, nature-based tourism and outdoor recreation, transboundary conservation, livelihoods, and well-being.

His experience spans 15 years in tourism planning and development, protected area management, and community development, both in the public and private sectors. For his innovative efforts in integrating conservation and community development for Indigenous communities adjacent to Mountain Gorilla habitat, he is the recipient of numerous awards. In addition to receiving the 2008 Young Conservationist of the Year Award as well as the 2015 UN-GRASP-Ian Redmond Conservation Award, he also received the 2016 Achievement Award. In 2010, he was selected and invited to participate in the Young African Leaders Forum by President Obama. CNN African Voices featured his work with mountain gorillas and indigenous people in 2015, and he was recognized as a CNN Hero.

Ngoni C. Shereni is a Postdoctoral Research Fellow in the School of Tourism and Hospitality at the University of Johannesburg in South Africa. He holds a PhD in Tourism and Hospitality Management from the University of Johannesburg. His research interests are in sustainable tourism, sustainable development goals (SDGs), climate change, tourism education, disruptive technology, tourism exhibitions, nature-based tourism, community-based tourism (CBT), and community-based natural resource management (CBNRM) practices.

Tembi M. Tichaawa is the Academic Head of the Tourism Department and a Full Professor in Tourism at the School of Tourism and Hospitality, University of Johannesburg. He holds a PhD in Geography and Environmental Management from the University of KwaZulu-Natal. He conducts research on a range of developmental issues linked to tourism with specific focus on the developing country context. In this area, he has supervised master's and doctoral students.

Kgosietsile Velempini is a Senior Lecturer in Environmental Education unit at the University of Botswana. He has a PhD in Curriculum and Instruction (Environmental Education) from the Patton College of Education at Ohio University, USA. His research focuses on environmental and sustainability education, sustainable development goals, tourism, community-based natural resource management, solid waste management, climate change, and local knowledge systems. He has published in various international journals. He has master's degrees in International Affairs and Environmental Sciences. He has a graduate certificate in Environmental Sustainability, a BA degree (Environmental Sciences), and a postgraduate diploma in Education.

Zibanai Zhou is a Lecturer in the Department of Tourism, Hospitality and Leisure Sciences at Midlands State University, Zimbabwe. His main research interests span tourism development, event tourism, and community-based tourism.

Abbreviations

ADMADE	Administrative Management Design for Game Management Areas
AWF	African Wildlife Foundation
AWPs	Artificial Water Points
BHAPU	Bumi Hills Anti-Poaching Unit
BTO	Botswana Tourism Organisation
CAMPFIRE	Communal Areas Management Programme for Indigenous Resources
CBD	United Nations Convention on Biological Diversity
CBNRM	Community-Based Natural Resource Management practices
CBO	Community-Based Organizations
CBT	Community-Based Tourism (CBT)
CECT	Chobe Enclave Community Trust
CEGs	Community Escort Guides
CHAs	Controlled Hunting Areas
CITES	International Trade in Endangered Species of Fauna and Flora
DNPW	Department of National Parks and Wildlife
DNPWM	Department of National Parks and Wildlife Management
DRC	Democratic Republic of Congo
DWNP	Department of Wildlife and National Parks
ENGOs	Environmental Non-governmental Organizations
FDI	Foreign Direct Investment
FMD	Foot and Mouth Disease
GDP	Gross Domestic Product
GNP	Gross National Product
GIZ	German Agency for International Cooperation
GMAs	Game Management Areas
GoZ	Government of Zimbabwe
GVTC	Greater Virunga Transboundary Collaboration
HWCs	Human Wildlife Conflicts
ITA	The International Trade Administration

IUCN	International Union for Conservation of Nature
JVPs	Joint Venture Partnerships
KAZA	Kavango–Zambezi Transfrontier Conservation Area
KTP	Kgalagadi Transfrontier Park
LNP	Liwonde National Park
MCPA	Makuleke Communal Property Association
MET	Ministry of Environment, Forestry and Tourism
MFAIT	Ministry of Foreign Affairs and International Trade
MZCDT	Mababe Zokotsama Community Development Trust
NACSO	Namibian Association of CBNRM Support Organisations
NGO	Non-governmental Organization
OECD	Organisation for Economic Co-operation and Development
PAs	Protected Areas
PPP	Public-Private Partnership
RDCs	Rural District Councils
RETOSA	Regional Tourism Organisation of Southern Africa
SADC	Southern African Development Community
SAN Parks	South African National Parks
SDGs	Sustainable Development Goals
STMT	Sankuyo Tshwaragano Management Trust
STCRC	Sustainable Tourism Cooperative Research Centre
TBCA	Transboundary Conservation Areas
TBNRM	Transboundary Natural Resources Management
TBPA	Transboundary Protected Area
TFCA	Transfrontier Conservation Area
TFCDA	Transfrontier Conservation and Development Area
TFCMA	Transfrontier Conservation Marine Area
TFCRA	Transfrontier Conservation Resources Area
TFP	Transfrontier Park
TMP	Transfrontier Marine Park
UNCCD	United National Convention to Combat Desertification
UNDP	United Nations Development Programme
UNEP	United Nations Environment Programme
UNESCO	United Nations Educational, Scientific and Cultural Organization
UNFCCC	United Nations Framework Convention on Climate Change
UNWTO	United Nations World Tourism Organization
USAID	United States Agency for International Development
VCA	Virunga Conservation Area
VCFs	Veterinary Cordon Fences
WINDFALL	Wildlife Industries New Development for All
WSSD	World Summit on Sustainable Development
WTTC	World Travel and Tourism Council
WWF	World Wide Fund for Nature
YODEP	Youth for Development and Productivity
ZTA	Zimbabwe Tourism Authority

List of Figures

List of Tables

Part I
Introduction: An Overview of Wildlife Tourism in Southern Africa

Chapter 1
Wildlife Tourism in Southern Africa: Prospects for Sustainability

Lesego S. Stone and Moren T. Stone

In many regions of the world, wildlife tourism is a popular form of ecotourism (Mangachena & Pickering, 2021; Newsome & Rodger, 2013). According to Novelli et al. (2006), wildlife-based tourism flourished in the mid-1960s due to the affordability of travel, more interest in nature and wildlife conservation, and Western tourists getting access to remote, unspoiled areas. This encouraged the development of a diverse African wildlife tourism sector. Wildlife tourism is a form of nature- based tourism based on encounters with wild animals; it can be consumptive or non-consumptive (Lovelock, 2008). It is a global activity of significance which forms a fundamental part world's massive nature-based tourism industry (Cong et al., 2017). Wildlife tourism is one of the prominent foreign exchange earners in numerous countries, especially those from the African continent (Novelli et al., 2006; Reynolds & Braithwaite, 2001). The UNWTO (2015) indicates the significance of wildlife tourism, with 80% of international travel to the sub-Saharan Africa being wildlife related.

For many Southern African countries, wildlife tourism is expanding rapidly. The region has one of the greatest endowments of wildlife in the world, with the 'Big Five' being abundant in the region (Chimuka, 2022). In countries such as South Africa, the 'Big Five' are the basis for wildlife tourism (Lubbe, 2022). Most Southern African countries use wildlife tourism as a conservation tool and as part of

L. S. Stone (✉)
Faculty of Business, Department Tourism and Hospitality Management,
University of Botswana, Gaborone, Botswana
e-mail: stonel@ub.ac.bw

M. T. Stone
Department of Environmental Science, University of Botswana, Gaborone, Botswana
e-mail: stonemt@ub.ac.bw

© The Author(s), under exclusive license to Springer Nature Switzerland AG 2024

L. S. Stone, M. T. Stone (eds.), *Wildlife Tourism Dynamics in Southern Africa*,
Geographies of Tourism and Global Change,
https://doi.org/10.1007/978-3-031-57252-4_1

integrated conservation and development programs (ICDP) (Stone, 2017). The Southern African region has a vast number of protected areas, including game reserves, transfrontier parks and national parks (Lekgau & Tichaawa, 2019; Tichaawa & Lekgau, 2020) The region is well known for its wildlife and nature-based tourism. According to Saayman and Saayman (2017), Southern African countries are dependent on the wildlife and their natural beauty for tourism. Wildlife tourism takes place mainly in the protected areas found in the region. Protected areas are acclaimed for their conservation efforts and achievements (Zafra-Calvo & Moreno-Penaranda, 2018). They play a significant role in community development and the improvement of local livelihoods. As a result, community-based projects are often encouraged (Mbaiwa, 2005). This is to ensure that despite losing the use of and access to resources found in protected areas, communities can derive benefits from tourism and in the process improve their livelihoods while at the same time conserving resources. At a global level, protected areas cover 15.4% of the world's land (Baghai et al., 2018). They are considered as being critical to the sustainable growth of countries within the Southern African region, due to their competitive advantage in wildlife and their capacity to generate revenue for tourism stakeholders (Lekgau & Tichaawa, 2019).

1.1 Wildlife Tourism and Sustainable Development

In many parts of the world where wildlife tourism is promoted, the sustainable and inclusive use of wildlife tourism resources is advocated for, to contribute to the attainment of the 2030 sustainable development goals (SDGs) (Zhou, 2023). The 2030 sustainable development goals (SDGs) call for the increased participation and involvement of communities. Wildlife tourism, that is anchored on the principles of sustainable development, community inclusion and conservation has been promoted as a significant tool for poverty alleviation, job creation, local economic development for communities residing adjacent protected areas and sustainable livelihoods (Mbaiwa, 2017; Snyman, 2017; Kimbu & Tichaawa, 2018; Lekgau & Tichaawa, 2019; Zanamwe et al., 2018). This is aimed at empowering local communities and eradicating inequality thereby promoting sustainable tourism development (Qin et al., 2020; Zhou, 2023). As a vital component of sustainable tourism development, local communities need to fully support wildlife tourism taking place amongst them. Wildlife tourism can provide significant financial benefits to local communities if it is done responsibly while also contributing to conservation efforts (Ballantyne & Packer, 2013; O'Malley et al., 2013; Trave et al., 2017). Many local and national economies can generate revenue and wealth directly and indirectly from wildlife resources (Naidoo et al., 2016).

For many countries in the region, sustainable tourism development is practised through community-based natural resource management (CBNRM). CBNRM has been promoted in Southern Africa since the late 1980s as a strategy to reverse environmental injustice linked to fortress conservation characterised by the displacement of indigenous populations and creating hostile relations between governments and local communities in conservation areas (Brockington, 2002; Dowie, 2009; Mbaiwa, 2018). In Botswana, community-based tourism takes place within the CBNRM programme; with communities engaging in both consumptive and non-consumptive tourism. Wildlife concessions are leased to local communities who are part of the CBNRM programme or private tourist safari operators by the Land Board (Mbaiwa & Mogende, 2022). Communities are able to derive benefits from the wildlife through renting their concessions and selling their hunting quotas to private tourism operators (Cassidy, 2021). Namibia also promotes CBNRM whose main goal is to conservation wildlife and to bring about socio economic development (Meyer et al., 2021). Through the Nature Conservation Amendment Act of 1996 wildlife and tourism rights are devolved to rural communities through partnerships (Mufune, 2015). The act recognises the role of the private sector and NGOs in contributing to rural development, hence it recognises their need to form partnerships with local communities to ensure natural and cultural assets are maintained within their areas of influence (Mufune, 2015).

Sustainable wildlife tourism in Southern African countries such as Namibia, Botswana and Zimbabwe that have communities participating in the CBNRM programme have experienced increases in wildlife, tourism-related jobs, improvements in rural communities' livelihoods, revenue generation through joint venture partnerships (JVPs) with the private sector and the co-management of natural resources (Zhou, 2023). Sustainable wildlife tourism is often promoted because it is believed when communities get direct benefits from thriving wildlife, they are more compelled to protect them (Mbaiwa, 2017; Zhou, 2023). Local support for wildlife management depends profoundly on benefit sharing (Mutanga, 2022; Tessema et al., 2010), with communities that receive more wildlife-related benefits being more likely to support conservation, while those that receive less benefits are more likely not to (Gandiwa et al., 2013).

1.2 Challenges of Wildlife Tourism in Southern Africa

Despite the various benefits that countries and communities derive from wildlife tourism, many challenges have also been reported. According to Zhou (2023), the failure to attain a balance between wildlife tourism resource utilisation and local community livelihoods needs often exacerbates human wildlife conflicts. Many countries in the region including Botswana, Namibia and Zimbabwe are experiencing increasing conflict between the humans and wildlife (Hitchcock, 2016). Authors such as Zhou & Chibaya (2022), Challenger et al. (2020) and Gandiwa et al. (2013) indicate that low community participation in wildlife tourism enterprises, insignificant community benefits and limited empowerment opportunities create problems such as human wildlife conflict.

In Botswana, Namibia and Zimbabwe, farmers struggle to coexist with elephants, which threaten livelihoods by destroying their crops (Cox, 2022; DeMotts, 2010).

Furthermore, many people residing adjacent to protected areas in Botswana have lost their lives to elephants (Madzwamuse, 2020; Madzwamuse et al., 2020). Thouless et al. (2016) and Cox (2022) report that over 70% of elephants in Botswana roam outside the country's protected areas. The country indicated that due to the growing human–elephant conflict and habitat destruction, it lifted the hunting ban in 2019 and supported Zimbabwe and Namibia in their application to have CITES lift the ban on the international commercial trade in ivory. According to Cox (2022), there is widespread human-elephant conflict across Southern Africa where people and elephants live close to each other, as a result, the author cautions us to 'forget the idealised 'Out of Africa' landscape of an unpopulated wilderness. The reality for African conservation is much more complicated. It takes place in human-dominated landscapes which brings with it questions of citizens' rights, ownership of land and political power which are increasingly overlooked in international public conservation discussions and policy processes' (p. 182).

The conflicts are exarcebated by the increasing human population growth in the region leading to elephants' traditional range areas reducing to accommodate human activity (Madzwamuse, 2020; Madzwamuse et al., 2020). Consequently, two competing philosophical approaches have emerged in the African wildlife conservation landscape. The first approach is the exclusionary approach that emphasises the legal protection of animals from human consumptive use and is primarily supported by Western conservation and animal protection movements. The second approach, the inclusive, rights-based approach, takes into account the sentiments and rights of citizens and is adopted by many African states, including Botswana, Namibia and Zimbabwe, it sees governments supporting the rights of their citizens (Cox, 2022; Greiber et al., 2009; Madzwamuse, 2020; Schramlc, 2018). This exclusionary approach seeks to strip rural communities of their rights, diminishes their capacity to engage with and influence policy, eliminates the economic value of their resources and aims to return authority over wildlife to a national and even international level (Madzwamuse et al., 2020). Madzwamuse et al. (2020) believe that exclusionary approaches to conservation shape the Western worldview of African people and wildlife found on the African continent and fail to consider different worldviews resulting from the different lived experiences of African citizens with different cultural lenses and ways of relating to wildlife. Similar sentiments are echoed by Chimuka (2022) who expressed that the Western media did not make any attempt to understand the viewpoint of Southern Africa regarding the Cecil, the Lion debate.

Moreover, authors such as Madzwamuse (2020) and Madzwamuse et al. (2020) are of the belief that what is missing in discussions about wildlife is the recognition that African countries own the wildlife and that they have been notable achievements in conservation and contributions to the livelihoods of rural Africans. Instead, wildlife is depicted as a global resource whose survival depends on people from the West and Africans are portrayed as 'trophy hunters', impoverished, corrupt 'others', who are receivers of the wisdom of lobbying organizations and their supporters in the west (Madzwamuse et al., 2020). Chimuka (2022) asserts that sustainable wildlife conservation in Southern Africa and other Global South destinations can succeed if the international community and the media include the views of local people and institutions who were historically and continue to be excluded from managing and benefiting from wildlife.

By ignoring African voices in the international environmental policy arena, Madzwamuse et al. (2020) believe this leads to the alienation, the criminalisation of rural African citizens regarding wildlife and impedes their ability to apply their agency within the policy space, disenfranchises and undermines the significant role that CBNRM plays in improving livelihoods, empowering rural citizens to call for inclusive and accountable governance and conserving wildlife.

Evidence from countries such as Tanzania and Botswana indicate that another challenge faced by the region is the foreign ownership of the wildlife tourism industry (safari hunting outfitters) leading to large revenue leakages (Baker, 1997; Mbaiwa, 2008). Furthermore, according to Mutanga (2022), the sustainability of numerous protected areas is threatened due to human–wildlife conflicts (HWC), crime and poverty among indigenous communities, habitat encroachment and destruction and increased illegal hunting.

1.3 Structure of the Book

The chapters in this volume indicate the importance of wildlife to many countries in the Southern African region. The volume is divided into six (6) parts. Part I titled '*Introduction—An overview of wildlife tourism in Southern Africa*' contains one chapter by Stone and Stone that provides an overview of wildlife tourism in Southern Africa. Part II, titled '*Resource use opportunities, contests, and disputes*' has four chapters. Zhou's chapter examines human-wildlife tourism conflicts (HWCs) in local communities adjacent to Matobo national park, Zimbabwe. The chapter determined that HWCs in the area are caused by interference with community life, damage to community residents' source of wealth and threats on human life. Stone's chapter examines different stakeholders' perspectives of the impacts of veterinary cordon fences (VCFs) on wildlife and tourism. The chapter determined that VCFs have led to the fragmentation of ecosystems and negative impacts on wild animals' migratory routes which have declined in both number and size over many years, with those that are in place being perceived to be under threat. Tichaawa and Lekgau's chapter uses Kgalagadi Transfrontier Park to explore the benefits and challenges of wildlife tourism for the Askham community in South Africa. The chapter determined that wildlife tourism promotes the economic development of surrounding communities. However, it also leads to the community relying largely on wildlife tourism, the main employers in the region. The last chapter in part II is Shereni's which analyses the evolution of CAMPFIRE and draw conclusions on its future developments. The author determined that the success of biodiversity conservation is dependent on local communities' attitudes towards wildlife. Furthermore, the positive attitudes of community members are influenced by the attainment of economic and material benefits.

Part III entitled '*The dynamics of tourism-based wildlife economies: Implications for sustainability*' comprises four chapters. Kamuti's chapter focuses on the state of wildlife tourism in the Southern African Development Community (SADC) region. The chapter reveals the benefits of wildlife tourism, including the conservation efforts and education but also the negative aspects such as ecological disruptions that are not implemented responsibly. Gohori's chapter discusses factors that influence and deter wildlife-based tourism in southern Africa. Results from the chapter indicate that the drivers of wildlife-based tourism in the region include amongst others the formation of transfrontier conservation areas, good conservation strategies, abundant wildlife and local communities. Deterrents on the other hand include the Convention on International Trade in Endangered Species [CITES] regulations, political instability, human-wildlife conflicts and poaching. The chapter also advances recommendations on how the region can amplify the drivers of wildlife-based tourism and minimise the barriers. Stone's chapter that follows discusses the negative impacts associated with countries relying heavily on the tourism sector using Botswana as a case study. Results indicate that the over reliance on wildlife tourism has led to tourism dependent communities, leading to doubts on the sustainability of the sector. The chapter calls for the need to develop destination adaptation and resilience strategies and the diversification of income earning activities to reduce the risk for tourism dependent communities especially during periods of uncertainty. Lastly, using the cases of Majete Wildlife Reserve and Liwonde National Park in Malawi, Bello and Lambulira's chapter discusses how sustainable tourism development is promoted in communities adjacent to protected areas. The chapter indicates that due to the benefits derived, communities appreciate natural resources in the protected areas and their culture.

Part IV entitled '*Hunting versus photographic tourism: Sustainability issues*' consists of four chapters. Lyakurwa & Sabuhoro's chapter gives an overview of consumptive wildlife tourism in East and Southern Africa. It discusses the current argument regarding utilitarian and ontological perspectives revolving around the trophy hunting industry and calls for the need to better understand each side of the debate (pro and anti-hunters). The authors conclude that with proper management, trophy hunting, conservation, and rationality of ethics can take place if the interests of pro-hunter and anti-hunters are harmonised. Mwakiwa's chapter assesses non-consumptive wildlife tourism use dynamics using case studies from Botswana, Namibia, South Africa, and Zimbabwe. Results from this chapter indicate the significance of non-consumptive tourism as a huge contributor to revenue generation and improving the livelihoods of people in southern Africa. This chapter discusses economic, socio-cultural and environmental benefits and challenges of non-consumptive tourism in the region. Mutanga and Gandiwa's chapter uses the case of COVID-19 in Botswana and cyclone Idai in Zimbabwe to assess the crises and disasters that have affected wildlife tourism in Southern Africa. Results indicate that natural disasters and the COVID-19 crisis have had negative impacts in the region including reduced conservation capacity and economic activity, destruction of infrastructure, less revenue for wildlife management, biodiversity loss and disturbance and negative tourist perceptions. The authors make recommendations that include addressing risk perceptions and behaviours of tourism stakeholders and creating emergency wildlife funds. They also call for tourism destinations to prepare in advance for crises and disasters to minimise negative impacts on visitors and the tourism industry.

The last chapter in part IV is one by Gumbo who examines the tourism in Botswana by examining the role of indigenous communities in the wildlife tourism development trajectory and exploring the conservation-local communities development nexus in wildlife tourism development. The author argues that local communities have been excluded and alienated from tourism spaces, furthermore, he highlights that sustainable wildlife tourism can only be attained with a harmonious relationship between wildlife conservation and the economic development of indigenous communities.

Part V includes four chapters and is entitled '*Protected areas and wildlife-based tourism*'. In the first chapter Lendelvo et al discuss the complex relationship between international law and wildlife tourism in Southern Africa. They discuss several key international environmental agreements guiding conservation in the region. The authors propose a framework consisting of four key areas to assess the legislation's influence on wildlife tourism. The chapter by Stone discusses transboundary conservation areas in Southern Africa. The chapter highlights the growth of transboundary conservation areas in southern Africa facilitated by several international and regional policy frameworks. Furthermore, it discusses the benefits and challenges experienced and their implications on tourism development in transfrontier conservation areas (TFCAs). Following that is Velempini and Garekae's chapter which assesses the relocation of communities from Kgalagadi Transfrontier Park (KTP) and the impact that the relocations had on the livelihoods of local communities. The authors interviewed KTP land claimants, some of whom were relocated from their ancestral land to make way for the establishment of the KTP. Results show that the relocation disrupted the locals' livelihoods and most lost their assets, which they had invested in within their ancestral land. Furthermore, local communities were given hope that they will return to their ancestral lands, something that never took place. The last chapter in this section is by Dibotlhale, Garekae and Stone who assess tourism joint venture partnerships (JVPs) in Southern Africa. The authors discuss economic benefits local communities in Southern Africa derive as a result of JVPs. Furthermore, the chapter discusses challenges experienced by communities with JVPs, including the passive participation of local communities in tourism development and the management of wildlife resources. The authors highlight the need for trust, communication and transparency between partners to ensure the success of JVPs in their role to build capacity and transfer skills to local communities.

Part VI, the 'conclusion' has one chapter by Stone and Stone. The chapter concludes the volume and synthesises the book chapters' findings. It summarises a few selected chapters with the main goal of contributing to knowledge on wildlife tourism dynamics in the Southern African region.

The book provides several case studies across the Southern African region. These cases illustrate rich insight into wildlife conservation and the wildlife tourism industry in the region.

References

Baghai, M., Miller, J. R. B., Blanken, L. J., Dublin, H. T., Fitzgerald, K. H., Gandiwa, P., Laurenson, K., Milanzi, J., Nelson, A., & Lindsey, P. (2018). Models for the collaborative management of Africa's protected areas. *Biological Conservation, 218*, 73–82.

Baker, J. E. (1997). Development of a model system for touristic hunting revenue collection and allocation. *Tourism Management, 18*(5), 273–286.

Ballantyne, R., & Packer, J. (2013). *International handbook on ecotourism*. Edward Elgar Publishing.

Brockington, D. (2002). *Fortress conservation: The preservation of the Mkomazi Game Reserve, Tanzania*. Indiana University Press.

Cassidy, L. (2021). Power dynamics and new directions in the recent evolution of CBNRM in Botswana. *Conservation Science and Practice, 3*(1), e205.

Challenger, D., Hinsley, A., Verissimo, D., & Sas Rolfes, M. (2020). *Why a blanket ban on wildlife trade would not be the right response*. Retrieved 5 December 2021, from https://theconversation.com/coronavirus.why-a-blanket-ban-onwildlife-trade-wouldnot-be-the-right-response-135745

Chimuka, G. (2022). The role of the media and the international community in recent conservation issues in Southern Africa: The case of Cecil the lion. In L. S. Stone, M. T. Stone, P. K. Mogomotsi, & G. E. J. Mogomotsi (Eds.), *Protected areas and tourism in Southern Africa: Conservation goals and community livelihoods* (pp. 146–160). Routledge.

Cong, L., Newsome, D., Wu, B., & Morrison, A. M. (2017). Wildlife tourism in China: A review of the Chinese research literature. *Current Issues in Tourism, 20*(11), 1116–1139.

Cox, C. (2022). International organisations and the ivory sales ban debate: The case of Zimbabwe, Namibia and Botswana. In L. S. Stone, M. T. Stone, P. K. Mogomotsi, & G. E. J. Mogomotsi (Eds.), *Protected areas and tourism in Southern Africa: Conservation goals and community livelihoods* (pp. 175–187). Routledge.

DeMotts, R. (2010). Whose elephants? Conserving, compensating, and competing in Northern Botswana. *Society and Nature Resources, 25*(9), 837–851.

Dowie, M. (2009). *Conservation refugees: The 100-year conflict between global conservation and native peoples*. MIT Press.

Gandiwa, E., Heitkönig, I. M. A., Lokhorst, A. M., Prins, H. H. T., & Leeuwis, C. (2013). CAMPFIRE and human–wildlife conflicts in local communities bordering northern Gonarezhou National Park, Zimbabwe. *Ecology and Society, 18*(4), 7–23.

Greiber, T., Janki, M., Orellana, M., Savaresi, A., & Shelton, D. (2009). *Conservation with justice: A rights-based approach*. IUCN Publications.

Hitchcock, R. K. (2016). African wildlife: Conservation and conflict. In *Life and Death Matters* (pp. 81–99). Routledge.

Kimbu, A., & Tichaawa, T. M. (2018). Sustainable development goals and socio-economic development through tourism in central Africa: Myth or reality? *Geojournal of Tourism and Geosites, 23*(3), 780–796.

Lekgau, R. J., & Tichaawa, T. M. (2019). Effects of institutional arrangements and policies on community participation in wildlife tourism in Africa. *Geojournal of Tourism and Geosites, 27*(4), 1280–1295.

Lovelock, B. (2008). *Tourism and the consumption of wildlife: Hunting, shooting and sport fishing*. Routledge.

Lubbe, B. (2022). The impact of rhino poaching on the economic dimension of sustainable development in wildlife tourism. In J. Saarinen, B. Lubbe, & N. N. Moswete (Eds.), *Southern African perspectives on sustainable tourism management* (pp. 187–198). Springer.

Madzwamuse, M. (2020). Contested conservation: Implications for rights, democratization. *Journal of the Society for International Development, 63*, 67–73.

Madzwamuse, M., Rihoy, E., & Louis, M. (2020). Contested conservation: Implications for rights, democratization, and citizenship in Southern Africa. *Development, 63*(1), 67–73.

Mangachena, J. R., & Pickering, C. M. (2021). Why are some animals popular with wildlife tourists: Insights from South Africa. *Journal of Ecotourism*, 1–17.

Mbaiwa, J. E. (2005). Wildlife resource utilisation at Moremi Game reserve and Khwai community area in the Okavango Delta, Botswana. *Journal of Environmental Management, 77*(2), 144–156.

Mbaiwa, J. E. (2008). The success and sustainability of consumptive wildlife tourism in Africa. In B. Lovelock (Ed.), *Tourism and the consumption of wildlife hunting, shooting and sport fishing*. Routledge.

Mbaiwa, J. E. (2017). Poverty or riches: Who benefits from the booming tourism industry in Botswana? *Journal of Contemporary African Studies, 35*(1), 93–112.

Mbaiwa, J. E. (2018). Effects of the safari hunting tourism ban on rural livelihoods and wildlife conservation in northern Botswana. *South African Geographical Journal, 100*, 41–61.

Mbaiwa, J. E., & Mogende, E. (2022). The state, community-based tourism, and wildlife user rights in tourism concessions in Botswana. In R. Musavengane & L. Leonard (Eds.), *Conservation, land conflicts, and sustainable tourism in Southern Africa*. Routledge.

Meyer, M., Klingelhoeffer, E., Naidoo, R., Wingate, V., & Börner, J. (2021). Tourism opportunities drive woodland and wildlife conservation outcomes of community-based conservation in Namibia's Zambezi region. *Ecological Economics, 180*, 106863.

Mufune, P. (2015). Community based natural resource management (CBNRM) and sustainable development in Namibia. *Journal of Land and Rural Studies, 3*(1), 121–138.

Mutanga, C. N. (2022). Tolerance for wildlife resources through community wildlife-based tourism: Implications for sustainability. In L. S. Stone, M. T. Stone, P. K. Mogomotsi, & G. E. J. Mogomotsi (Eds.), *Protected areas and tourism in Southern Africa: Conservation goals and community livelihoods* (pp. 56–69). Routledge.

Naidoo, R., Weaver, L. C., Diggle, R., Matongo, G., Stuart-Hill, G., & Thouless, C. (2016). Complementary benefits of tourism and hunting to communal conservancies in Namibia. *Conservation Biology, 30*(3), 628–638.

Newsome, D., & Rodger, K. (2013). *Wildlife tourism*. In A. Holden & D. Fennell (Eds.), *A handbook of tourism and the environment* (pp. 345–358). Routledge.

Novelli, M., Barnes, J. I., & Humavindu, M. (2006). The other side of the ecotourism coin: Consumptive tourism in Southern Africa. *Journal of Ecotourism, 5*(1-2), 62–79.

O'Malley, M. P., Lee-Brooks, K., & Medd, H. B. (2013). The global economic impact of manta ray watching tourism. *PLoS One, 8*(5), e65051.

Qin, H., Bass, M., Ulrich-Schad, J. D., Matarrita-Cascante, D., Sanders, C., & Bekee, B. (2020). Community, natural resources and sustainability: Overview of an interdisciplinary and international literature. *Sustainability, 12*, 1061.

Reynolds, P. C., & Braithwaite, D. (2001). Towards a conceptual framework for wildlife tourism. *Tourism Management, 22*, 31–42.

Saayman, M., & Saayman, A. (2017). Is the rhino worth saving? A sustainable tourism perspective. *Journal of Sustainable Tourism, 25*(2), 251–264.

Schramlc, O. S. (2018). Making sense of protected area conflicts and management approaches: A review of causes, contexts and conflict management strategies. *Biological Conservation, 222*, 136–145.

Snyman, S. (2017). The role of private sector ecotourism in local socio-economic development in Southern Africa. *Journal of Ecotourism, 16*(3), 247–268.

Stone, L. S. (2017). Wildlife Tourism. In L. Lowry & J. G. Golson (Eds.), *The SAGE international encyclopedia of travel and tourism, sage* (pp. 1437–1439). Sage Publishers.

Tessema, M., Lilieholm, R., Ashenafi, Z., & Leader-Williams, N. (2010). Community attitudes toward wildlife and protected areas in Ethiopia. *Society and Natural Resources, 23*(6), 489–506.

Thouless, C., Dublin, H. T., Blanc, J., Skinner, D. P., Daniel, T. E., Taylor, R., ... & Bouché, P. (2016). African elephant status report 2016. *Occasional paper series of the IUCN Species Survival Commission, 60*, 309.

Tichaawa, T., & Lekgau, R. J. (2020). Nature and characteristics of wildlife tourism in the Kgalagadi Transfrontier Park. *Acta Universitatis Danubius. Œconomica, 16*(5), 69–88.

Trave, C., Brunnschweiler, J., Sheaves, M., Diedrich, A., & Barnett, A. (2017). Are we killing them with kindness? Evaluation of sustainable marine wildlife tourism. *Biological conservation, 209*, 211–222.

United Nations Word Tourism Organisation (UNWTO). (2015). *Towards measuring the economic value of wildlife watching tourism in Africa* (Briefing paper). UNWTO.

Zafra-Calvo, N., & Moreno-Peñaranda, R. (2018). Exploring local people's views on the livelihood impacts of privately versus community managed conservation strategies in the Ruvuma landscape of North Mozambique-South Tanzania. *Journal of Environmental Management, 206*, 853–862.

Zanamwe, C., Gandiwa, E., Mukobo, N., Kupika, O. L., & Mukamuri, B. B. (2018). Ecotourism and wildlife conservation-related enterprise development by local communities within Southern Africa: Perspectives from the Greater Limpopo Transfrontier Conservation, South-Eastern Lowveld, Zimbabwe. *Cogent Environmental Science, 4*, 1–13.

Zhou, Z. (2023). Human–wildlife tourism conflict: Towards an inclusive wildlife resources utilisation approach in Southern Africa. *GeoJournal, 88*, 931–949.

Zhou, Z., & Chibaya, T. (2022). Towards gender and disability inclusive tourism development in Zimbabwe: insights through sustainable development goals. In *Gender, Disability, and Tourism in Africa: Intersectional Perspectives* (pp. 137–154). Springer International Publishing.

Part II
Resource Use Opportunities, Contests, and Disputes

Chapter 2
Wildlife Tourism and Human-Wildlife Conflicts in Zimbabwe: Causes, Compensation, and Community Perceptions

Zibanai Zhou

2.1 Introduction

Wildlife tourism has morphed into a multi-trillion-dollar sector widely known for its favourable socioeconomic transformative prowess on local communities across diverse geographical contexts (Mbaiwa & Stronza, 2011; Makumbe et al., 2022; Matseketsa et al., 2019; UNWTO, 2019). However, wildlife tourism has always been associated with an equal measure of adverse effects on communities adjacent to wildlife reserves such as forced displacements and relocations, and interference with communities' traditional sources of livelihood (Zhou, 2019a, 2023). Consequently, human-wildlife conflicts (HWCs) are inevitable in the wildlife tourism value chain. To date, HWCs have become an issue of intense debate among policy makers, tourism practitioners, wildlife authorities, and academia with a view of finding a lasting solution to peaceful co-existence. However, given the intricate dynamics around HWCs, attaining this objective remains elusive.

At a global scale, HWCs have been attributed to increasing competition on shared natural resources between wildlife and humankind (Merkebu & Yazezew, 2021). Past studies show that local communities around wildlife tourism spaces are exposed to a plethora of wildlife-induced devastating effects encompassing crop destruction, loss of livestock through predation or disease, and human mortality (Andersson et al., 2013; Kahuni et al., 2014). Wildlife plays a dual role as the mainstay of tourism, and a source of local communities' livelihood. Consequently, in circumstances where wildlife resources are exclusively used for tourism, conflicts with communities are inevitable (UNEP, 2019, Zhou, 2023). To this end, HWC has

Z. Zhou (✉)
Tourism, Hospitality & Leisure Sciences, Midlands State University, Gweru, Zimbabwe
e-mail: zhouz@staff.msu.ac.zw

© The Author(s), under exclusive license to Springer Nature Switzerland AG 2024

L. S. Stone, M. T. Stone (eds.), *Wildlife Tourism Dynamics in Southern Africa*, Geographies of Tourism and Global Change,
https://doi.org/10.1007/978-3-031-57252-4_2

become an issue of huge concern in developing countries particularly in communities bordering wildlife spaces (Shilongo et al., 2018; Nyhus, 2016; Seoraji-Pillai & Pillay, 2016).

While HWC is a global issue, southern Africa in particular is under the spotlight as it has in excess of 300 protected areas (PAs). It has the largest reservoirs of wildlife population in terms of density and diversity as compared to the rest of the world (DeGeorges & Reilly, 2008). Most local communities in southern Africa depend to a greater extent on subsistence agriculture, small scale animal husbandry, and direct exploitation of natural resources, inclusive of wildlife. To this end, crop raiding by elephants, baboons and monkeys; wild-pigs, birds and livestock depredation by lions, and hyenas have been reported across southern Africa, and Zimbabwe is no exception. According to Spenceley (2005), the causes of HWCs are multi-dimensional. Spenceley (2005) frames the multi-dimensional aspect of the causes of HWCs as a significant conservation and socio-economic issue that warrants further investigation. Similarly, Redpath et al. (2013) observed that despite the existence of extensive research on HWCs, there is still little consensus about how the drivers of HWCs can be mitigated. This is further strengthened by Seoraji-Pillai and Pillay (2016) who opined that systematic and in-depth examination of the most vulnerable groups of people such as local communities bordering wildlife in protected areas, should be the focus of future HWCs research avenues.

The present study builds on the literature gap highlighted by Mutanga et al. (2015) suggesting that empirical studies which focus on the relationships between wildlife and humans in Zimbabwe are yet to permeate the academic spectra. This is even more relevant in the Zimbabwean context where human settlements continue to expand in the direction of wildlife tourism spaces inspired by the land reform programme, thus exacerbating direct conflict. The ever-increasing competition over dwindling resources, coupled with limited progress by relevant wildlife authorities in preventing antagonism between local communities and wild animals makes the present study significant.

Zimbabwe has the second highest level of biodiversity worldwide, and therefore presents an ideal study setting to investigate human-wildlife tourism conflict. Furthermore, the prevalence of many local communities sharing borders with wildlife reserves, which at the same time compete with wild animals for critical natural resources for sustenance is even more compelling. The present study acknowledges the existence of extensive research around HWCs in Zimbabwe (Gandiwa et al., 2013, 2014; Makumbe et al., 2022; Mudimba & Tichaawa, 2019), whose focus has been predominantly on wildlife conservation, attitudes and perceptions themes. Such themes were however loosely tied to other critical aspects of wildlife tourism and HWCs such as causes of HWCs; available wildlife damage compensation schemes; and community perceptions. Additionally, these studies are also fraught with a number of limitations. Most notably, the analyses of the various sub-themes around HWCs were majorly done independent of other aspects, hence glossing over the existing interdependence. It is pertinent to highlight that a deep understanding of the intricate relationship around the causes of HWCs, wildlife damage compensation, and the overall perception of local communities on wildlife tourism and compensation frameworks is only possible if such sub-themes are investigated holistically and not in isolation. It is against this backdrop that the present study argues that interventions hastily prepared by central government, non-governmental organisations (NGOs), and wildlife authorities without deep contextual knowledge of the intricate nexus around the causes of HWCs, available compensation frameworks, and communities' perception toward wildlife tourism would invariably end badly. Additionally, the studies are overly restricted to Save Valley Conservancy and communities bordering Gonarezhou at the expense of communities in other regions, especially south-western Zimbabwe.

There is growing concern that the south-western region of Zimbabwe is marginalised. Further still, it is apparent that the theme of wildlife damage compensation schemes is sparsely represented or mentioned in passing in most previous studies. Finally, less emphasis was placed on the interlinkage among the various subthemes; and overall implication on the conservation agenda thus providing a rather limited HWCs perspective. By integrating the themes, the present study addresses the aforementioned literature gaps focusing on communities around Matobo national park, located in the southern region of Zimbabwe.

The study aimed to examine the main causes of HWCs in local communities around Matobo national park; analyse the compensation measures available to local communities; and determine the perceptions of local communities on wildlife tourism in general and compensation schemes in particular. The study amplifies the voices of local communities in wildlife tourism-human conflicts discourse. The study adopted a qualitative approach where semi-structured interviews were used for data collection. The study is significant as it provides insights, which can be employed as a basis to review the contemporary wildlife damage compensation framework; reduce HWCs and enhance co-existence. This may improve local communities' perception on wildlife tourism thereby reducing wide scale retaliatory killing of wildlife. The chapter is organised as follows. The next section discusses wildlife tourism-human conflicts in the wider tourism discourses. The methodology that was adopted is described in the subsequent section. The last section of the chapter discusses the major findings of the study and provides a conclusion.

2.2 Literature Review

2.2.1 Wildlife Tourism-Human Conflicts, Compensation and Community Perception Discourses

Historically, people have always competed with wildlife for food and resources (Nyhus, 2016). HWCs occur when the needs of wildlife encroach on those of human populations (Mukeka et al., 2019). Woodroffe et al. (2005) describe HWCs as scenarios where the resource demands of wild animals and humans overlap, fuelling competition for food, space and water, and the ensuing tension between people and wildlife authorities. HWCs are rife around the edges of wildlife tourism spaces where there is high wildlife and humankind interaction (Matseketsa et al., 2019). Furthermore, Efio et al. (2018), Eustace et al. (2018), and Zhou (2023) assert that conflicts are more severe in spaces where both wildlife and humans live and share resources. HWC manifests through loss of livestock to predators, crop-raiding by elephants, physical attacks on humans resulting in injury or death, transmission of diseases, and competition over resources. Past research point to habitat disturbances and population growth, leading pressure on natural resources, urbanisation and industrialisation as drivers of HWCs. Acharya et al. (2016) assert that the unavailability of fodder and the presence of human habitat in the vicinity of most forests compel wild animals to attack humans, livestock and their crops. However, the paradox is that local communities around wildlife tourism reserves are limited by law and regulation to defend themselves from the wild animals. According to Shilongo et al. (2018) this is partly attributed to ecocentric and protectionist paradigms' failure to fully accommodate the needs of people living with wildlife. Shilongo et al. (2018) further outlined that conflict between wildlife and humans undermines coexistence and threatens biodiversity conservation and livelihoods.

Globally, it has become common practice to compensate communities for wildlife-induced damages or losses (Ravenelle & Nyhus, 2017). Compensation schemes as control measures are designed to mitigate the effects of wildlife conflict once damages have been incurred by making payments to cover losses. Ayivor et al. (2020) assert that compensating communities for wildlife related damages reduces hunting pressures. However, Dublin and Hoare (2004) argue that despite compensation efforts, biodiversity worldwide continues to decline while HWCs remain a huge threat to humankind livelihood. This contradicts the widely held view that compensation, reimbursement to communities and people who have been injured, 'killed' or physically threatened by wildlife can incentivise coexistence. The aim to reimburse costs of lost property or life is a way of nurturing tolerance for wildlife by communities, and conservation policies (Demotts & Hoon, 2012; Mckenzie et al., 2018). However, as already pointed out by Shilongo et al. (2018) compensation programs have been criticised for being riddled with corruption and inefficiency.

Zimbabwe like other southern Africa countries grapples with HWCs (Gandiwa et al., 2013; Zhou, 2023; ZTA, 2019). Wildlife authorities and government officials worldwide recognise that HWC is an issue that has the potential to undermine biodiversity conservation. To achieve biodiversity conservation, the involvement of local communities is inevitable. Pushing the conservation agenda of fauna to local communities ensures success of such a programme. In support, Thondhlana et al. (2015) and Chiutsi and Saarinen (2019) pointed out that the inclusion of community livelihoods in protected areas management was recognised by governments after the failure of previous exclusionary approaches to conservation, which had removed communities from the natural asset rich areas. Similarly, Ghosal et al. (2015) called for more nuanced approaches toward navigating HWCs in different geographical contexts. Wildlife tourism is popular in Zimbabwe and remain the largest foreign currency earner (ZTA, 2019). However, the widespread and incessant conflict between humans and wildlife across Zimbabwe undermines wildlife tourism and creates policy dilemmas for conservationists and wildlife authorities. This is against a backdrop where wildlife tourism spaces in developing countries have become major tourist attractions offering a variety of wildlife viewing and outdoor recreational activities (Mbaiwa & Hambira, 2020).

Just like other southern African countries, there is growing recognition in Zimbabwe that PAs and subsequently wildlife tourism hinges on co-existence with local communities who depend on such resources for their sustenance (Sene-Harper et al., 2018). Wildlife tourism holds significant potential in generating much needed economic relief in such communities through its contribution toward national economies (Mbaiwa & Stronza, 2011). This is particularly so in many sub-Saharan Africa countries, where community livelihoods are tied to wildlife resources within the local vicinity (Pour et al., 2019; Tichaawa & Lekgau, 2020; Zhou, 2018). Wildlife tourism has been embraced as a vector of achieving sustainable and inclusive development in adjacent host communities as well as supporting conservation activities (Moswete et al., 2020; Zhou, 2019b). Chilembwe (2020) opined that local communities' perceptions and attitudes towards the socioeconomic impacts of wildlife tourism are instrumental toward fostering community participation and managing efficient and sustainable conservation. The host communities are hampered from benefiting fully from wildlife tourism by constraints such as institutional arrangements, imperial policies and legal systems, marginalisation, and autocratic approaches (Mbaiwa & Hambira, 2020; Moyo & Tichaawa, 2017; Zhou, 2023).

Sharma et al. (2021) cautioned that while efforts are being made to tackle HWCs, most initiatives are limited in scope as they remain country and location specific. Similarly, Wester et al. (2019) observed that most countries suffer from inadequate and scattered knowledge generation, which is a hindrance to understanding the underlying drivers and effects of HWCs. This undermines informed collaborative natural resource governance (Davies & White, 2012). To this end, Sharma et al. (2021) emphasised the importance of understanding the dynamics that drive the relationship between humans and wildlife on the basis that the factors are area specific and highly complex. This therefore implies that while viewing existing HWCs through a universal lens in a case-specific context is important, in practice it does not always comply with the complexity of the underlying phenomenon.

2.3 Methodology

2.3.1 Study Context

HWCs have reached alarming levels in several districts in Zimbabwe, with statistics increasing annually (ZTA, 2019). Communities in Matobo district grapple with high rates of reported cases of HWCs every year such as loss of life, crop destruction, killing of livestock, and retaliatory killings have become widespread. This has created a dilemma for Zimbabwean wildlife authorities, Government of Zimbabwe

[GoZ], and local communities. Matobo national park is located in the south-western part of Zimbabwe, approximately 40 km outside Bulawayo. The national park is famous for its diverse range of wild animals which form the basis for a thriving wildlife tourism sector. Popular tourism activities encompass wildlife viewing, trailing, elephant rides, photographing, bird watching, and heritage tourism. Surrounding communities rely on wildlife resources for their livelihoods.

2.3.2 Data Collection and Sample

The interviews for this study took place in communities adjacent to Matobo national park. Most of the surrounding land is primarily for subsistence agriculture. While there is thriving wildlife tourism enterprising, reports of HWCs were plentiful, making it an ideal location for the study. Local community participants were drawn from Ward 24, Malunga village. The community was chosen on the basis of having the highest reported cases of HWCs. A total of 21 participants living in the community selected through purposive and snow-ball sampling were interviewed during February and March 2023 (Table 2.1). Initially, the first primary contacts were

Table 2.1 Interviewee profiles

Interviewee code	Gender	Age	Designation
Participant 1	F	24	Unemployed youth graduate
Participant 2	F	40	Community resident
Participant 3	M	46	Area manager
Participant 4	M	50	Victim 1
Participant 5	F	62	Livestock owner
Participant 6	M	58	Community resident
Participant 7	F	36	Resident
Participant 8	M	41	NGO
Participant 9	F	49	Victim 2
Participant 10	M	55	Victim 3
Participant 11	F	25	Youth
Participant 12	M	22	Youth community resident
Participant 13	F	60	Community cooperative leader
Participant 14	F	46	Subsistence farmer
Participant 15	M	65	Community headman
Participant 16	M	70	Political community leader
Participant 17	F	58	Community resident
Participant 18	F	66	Elderly community resident
Participant 19	M	44	Zimparks official
Participant 20	M	56	Matobo Rural district council official
Participant 21	F	61	Wildlife business proprietor

Source: Author (2023)

purposively selected through a local NGO with interest in wildlife, conservation and community development, which provided the names of local community residents and wildlife parks officials to interview. The study targeted participants with a granular understanding of dynamics around the causes of HWCs, compensation schemes and community perceptions. Subsequently, snowball sampling was used to identify additional participants. The recommended participants were then contacted individually by phone and asked if they would be willing to talk about the causes of HWC, compensation schemes; all but one person agreed.

The objectives of the study were explained to participants and their anonymity was assured. The semi-structured interviews with local residents and conservation practitioners lasted about 2 h, with the researcher visiting respondents at their homes after calling to set up an appointment. Cursory notes were taken during interviews to put respondents at ease. A semi-structured interview guide was designed informed by a literature review and previous studies (Gandiwa et al., 2014; Makumbe et al., 2022).

Thematic content analysis was employed to analyse data from interviews. Braun and Clarke's (2006) framework of analysing data was adopted. The stages of the analytical framework encompassed familiarisation with data; reading interview data repeatedly; generating preliminary codes; theme searching; reviewing themes; naming themes, and report writing. Microsoft excel was used to categorise the codes and themes.

2.3.3 *Findings and Discussion*

This section presents and discusses the study's findings. The findings are presented in themes. The themes derived from the data are causes of HWCs in communities adjacent to Matobo national park; wildlife damage compensation framework; and perception of local communities on compensation measures and wildlife tourism. These are discussed below.

2.4 Causes of HWCs in Communities Adjacent Matobo National Park

Findings show that the major causes of wildlife tourism and human conflict in local communities surrounding Matobo national park are varied and revolve around competition over the use of finite faunal and floral resources among different actors in the wildlife tourism resource value chain. The main actors encompassed residents of surrounding communities, wildlife park authorities; wildlife tourism business operators; wildlife conservation organisations and wild animals. The conflicts were attributed to the stakeholders' diverse interest around wildlife tourism resources.

Participants outlined five main causes of HWCs notably: (i) threats on human life; (ii) interference with community life; (iii) damage on community residents' source of wealth; (iv) damage on communities' traditional sources of livelihood; and (v) competition in sharing of natural resources. The following interview excerpts summarise the major causes of HWCs in communities adjacent Matobo national park:

> There are widespread attacks on people at dusk, residents working in the fields, on night prayers, fetching firewood and water, catching buses, and visiting service centres such as grinding mills and clinics. In fact, wild animals interfere with every aspect of community life. (Participant 19)

> Our livestock herds have depleted as they are being devoured by hyenas and lions. Every year we lose cattle, goats, sheep, and donkeys to marauding wild animals. (Participant 15)

> A lifetime investment in form of fruit trees is destroyed overnight by elephants. (Participant 5)

> Hippos destroyed my sugar beans which were at flowering stage, elephants trampled my mangoes and bananas. My entire fields of millet and rapoko were shelled by birds. (Participant 6)

The interview narratives are consistent with past studies, which found that HWCs emanate from multiple causes, such as competition for resources due to habitat encroachment, destruction of crops, and predator-livestock conflicts (Gandiwa et al., 2013; Ravenelle & Nyhus, 2017). In Zimbabwe's communal areas, livestock is not only a source of wealth but also a status symbol and a source of draught power. Destruction of crops and orchards is a significant threat to livelihoods which undermine household food security. Similarly, Quirin and Dixon (2012) found that baboons caused extensive damage to crops in Uganda and Ethiopia posing a great threat to crop yields. Gandiwa et al. (2013) established that hyenas and elephants were the most troublesome wild animals in communities around Save Valley conservancy, and a similar pattern is emerging in Matobo national park. The preceding findings are also supported by Khumalo and Yung (2015) who outlined that HWCs cause local communities to lose investments in the form of fruit trees. Furthermore, HWCs cause emotional stress and trauma on victims in local communities especially children and women and the elderly who always live in constant fear of physical harm. Residents of local communities around Matobo national park no longer feel safe even outside their homesteads, and when they go about on daily chores as revealed by the study.

2.5 Wildlife Damage Compensation Framework

The analysis of data showed that on paper there is a wildlife damage compensation framework in place, which was restrictive and pays varying amounts of money in local currency to victims of HWCs. These views were summed up by the following excerpts:

> Funds paid out in local currency to offset the loss are symbolic. (Participant 11)

> In cases of death caused by elephants and buffaloes families are robbed of their sole bread winners. This leaves the deceased's families, dependents, and children vulnerable. Such families get nothing beyond the once off payment. Why not paying school fees for the victims' children, or prioritising provision of employment to the dependents? This would be far more significant than a once off payment. (Participant 4)

> Compensation for poultry lost through wild dogs and loss of small grain crops like rapoko and millet to birds is not covered. (Participant 15)

Victims of wildlife tourism conflicts are reimbursed for every hectare destroyed by elephants, whereas livestock, goats, sheep, cattle, and donkeys have varying amounts of compensation. The tragedy in Zimbabwe is that compensation is given in local currency which is very volatile. More so, it is a requirement that the money is deposited into one's bank account or mobile wallet which most people in local communities surrounding Matobo national park do not have. This creates further challenges for the victims of wild animals to access the funds on time. HWCs become even more challenging for local communities in a non-functional economic environment characterised by currency volatility. In communal Zimbabwe livestock is sold strictly in South African Rands and or USD. However, Zimparks argues that they pay the equivalent of the local currency using the interbank exchange rate. Nyhus (2016) notes that in the global South, the debate around the benefits and costs of biodiversity conservation continue to rage unabated particularly for local communities living near or within wildlife spaces. A similar pattern unfolds in Matobo national park.

However, from the perspective of local communities the major issue was not the availability of such a framework rather most respondents were concerned about fairness and justice, which seem to be lacking. Most participants were concerned that their rights to recourse and justice were trampled upon by the very same government which was supposed to protect them, and they had limited options except to retaliate by killing wild animals. Some residents in local communities are less able to recover from crop damage, livestock loss and human injury. Consistent with the compensation policies for southern Africa countries, compensation is considered for damage caused by certain species, which means damages by birds, bush pigs, and warthogs are not covered. Furthermore, communities around Matobo national park plant drought resistant crops, which are shelled by birds, yet this is not covered under the current compensation framework. Similarly, stray wild dogs eat poultry which is not covered in the compensation scheme. Overall, that makes the wildlife damage compensation framework largely symbolic and exclusionary. The amount of money given as compensation is neither proportional to the value of livestock nor crop yield lost. This finding mirrors compensation trends worldwide as highlighted in literature (Ravenelle & Nyhus, 2017). Compensation measures largely provide short term relief and are broadly branded as stop gap measures while damage caused by wild animals on crops and livestock is permanent.

Overall, the study unravelled that the current compensation mechanism partially and selectively addresses the concerns of local communities' issues. This is in consonance with Mckenzie et al. (2018) who outlined that in most developing countries, the compensation process is controlled by a cumbersome bureaucratic process, using a primary neoliberal model of cost benefit analysis. Such a model is out of synch with the expectations of local people and worsens the discontent between the contemporary compensation mechanism and the needs of local communities, thereby worsening the disconnect between policy and practice. The compensation model disenfranchises villagers at household level, it also disempowers local communities by stripping them of their bargaining power.

2.6 Perceptions of Local Communities on Compensation Measures and Wildlife Tourism

The perceptions of local communities towards compensation measures in particular and wildlife tourism in general was found to be overly negative. The responses below illustrate this view.

> Despite the few employment opportunities offered by wildlife tourism, it has hugely brought more problems than jobs for us. Only a few are employed however everyone is affected by the destruction of crops and killing of livestock. (Participant 14)

> Wildlife tourism benefits are on paper, practically they are superficial. (Participant 11)

> We are being stripped of our assets under the cover of wildlife tourism as we are becoming poorer with each passing day. We are better off without wild animals. (Participant 18)

These findings are in line with Acha and Temesjen (2015) and Biru et al. (2017) who found that local communities had negative perceptions of wildlife tourism and the compensation measures put in place. Crop damage, livestock depredation and attacks on humans are the main issues causing negative perceptions. The involvement and empowerment of local communities in natural resource management, linked to skills development and capacity building are critical to influencing positive perceptions of local communities toward wildlife tourism. According to residents interviewed, negative perceptions about the compensation measures stemmed from the fact that after application for compensation, it took too long to obtain the claim. In contrast, NGO staff and government officials stated that the processing of compensation claims and disbursement is standard as they have to do verification. This disconnect in perspectives fuels negative perceptions. Despite disagreement with local communities on compensation measures, government officials acknowledged broader failures in their responses to HWCs. The negative perception, inconsistency, and mistrust created an unhealthy environment between local communities and wildlife authorities.

According to Treves (2009), access to conservation-related benefits and the involvement of local communities in decision making for resource management positively influences local communities' attitude towards wildlife, PAs and conservation. The exclusion of local communities from wildlife tourism, opaque compensation schemes, incessant destruction of crops and the killing of livestock continue to influence the communities' perceptions of wildlife conservation negatively. Abukari and Mwalyosi (2020) posit that perceptions provide an important means of assessing the influence of conservation initiatives so that better policies may be developed for effective biodiversity protection and the well-being of people living near wildlife tourism spaces. The impacts of wildlife and compensation schemes on local communities' livelihoods can be major determinants of the perceptions of local communities.

2.7 Conclusion

This study examined the causes of human-wildlife tourism conflicts in communities surrounding Matobo national park, available wildlife damage compensation schemes and perceptions of communities towards wildlife tourism. The study has revealed that the major causes of HWCs in communities adjacent to Matobo national park revolved around wild animals' threats on human life, interference with community life, and damage on community residents' source of wealth among others.

The study confirmed the availability of a wildlife damage compensation framework despite its shortcomings in meeting the expectations of the victims of HWCs. The compensation framework was overall condemned for being too insensitive to the specific needs of local communities, and peculiar individual circumstances. Most participants loathed the one size fits all approach when compensating for losses incurred instead of treating each case differently. Most previous studies across diverse contexts in southern Africa concluded that the cost of biodiversity conservation is borne directly by local communities in rural households (Zhou, 2023; De Motts & Hoon, 2012). The efficacy of compensation policies to address HWCs costs has been questioned in previous studies (Barua et al., 2013). This has also contributed to the general negative perceptions on wildlife tourism.

The study revealed that communities have a negative perception on both wildlife damage compensation schemes and wildlife tourism. The perception is largely influenced by wildlife tourism costs borne by communities. It is critical that policy makers and wildlife authorities increase wildlife tourism benefits to communities to influence positive perceptions. The study contributed to three areas in the broad area of HWCs precinct. Firstly, it contributes to the evolving discourse on the causes of HWCs focusing on communities surrounding Matobo national park, which is underrepresented in the current Zimbabwean HWCs discourse. Secondly, it adds to the literature by integrating three important HWCs sub-themes largely discussed in isolation in previous studies. Finally, the study expands on the literature pertaining to the perceptions of local communities towards HWCs, and wildlife tourism in the context of Zimbabwe. By addressing these gaps in the Zimbabwean wildlife tourism context, the study may help in the formulation of inclusive and sustainable HWCs policies that are responsive to the needs of local communities, who bear the brunt of HWCs. This would ensure co-existence and promote biodiversity conservation.

One important limitation of this research is its small sample size. Given that the study did not seek generalisability of findings, the sample size of 21 was sufficient. The findings from those interviews reflect the existing literature related to HWCs. In addition, a large body of social science research supports the assumption that for qualitative research a large sample size or interview pool is not always necessary (Robinson, 2013; Fuji, 2018; Saldana & Omasta, 2017). The study is also limited to communities in Zimbabwe. Future studies may focus on other communities in other geographical locations and increase the sample size to compare findings.

References

Abukari, H., & Mwalyosi, R. B. (2020). Local communities' perceptions about the impact of PAs on livelihoods and community development. *Global Ecology and Conservation, 22*. https://doi.org/10.1016/j.gecco.2020.e009

Acha, A., & Temesjen, M. (2015). Approaches to HWC management in and around Chebera-Churchara national park, southern Ethiopia. *Asian Journal of Conservation Biology, 4*, 136–142.

Acharya, K. P., Paudel, P. K., Neupane, B. P. R., & Koehl, M. (2016). Human-wildlife conflict in Nepal: Patterns of human fatalities and injuries caused by large mammals. *PLoS ONE, 11*(9), e0161717. https://doi.org/10.1371/journal.Pone-0161717

Andersson, J. A. D., Garine-Wichatitsky, M., Cumming, D. H., Dzingirai, V., & Giller, K. E. (2013). *Trans-frontier conservation areas: People living on the edge*. Earthsca Publications.

Ayivor, J. S., Nyameto, J. K., & Ayivor, S. (2020). Protected area governance and its influence on local perceptions, attitudes and collaboration. *Land, 9*(9), 310.

Barua, M., Bhagwat, S. A., & Jadhav, S. (2013). The hidden dimensions of human-wildlife conflict: Health impacts, opportunity and transaction costs. *Biological Conservation, 157*, 309–316.

Biru, Y., Tessema, Z. K., & Uige, M. (2017). Perceptions and attitudes of pastoralists on livestock-wildlife interactions around Awash national park. Ethiopia: Implications for biodiversity. *Conservation Ecological Process, 6*, 1–13.

Braun, V., & Clarke, V. (2006). Using thematic content analysis in psychology. *Qualitative Research in Psychology, 3*(2).

Chilembwe, J. M. (2020). Nature tourism, wildlife reserves and community-based conservation. In M. T. Stone, M. Lenao, & N. Moswete (Eds.), *Natural resources, tourism and community livelihoods in southern Africa: Challenges to sustainable development* (pp. 26–38). Routledge.

Chiutsi, S., & Saarinen, J. (2019). The limits of inclusivity and sustainability in Trans-frontier peace parks: Case of Sengwe community in the Great Limpopo Trans-Frontier Conservation area, Zimbabwe. *Critical African Studies, 11*(3), 348–360.

Davies, A. L., & White, R. M. (2012). Collaboration in natural resources governance. Reconciling stakeholder expectations in deer management in Scotland. *Journal of Environmental Management, 112*, 160–169. https://doi.org/10.1016/j.jenvman.2012.07.032

DeGeorges, P. A., & Reilly, B. K. (2008). *A critical evaluation of conservation and development in sub-Saharan Africa: Last chance Africa* (pp. 10–572). Edwin Mellen Press.

Demotts, R., & Hoon, P. (2012). Whose elephant? Conserving, compensating and competing in northern Botswana. *Society and Natural Resources, 25*(9), 837–851.

Dublin, H. T., & Hoare, R. C. (2004). Searching for solutions: the evolution of an integrated approach to understanding and mitigating human-elephants conflict in Africa. *Human Dimensions of Wildlife, 9*(4), 271–278.

Efio, S., Sogbohossoy, E. A., Magnon, Z. Y., Houinato, M. R. B., Habiyaremye, M., Sin, B. A., & Tosson, C. R. (2018). Human-wildlife conflicts and mitigation measures in Pendjari Biosphere Reserve, northern Benin. *Annals of Agronomic Sciences, 22*, 15–31.

Eustace, A. A., Kisingo, W., & Mbwiliza, J. (2018, May 1). Wildlife damage in villages surrounding Serengeti ecosystem. *Parks, 24*.

Fuji, L. A. (2018). *Interviewing in social science: A rational approach: 1. Routledge Series on interpretive methods*. Routledge, Taylor and Francis Group.

Gandiwa, E., Heitkonig, I. M. A., Lokhorst, A. M., Prins, H. H. T., & Leewis, C. (2013). CAMPFIRE and human-wildlife conflicts in local communities bordering northern Gonarezhou national park, Zimbabwe. *Ecology and Society, 18*(4), 7.

Gandiwa, E., Zisadza-Gandiwa, P., Muboko, N., Libombo, E., Mashapa, C., & Gwazani, R. (2014). Local people's knowledge and perceptions of wildlife conservation in south eastern Zimbabwe. *Journal of Environmental Protection, 5*, 475–481.

Ghosal, S., Skogen, K., & Krishnan, S. (2015). Locating human-wildlife interactions: Landscapes contestations and responses to large carnivore conservation in India and Norway. *Conservation and Society, 13*(3), 265–274.

Kahuni, T., Zisadza-Gandiwa, P., Mhlanga, W., Libombo, E., Mashapa, C., Muboko, N., & Gandiwa, E. (2014). Are fences effective in combating human-wildlife conflicts around protected areas in Zimbabwe? In C. G. Mararike (Ed.), *Land: An empowerment asset for Africa: the human factor perspective* (1st ed., pp. 278–287). University of Zimbabwe Publications.

Khumalo, K. E., & Yung, L. A. (2015). Women, human-wildlife conflict and CBNRM. Hidden impacts and vulnerabilities in Kwanda Conservancy, Namibia. *Conservation and Society, 13*(3), 232–243.

Makumbe, P., Mapurazi, J., Jaravani, S., & Matsilele, I. (2022). Human-wildlife conflict in Save Valley Conservancy: Residents' attitude toward wildlife conservation. *Scientifica*. https://doi.org/10.1155/2022/2107711

Matseketsa, G., Muboko, N., Gandiwa, E., Kombora, D. M., & Chibememe, G. (2019). An assessment of human-wildlife conflicts in local communities bordering the western part of Save Valley conservancy, Zimbabwe. *Global Ecology and Conservation, 20*, e0007, 37. https://doi.org/10.1016/j-gecco.2019.eoo737

Mbaiwa, J., & Hambira, W. L. (2020). Enclave and shadow state tourism in the Okavango Delta, Botswana. *South African Geographical Journal, 102*(1), 1–21.

Mbaiwa, J. E., & Stronza, A. (2011). Changes in resident attitudes toward tourism development and conservation in the Okavango Delta, Botswana. *Journal of Environmental Management, 92*(8), 1950–1959.

Mckenzie, F. J., Karanth, K. K., & Weinthal, E. (2018). Compensation as a policy for mitigating human-wildlife conflict around four protected areas in Rajasthan, India. *Conservation and Society, 16*(3), 305–319.

Merkebu, S., & Yazezew, D. (2021). Assessment of HWC and the attitude of local communities towards wild animal conservation and Borena Sayint national park. *Ethiopia. International Journal of Ecology*, 1–8. https://doi.org/10.1155/2021/66197-57

Moswete, N., Thapa, B., & Darley, W. K. (2020). Local communities' attitude and support of the Kgalagadi Transfrontier Park in southwest Botswana. *Sustainability, 12*(4), 15–24.

Moyo, S., & Tichaawa, T. M. (2017). Community involvement and participation in tourism development. A Zimbabwe study. *African Journal of Hospitality, Tourism and Leisure, 6*(1), 1–15.

Mudimba, T., & Tichaawa, T. M. (2019). Perceptions of local residents and authorities on human-wildlife conflicts in Zimbabwe. *African Journal of Hospitality, Tourism and Leisure, 8*(4), 1–17.

Mukeka, J. M., Ogutu, J. O., Kanga, E., & Roskaft, E. (2019). Human-wildlife conflicts and their correlates in Narok, country, Kenya. *Global Ecology and Conservation, 18*, 1–22. https://doi.org/10.1016/j.gecco.2019.e00620

Mutanga, C. N., Vengesayi, S., Gandiwa, E., & Muboko, N. (2015). Community perceptions of wildlife conservation and tourism: A case study of communities adjacent to four protected areas in Zimbabwe. *Tropical Conservation Science, 8*(2), 564–585.

Nyhus, P. J. (2016). Human-wildlife conflicts and co-existence. *Animal Review of Environmental Resources, 41*, 143–171.

Pour, M. D., Barati, A. A., Azadi, H., & Scheffran, J. (2019). Revealing the role of livelihood assets in livelihood strategies: towards enhancing conservation and livelihood development in Haru Biosphere Reserve in Iran. *Ecological Indicators, 94*, 336–347.

Quirin, C., & Dixon, A. (2012). Food security, politics and perceptions of wildlife damage in Western Ethiopia. *International Journal of Pest Management, 58*, 101–114.

Ravenelle, J., & Nyhus, P. J. (2017). Global patterns and trends in human-wildlife conflict compensation. *Conservation Biology, 31*(6), 1247–1256.

Redpath, S. M., Young, J., Evelyn, A., Adams, W. M., Sutherland, W. J., Whitehouse, A., & Amar, A. (2013). Understanding and managing conservation conflicts. *Trends in Ecology & Evolution, 28*(2), 100–109.

Robinson, O. C. (2013). Sampling in interview based qualitative research: A theoretical and practical guide. *Qualitative Research in Psychology, 11*, 25–41.

Saldana, J., & Omasta, M. C. (2017). *Qualitative research: Amazing life* (2nd ed.). Sage.

Sene-Harper, A., Matarrita-Cascante, D., & Larson, L. R. (2018). Leveraging local livelihood strategy to support conservation and development in West Africa. *Environmental Development, 29*, 16–28.

Seoraji-Pillai, N., & Pillay, N. (2016). A meta-analysis of human-wildlife conflict: South African and global perspectives. *Sustainability, 9*, 34–24. https://doi.org/10.3390/su9010034

Sharma, P., Chettri, N., & Wangchuk, K. (2021). Human-wildlife conflict in the roof of the world: Understanding multi-dimensional perspectives through a systematic review. *Ecology and Evolution*, 1–18. https://doi.org/10.1002/ece3.7980

Shilongo, S. M., Sam, M., & Simuela, S. (2018). Using incentives as mitigation measures for human-wildlife conflict management in Namibia. *International Journal of Scientific and Research Publications, 8*(11), 677–682.

Spenceley, A. (2005). Nature-based tourism and environmental sustainability in southern Africa. *Journal of Sustainable Tourism, 13*, 136–170.

Tichaawa, T. M., & Lekgau, R. J. (2020). Nature and characteristics of wildlife tourism in the Kgalagadi Trans-frontier park. *Acta Universitatis Danubius Economica, 16*(5), 69–88.

Tondhlana, G., Shackleton, S., & Blignaut, J. (2015). Local institutions, actors and natural resources governance in Kgalagadi Transfrontier park and surrounds, South Africa. *Land Use Policy, 47*, 121–129.

Treves, A. (2009). The human dimensions of conflicts with wildlife around protected areas. In M. J. Manfredo & J. J. Vaske (Eds.), *Wildlife and society: The science of human dimensions* (pp. 214–228). Island Press.

United Nations Environment Programme (UNEP). (2019). *Wildlife, wild livelihoods: Involving communities in sustainable wildlife management and combatting the illegal wildlife trade*. UNEP.

United Nations World Tourism Organisation (UNWTO). (2019). *The economic impacts of global wildlife tourism.*

Wester, P., Mishra, A., Mukherji, A., & Shrestha, R. B. (2019). *The Hindu_kush Himalaya assessment: Mountains, climate change, sustainability and people*. Springer.

Woodroffe, R., Thirgood, S., & Robinowitz, A. (2005). *People and wildlife: Conflict or co-existence?* Cambridge University Press.

Zhou, Z. (2018). The tourism sector: A bright light in Zimbabwe's depressed economic environment. *African Journal of Hospitality, Tourism and Leisure, 7*(1), 1–19.

Zhou, Z. (2019a). Trans-frontier parks as vehicles for tourism development and poverty alleviation. Lessons from southern Africa. *International Journal of Hospitality and Tourism Systems, 12*(2), 67–81.

Zhou, Z. (2019b). Tourism progress in the SADC region: Postcolonial era milestones. In M. Mkono (Ed.), *Positive tourism in Africa* (pp. 137–146). Taylor & Francis.

Zhou, Z. (2023). Human-wildlife tourism conflict: Towards an inclusive wildlife resources utilisation approach in southern Africa. *GeoJournal, 88*, 931–949. https://doi.org/10.1007/s10708-022-10667-8

ZTA. (2019). *Zimbabwe Tourism Authority annual report.*

Chapter 3
Cordon Fences and Wildlife Tourism: Impacts on Wildlife

Moren T. Stone

3.1 Introduction

Foot and Mouth Disease (FMD) is a globally significant infectious disease affecting cloven-hoofed animals. It is most frequently transmitted to cattle from reservoir infections in wildlife such as buffalo, kudu and impala (Jori & Etter, 2016). While it has a limited effect on the production and mortality of cattle, the disease accounts for dramatic revenue losses for export beef producers (Bunn et al., 2022). In southern Africa, FMD control has principally led to the deployment of veterinary cordon fences (VCFs) that isolate wildlife such as the African buffalo (*Syncerus caffer*), known to transmit FMD virus to livestock (Osofsky, 2020). Over the many decades of using fencing as a focus of control, hundreds of thousands, if not millions, of wild animals have died due to their being impeded from undertaking seasonal migrations critical for accessing grazing and water resources (Gadd, 2012; Osofsky, 2019). For many years the fences have taken a massive toll of animals whose traditional migration routes to grazing land and water have been sealed off (Williamson & Williamson, 1984).

The history of veterinary cordon fencing in the region is associated with significant external subsidies going back to the colonial era (Cumming et al. 2015; Taylor & Martin, 1987). One sector (livestock) has been supported at the expense of another (wildlife) (Osofsky, 2019). While this is the case, nature-based tourism now contributes as much or more to the gross domestic product of the Southern African Development Community (SADC) region compared to agriculture (including livestock), fisheries, and forestry combined (Scholes & Biggs 2004). With wildlife having moved into such a prominent position in terms of its contribution to SADC's

M. T. Stone (✉)
Department of Environmental Science, University of Botswana, Gaborone, Botswana
e-mail: stonemt@ub.ac.bw

© The Author(s), under exclusive license to Springer Nature Switzerland AG 2024

L. S. Stone, M. T. Stone (eds.), *Wildlife Tourism Dynamics in Southern Africa*, Geographies of Tourism and Global Change,
https://doi.org/10.1007/978-3-031-57252-4_3

economy through nature-based tourism and associated economic activities, the importance of rethinking a reliance on fencing becomes questionable, and even contested (Osofsky, 2019). What makes the debate on fencing (or not fencing) more contested, is the fact that the global impacts of FMD have been estimated to potentially exceed US$22.5 billion annually (Knight-Jones & Rushton 2013), making this disease arguably the most economically impactful animal disease in the world (Osofsky, 2020). Although the efficacy of fences has long been questioned (Owens & Owens, 1984; Ross, 2003; Osofsky, 2020), fences are still regarded as an essential component of disease control because they provide partial animal diseases control. Indications are that the rate of disease spread is proportional to the amount of animal traffic; fences facilitate this by reducing the spread of diseases (Gadd, 2012; Taylor & Martin, 1987).

In view of this and other factors, the earth is facing the greatest mass extinction in 65 million years, driven by the rapid growth in human consumption and population and related environmental effects, such as global warming, habitat loss, and fragmentation (Cushman & Huettmann, 2010). Therefore, the scientific community has a critical role to play in the urgent task of redirecting human civilization towards a sustainable trajectory. This chapter joins the debate on VCFs and their impacts on wildlife resources and wildlife-based tourism. The chapter does not question the efficacy of VCFs nor the impacts they have on conservation and development, rather, it outlines the context, reviews, and synthesizes literature to assess how VCFs have an impact on wildlife and wildlife-based tourism.

3.2 Veterinary Cordon Fences: Background in Southern Africa

Southern African rangelands support both livestock and a rich diversity of wildlife, with the latter also central to national economies and conservation efforts (Mogotsi et al., 2016). Veterinary cordon fences zigzag across the Southern African countries. They are mainly intended to separate disease-free livestock from infected livestock and their closest wild relatives the Cape buffalo, and to restrict the movement of antelope that could carry diseases of concern (Gadd, 2012). As a result, the physical separation of wildlife and livestock (a pre-requisite for access to the EU beef export market), movement controls and repeated vaccination of cattle herds exposed to wildlife have become basic to FMD disease control strategies (Maree et al., 2014). The veterinary cordon fences of southern Africa have become so structural that few policy makers dare to question their relevance, even though the subsidies that drive them have long had perverse consequences for both the beef and wildlife industries and sustainable rural livelihoods (Perkins & Ringrose, 1996). For Botswana, the British Protectorate Government saw cattle keeping as Botswana's only hope for development (Debenham, 1952) and at the time of independence in 1966 it was the number one beef exporter in Africa (MOD, 1965).

The erection of VCFs (e.g. Kuke Cordon Fence in Botswana in 1954) remained a spatially complex contradiction of policies that on the one hand have led the government to gazette large areas of wilderness for free ranging ungulates and their predators, and on the other the fence is seen to actively develop and expand the EU beef export market through enclosure and fencing for geographic disease control (Perkins, 2020). The Kuke fence caused enormous die-offs of more than half a million of blue wildebeest and red hartebeest in the 1980s, with continuous but largely undocumented losses occurring through further fence related mortality, lethal control of buffalo when they stray out of the FMD zone and the loss of many ungulates and predators through problem animal control (Perkins, 2020). While the veterinary cordon fences target buffaloes, the main vector agent for the FMD, Perkins (2020) argues that the buffalo, a key bulk grazer has consequently been removed from vast tracts of rangeland, with un-quantified impacts on ecosystem functioning and resilience. The policy is therefore one of segregation and surveillance of the buffalo herds.

Wildlife mobility is recognised as critical for the maintenance of the Kalahari as a major wild animal system (DHV, 1980) but it is compromised primarily by veterinary cordon fences for disease control, and also by ranch and other fences, such as those along main roads. Perkins (2020) argues that huge die-offs along these fences occurred in the 1980s with the key populations of blue wildebeest and red hartebeest never recovering. This has created a comparative advantage for eastern African countries as far as wildebeest migration tourism is concerned, as rich and spectacular wildebeest migration are only currently witnessed there. Perkins (2020) further argues that it is not only veterinary cordon fences that are a threat to Botswana's wildlife, but fragmentation has continued with new fences and extensive borehole drilling and ranch provision leading to the large animal biomass over extensive areas of the Kalahari to be increasingly dominated by domestic stock.

3.3 Wildlife Resources and Veterinary Fences Relationship Dynamics

There is a wide variety of wildlife found in southern Africa. These herds of wildlife, more especially herbivores, sometimes accumulate in an area and may number up to 40,000 animals or more (Campbell, 2004). Such species, to mention but a few, include hartebeest, wildebeest, zebras, springbok, impalas, gemsbok and eland, and are the major utilizers of the savannah environments and are said to be most affected by fences (Campbell, 2004). The highlighted species, obtain their moisture from the food they eat and if their grazing rangelands are interfered with, barriers like veterinary fences affect their rangeland dynamics. Of the mentioned species, the wildebeest is less well adapted than the other species and during very dry periods, which are frequent, needs to move to better pastures and surface water (Campbell, 2004), therefore, wildebeest are the widest ranging and it becomes a problem when their rangeland becomes fragmented by the veterinary fences.

Fences have adverse effects on wild mammals at the individual, population, and species levels, and alter community structure and ecosystem productivity; furthermore, VCFs disrupt individual wildlife daily movements and may lead to death by starvation, dehydration or entanglement (Gadd, 2012). The fence problems have brought dilemmas to both Southern African communities and governments on how to secure farmers livelihoods through the beef industry while at the same time safeguarding conservation of wild resources (Stone & Nyaupane, 2018). Fencing is seen as one method of controlling livestock diseases by directly halting host/ pathogen traffic, but inevitably protected areas are perceived as reservoirs of economically extremely harmful diseases that risk a spill-over into financially struggling communities (Reisen, 2010). Ferguson and Hanks (2010) argue that fencing aids the fragmentation of landscapes that surround protected areas and result in impassable barriers to the dispersal of highly mobile species. While these fences often fulfil their intention, both target and non-target species within the system may experience unintended side effects (Smith et al., 2020).

Most exclusion fences have been designed with target species in mind, but these designs can also affect other species (non-target species) which may experience net positive or negative effects (Peters et al., 2018; Smith et al., 2020). Southern Africa has semi-arid environments characterized by erratic rainfall and scattered primary forage production, migratory movements of wildlife are a key survival strategy for large herbivores to track resources over vast areas, but VCFs act as a barrier against wildlife migratory routes and survival movements (Hering et al., 2022). VCFs fragment habitats and shape the environment (Pirie et al., 2017), redirect wildlife to less suitable habitat (Hering et al., 2022), and cause injuries and mortality to wildlife (McKay et al., 2021). McInturff et al. (2020), argue that fences are globally ubiquitous, yet they are often omitted from discussions on anthropogenic impacts. As a result, fences are often framed as a management tool rather than a globally significant ecological feature, and they are a notable omission from efforts to map global infrastructure, including the human footprint (Sanderson et al. 2002).

3.4 Impact of Climate Change and Cordon Fences on Wildlife Resources Dynamics

The last century in the Kalahari system of southern Africa, especially Botswana, has been marked by increased climate desiccation and the substitution of livestock for wild animals over extensive areas (Perkins, 2020). Wildlife mobility was recognised as critical for the maintenance of the Kalahari as a major wild animal system (DHV, 1980) but was compromised primarily by veterinary cordon fences for disease control, but also by ranch and other fences, such as those along main roads (Ferguson & Hanks, 2010). It is a change that could well be duplicated throughout the subcontinent as the interior dries and the loss of water dependent wildlife from extensive areas on the one hand is seized as an opportunity to expand the beef sector and agriculture on the other. Range ecologists predict that agriculture is not going to fare well under the predicted climate change scenarios. For example, Perkins (2020) envisions that cattle-keeping over extensive areas of the Kalahari will not be possible due to heat stress and declining forage quantity and quality. Crops production will likewise be threatened by the high temperatures and extreme moisture stress.

Consequently, the maintenance of wildlife mobility across regional conservation landscapes provides one of the most critical adaptive strategies to climate change albeit the existence of cordon fences. It is therefore important to keep wild animals' corridors open; otherwise the impact of climate change will be worsened by the barriers of cordon fences. While the impact of cordon fences have dismally affected wild animals mobility, climate change predictions are particularly bleak for key strongholds of elephant, buffalo, zebra and other water dependent populations in southern Africa (Perkins, 2020). While others (National Research Council, 2005) are calling for artificial water points (AWPs) to be drilled on river beds and pumped all year round to mimic the 'natural' system, others (Coe et al., 1976) are cautious about the dearth of rainfall in the subcontinent and the direct linear correlation between rainfall and primary biomass. Perkins (2020) concludes by postulating that, even if water is supplied from AWPs, there will be an absolute lack of forage around them. This development will ultimately affect both consumptive and non-consumptive forms of tourism as they are dependent of the availability of wildlife resources.

Hedges and Durant (2014) argue that while VCFs may serve a purpose, they may, on the other hand, alter the ecosystem to the extent of rendering some animal species extinct and reduce their resilience in the face of climate change. Nevertheless, Hambira et al. (2020) are of the view that there is limited research on how the VCFs of Southern Africa contribute to the uncertainty or vulnerability posed by climate change. Perkins (2020) however, theorises that in the long run, due to the impacts of climate change aided by the existence of VCFs fragmentation, cattle-keeping over extensive areas of the Kalahari will not be possible due to heat stress and declining forage quantity and quality. Therefore, he concludes that climate change predictions are particularly bleak for key strongholds of elephant, buffalo, zebra and other water dependent populations in southern Africa.

3.5 The Veterinary Cordon Fences Debate

Foot-and-mouth disease (FMD) endemic regions contain three-quarters of the world's FMD susceptible livestock and most of the world's poor livestock keepers (Knight-Jones et al., 2017). While this is the case, food security in the form of livestock production is an important aspect of many African nations (where poor livestock keepers are found), economically and culturally, locally and nationally, for both subsistence and commercial producers. However, the introduction of VCFs purported to improve food production, do not discriminate between targets and non-targets, and create obstacles for several large wildlife, restricting their movements (Gadd, 2012). The VCFs may not necessarily be internal motivated, but externally. For example, European policies and tariffs have buoyed cattle export from some African countries and have subsidized further fence building; the 1976 Lomé Convention and its successor, the 2000 Cotonou Agreement, gave Botswana preferred trading status for beef, guaranteeing prices 25% higher than the global average (EIA, 2004; Gadd, 2012; Mapitse, 2008). Furthermore, it is the role of donors as indications have shown that donors funded fences under the guise of poverty alleviation and economic development, but evidence indicates that the majority of revenue from beef export is captured by elites (Perkins 1996; Scoones & Wolmer, 2008).

The debate on the usefulness of the erection of VCFs to protect wildlife or people and as a land management tool has been discussed; however, VCFs are viewed as being counterproductive, if one is to consider the division of opinions of interested parties, including scientists and decision-makers (Durant et al., 2015). Those who deny that the fences have a serious effect or feel their benefits (better control of foot-and-mouth disease out-breaks) outweigh other considerations, especially wildlife and wildlife-based tourism, usually prefer not to commit themselves to paper (Campbell, 1981). Those who castigate the fences generally base their conclusions on research carried out on observation without scientific basis or purely on sentiment (Campbell, 1981). Others, like Williamson and Williamson (1984) show that there are factors other than fences involved in population changes and that too little research has been carried out to be dogmatic about the harm fences cause. The authors further argue that the fences have enclosed much of that part of the Kalahari that lies within Botswana, blocking or severely disrupting wildlife movements both out of and into the region, therefore, in times of drought species that are unable to cope with the abnormally arid conditions need to move out of the Kalahari to perennial water sources such as the Okavango and Chobe system (Williamson & Williamson, 1984).

Nevertheless, Williamson and Williamson (1984) are of the view that the perishing of wildlife, especially the wildebeest and zebras, should not be solely blamed on veterinary fences since there are other factors such as the spread of pastoralism which was aided by the sinking of boreholes in otherwise waterless areas, the prevailing droughts, expanding human settlements, the proliferation in hitherto inaccessible areas of new roads and tracks made by mineral prospecting companies, the diversion of Boteti River to supply water for the Orapa diamond mine, the settlement of large numbers of people in the vicinity of mines in previously remote areas, and illegal hunting on a very large scale. The authors call for more research to examine how each of these highlighted developments has affected wildlife resources rather than advancing a blanket argument that blames veterinary fences in solo.

Williamson and Williamson (1984) also criticize the methods used to assess wildlife populations, challenging aerial surveys works that are unavoidably fraught with inaccuracies arising from limited manpower and equipment, therefore, concluding that it would be disingenuous to imply that estimates of large herbivore biomass should be accepted uncritically. For Campbell (1981) views is that the general contention appears to be, remove the fences and Kalahari wildlife populations will increase; but he contends that, even the removal of fences will not save the Kalahari wildlife populations, as it will continue to decrease because the fences are only a small part and not the root cause of the decline. Campbell (2004) further postulates that emphasises that place the blame for declining wildlife populations on the fences is simply using the fences as a whipping-boy, when the real blame should be placed elsewhere. For him, research should establish the impact of the veterinary fences on wildlife populations in relation to other causes. For Bunn et al. (2022) the VCFs have a dual function: they control the movement of wildlife and cattle, impose restrictions on local livelihoods; and they also enable an enhanced form of policing and population surveillance by dividing space into a series of articulated control zones. While others are for the VCFs, others oppose the development, but the essence of the fences does not necessarily stop the transmission of FMD. This is echoed by Knight-Jones et al. (2017) in their argument that when control measures are ineffective, farmers, society and wildlife may experience the burden of control without reducing disease burden.

Fencing can divide populations, prevent re-colonisation and render subpopulations prone to the risks faced by small populations (Gadd, 2012). Coe and Skinner (1993) show how the mammals of East and southern Africa exhibit patterns of distribution that are disjunct (10%), continuous (48%) or restricted (42%) to just one of these two regions. In view of Coe and Skinner (1993) findings, Perkins (2020) posits that many mammal species such as lions in the Ngorogoro Crater are fragmented and suffering from the effects of inbreeding and mutation accumulation. Others (e.g. Cumming et al., 2015) argue that while a primary conservation reason for developing Transfrontier Conservation Areas (TFCAs) is to re-establish ecological processes such as large mammal migrations and historical dispersal routes across environmentally artificial national boundaries, the presence of the VCFs work against this noble idea. Therefore, Perkins (2020) argues that the movement corridors for wildlife and the co-existence of people livestock and wildlife over contiguous landscapes is abhorrence to most disease control veterinarians.

3.6 The Implications of Veterinary Cordon Fences on Wildlife-Based Tourism

Participation in wildlife-based tourism within and around protected areas is seen as a tool to link biodiversity conservation and community livelihoods improvement (Stone & Nyaupane, 2018). While the adoption of tourism as a livelihood option has the potential to improve community livelihoods, cattle farmers bear the brunt of protecting wild animals for the tourism sector. Wildlife resources, in particular buffaloes, continue to frustrate farmers in their quest to sell their cattle to the lucrative European Union (EU) market. The buffalo carries foot and mouth disease (FMD) virus, which is highly contagious and easily transmitted to cattle, resulting in a dilemma for both Southern African communities and governments on how to secure farmers livelihoods through the beef industry while at the same time safeguarding the tourism sector (Stone & Nyaupane, 2018). The FMD outbreak heightened farmers' losses due to the disease killing cattle, and eradication exercises for FMD infected cattle and as a result restricted EU market access as the beef originating from the region could not meet the EU market standards. The introduction of VCFs aided the fragmentation of landscapes that surround protected areas and resulted in impassable barriers to the dispersal of highly mobile species (Ferguson & Hanks, 2010) affecting the tourism industry that is dependent on wildlife resources as main attractions.

Using the same land for both agriculture and tourism fuels the resultant conflicts. Tourism and agriculture both contribute to communities' livelihoods and contributes to a country's GDP and both are equally needed to diversify the economy. The difficulties emanate from how both could be practiced on the same piece of land without creating conflicts or adversities. The importance of conservation and wildlife-based tourism justified the economic rationale for developing TFCAs based on the realization that southern Africa's charismatic large mammal fauna provides a major local and international tourist attraction (Cumming et al., 2015), but the presence of the VCFs impede the TFCAs role. Nature-based tourism is an area in which southern Africa has a high comparative advantage and it contributes as much, if not more, to gross domestic product (GDP) than the livestock industry (Cumming, 2004). The VCFs are in no doubt working against the mandate of TFCAs. Cumming et al. (2015) posit that given southern Africa's long history of investment in fences as a means of separating wild and domestic animals to control disease, it is not surprising that shifting from closed to open landscapes and removing fences is a major issue in implementing TFCAs. The TFCAs are pro-wildlife-based tourism while the VCFs are not.

The VCFs and TFCAs symbolize the complex patterns of land tenure and land use, with overlapping jurisdictions governing both resources and human and animal health. Nevertheless, the move towards a return to open landscapes in Southern Africa (i.e., TFCAs) is challenged as it raises several ecological, social and economic questions. For instance, Cumming et al. (2015) raises key questions: are open landscapes and multispecies systems in arid and semi-arid rangelands likely to result in improved livelihoods and healthier people, healthier wild and domestic animals and healthier ecosystems than the VCFs fragmented ecosystems?, and how feasible is it to establish multispecies systems given present land tenure and land-use systems? This led to Perkins (2020) conclusion that the overwhelming message that we are the engineers of the unfolding environmental disaster and the need to totally overhaul our approach to agriculture and wildlife conservation is unpopular with many, including politicians and farmers who cling desperately to what they know.

3.7 Discussions

Though Botswana was projected to be one of the main beef mainstay industries prior and during independence, in 2015 Botswana was ranked 22nd in terms of cattle inventory in Africa and 73rd in the world (Cook, 2015). Brazil, India, Australia and the US are estimated to account for roughly two thirds of the world's beef exports and Botswana accounts for less than 0.01% (Cook, 2018). Therefore, in line with Perkins (2020), we argue that, in view of this development, there is little doubt that Botswana's comparative advantage globally lies in wildlife and wilderness, a position that climate change is likely to exacerbate, due to the dire consequences it holds for both beef production and rainfed arable farming.

Indifference to wildlife conservation has certain logic when one views the world through the eyes of rural people living at a meagre subsistence level. To such people wildlife is seldom more than a source of free meat and hides and, perhaps, of additional income and it is often a nuisance or a threat to their livelihoods and lives (Williamson & Williamson, 1984). It is also logical for people who see economic development as an overriding concern to accord a low priority to wildlife conservation, as economic benefits from wildlife may be very small compared to those derived, for example, from the cattle and mining industries in Botswana (Williamsons & Williamson, 1984).

The debate went on to devise mitigations that could ameliorate the situation, nevertheless, without contentions. Even if, when there is no programme that could be guaranteed to reverse present trends in wildlife dynamics due to the relationships with veterinary cordon fences, others believe that there are steps that can be taken to improve the situation. Such mitigation measures include a detailed assessment of the ecological status of southern Africa's ecosystem and veterinary cordon fences; formulation of an integrated land-use plan for the southern Africa through which orderly development of the region could occur; pilot projects for the provision of artificial water supplies for wildlife; the development of transfrontier conservation areas; vigorous promotion of the development of wildlife-based industries/tourism, with the object of convincing local people that the wildlife resource is a valuable asset with the potential to contribute significantly to economic development (Williamson & Williamson, 1984). While a significant number of mitigation measures are suggested, scholars like Williamson & Williamson (1984) are still sceptical that no one should make a claim that these suggestions are either comprehensive or definitive, nonetheless, remedial action is needed and increasingly agent.

While Boone and Hobbs (2003) are of the view that the declines in the mobility of people, livestock and wildlife resulting from the increasing fragmentation of rangelands by VCFs are likely to increase land degradation and environmental change, McGahey and McGahey (2008) feels that there is an urgent need for a thorough interdisciplinary investigation of the socio-environmental impacts of VCFs'. Perkins (2020) warns that concepts such as 'rewilding' are likely to become increasingly prevalent in southern Africa, probably more by accident than design, as large tracts of land become unsuitable for human occupation due to the impacts of VCFs and other factors like climate change.

3.8 Conclusion

In conclusion, I borrow from Osofsky (2019) who posits that it is long past the time when we should be recognizing that the potential impacts of animal health interventions focused on livestock may in fact be far from environmentally benign. Terrestrial wildlife migrations, once common, are now rare because of ecosystem fragmentation caused by VCFs in southern Africa, affecting the prospect of wildlife-based tourism. Ecosystem fragmentation has now restricted the practice of wildlife-based tourism in certain geographical areas and prevents the equitable re-distribution of wildlife animals across ecosystems. The fragmentation of ecosystems as a result of VCFs have impacted wild animals migrations; the migrations have declined dramatically in both number and size over many years, and many of those that still occur are believed to be under threat.

There should be renewed global efforts to make wildlife conservation the foundation for broad-based economic development. More generally, these bio-political divisions (having fences or not) are part of a broad contestation of wildlife and livestock value, organised around changing regimes of protected area enclosure and the spacing of human and non-human life. Demolishing fences and increasing habitat connectivity through TFCAs, challenging political territorialisation and ecological fragmentation should continue to be important issues that need to be debated by stakeholders to save wildlife resources that are directly and indirectly affected. Overall, it should be noted that there is no single FMD eradication strategy that is perfect or even appropriate for all circumstances, and thus, epidemiological factors, livestock husbandry systems, stakeholder lobbying, community acceptance, likely cost and other public health policies (e.g. health and environmental) interacting with animal health policy should be considered (FAO, 2002; Mogotsi et al., 2016; OECD, 2012).

References

Boone, R. B., & Hobbs, N. T. (2003). Lines around fragments: Effects of fencing on large herbivores. In N. Allsopp, A.R. Palmer, S. J. Milton, K. P. Kirkman, G. I. H. Kerley, C. R. Hurt, & C. J. Brown (Eds.), *Proceedings of the VIIth international rangelands congress* (pp. 580–591).

Bunn, D., Büscher, B., McHale, M. R., Cadenasso, M. L., Childers, D. L., Pickett, S. T., ... & Swemmer, L. (2022). Golden wildebeest days: Fragmentation and value in South Africa's wildlife economy after apartheid. *Journal of Southern African Studies, 48*(6), 1013–1035.

Campbell, A. (1981). A comment on Kalahari wildlife and the Khukhe fence. *Botswana Notes & Records, 13*(1), 111–118.

Campbell, A. C. (2004). Establishment of Botswana's National Park and game reserve system. *Botswana Notes and Records, 36*, 55–66. http://www.jstor.org/stable/40980360

Coe, M. J., & Skinner, J. D. (1993). Connections, disjunctions and endemism in the eastern and Southern African mammal faunas. *Transactions of the Royal Society of South Africa, 48*, 233–255. https://doi.org/10.1080/00359199309520273

Coe, M., Cumming, D. H. M., & Phillipson, J. (1976). Biomass and production of large African herbivores in relation to rainfall and primary production. *Oecologia, 22*, 341–354.

Cook, R. (2015). *World cattle inventory: Ranking of countries* (FAO). Published on August 2, 2015, 05:51 PM. https://www.drovers.com/article/world-cattle-inventory-ranking-countries-fao

Cook, R. (2018). *World beef exports: Ranking of countries*. Published on April 11, 2019. https://beef2live.com/story-world-beef-exports-ranking-countries-0-106903

Cumming, D. H. M. (2004). Performance of parks in a century of change. In B. Child (Ed.), *Parks in transition: Biodiversity, development and the bottom line* (pp. 105–124). Earthscan.

Cumming, D. H. M., Osofsky, S. A., Atkinson, S. J., & Atkinson, M. W. (2015). Beyond fences: Wildlife, livestock and land use in Southern Africa. In J. Zinsstag, E. Schelling, D. Waltner-Toews, M. Whittaker, & M. Tanner (Eds.), *One health: The theory and practice of integrated health approaches* (pp. 243–257). CABI.

Cushman, S. A., & Huettmann, F. (Eds.). (2010). *Spatial complexity, informatics, and wildlife conservation* (pp. 349–368). Springer.

Debenham, F. (1952, March). The Kalahari Today. *The Geographical Journal, 118*(1), 12–21. http://www.jstor.org/stable/1791232

DHV. (1980). *Countrywide animal and range assessment project*. 7 vols. European Development Fund and Ministry of Commerce and Industry.

Durant, S. M., Becker, M. S., Creel, S., Bashir, S., Dickman, A. J., & Pettorelli, N. (2015). Developing fencing policies for dryland ecosystems. *Journal of Applied Ecology, 52*, 544–551.

EIA. (2004). *A line in the sand: The unsustainable expansion of Botswana's beef industry at the expense of local communities and the Okavango Delta*.

FAO, Food and Agriculture Organization. (2002). *Preparation of foot and mouth disease contingency plans*. FAO Animal Health Manual No. 16.

Ferguson, K., & Hanks, J. (2010). *Fencing Impacts: A review of the environmental, social and economic impacts of game and veterinary fencing in Africa with particular reference to the Great Limpopo and Kavango-Zambezi Transfrontier Conservation Areas. Pretoria, South Africa* (p. 329).

Gadd, M. E. (2012). Barriers, the beef industry and unnatural selection: A review of the impact of veterinary fencing on mammals in Southern Africa. In M. J. Somers & M. W. Hayward (Eds.), *Fencing for conservation: Restriction of evolutionary potential or a riposte to threatening processes?* (pp. 153–186). Springer.

Hambira, W. L., Saarinen, J., & Moses, O. (2020). Climate change policy in a world of uncertainty: Changing environment, knowledge, and tourism in Botswana. *African Geographical Review, 39*(3), 252–266. https://doi.org/10.1080/19376812.2020.1719366

Hedges, S. W. R., & Durant, S. M. (2014). To fence or not to fence. *Science, 344*(6179), 46.

Hering, R., Hauptfleisch, M., Jago, M., Smith, T., Kramer-Schadt, S., Stiegler, J., & Blaum, N. (2022). Don't stop me now: Managed fence gaps could allow migratory ungulates to track dynamic resources and reduce fence related energy loss. *Frontiers in Ecology and Evolution, 10*, 907079.

Jori, F., & Etter, E. (2016). Transmission of foot and mouth disease at the wildlife/livestock interface of the Kruger National Park, South Africa: Can the risk be mitigated? *Preventive Veterinary Medicine, 126*(201), 19–29.

Knight-Jones, T. J. D., & Rushton, J. (2013). The economic impacts of foot and mouth disease – What are they, how big are they and where do they occur? *Preventive Veterinary Medicine, 112*, 161–173.

Knight-Jones, T. J., McLaws, M., & Rushton, J. (2017). Foot-and-mouth disease impact on smallholders-what do we know, what don't we know and how can we find out more? *Transboundary and Emerging Diseases, 64*(4), 1079–1094.

Mapitse, N. (2008). *Botswana's foot-and-mouth disease and beef trade policy*. Institute of Development Studies.

Maree, F. F., Kasanga, C. J., Scott, K. A., Opperman, P. A., Chitray, M., Sangula, A., Sallu, R., Sinkala, Y., Wambura, P. N., King, D. P., Paton, D. J., & Rweyemamu, M. M. (2014). Challenges and prospects for the control of foot and-mouth disease: An African perspective. *Veterinary Medicine: Research and Reports, 2014*(5), 119–138.

McGahey, D., & McGahey, D. J. (2008). *Maintaining opportunism and mobility in drylands: The impact of veterinary cordon fences in Botswana*. Doctoral dissertation, University of Oxford.

McInturff, A., Xu, W., Wilkinson, C. E., Dejid, N., & Brashares, J. S. (2020). Fence ecology: Frameworks for understanding the ecological effects of fences. *BioScience, 70*(11), 971–985.

McKay, T. L., Pigeon, K. E., Larsen, T. A., & Finnegan, L. A. (2021). Close encounters of the fatal kind: Landscape features associated with central mountain caribou mortalities. *Ecology and Evolution, 11*, 2234–2248. https://doi.org/10.1002/ece3.7190

MoD. (1965). *The development of the Bechuanaland economy*. Report of Ministry of Overseas Development Economic Survey Mission (November 1965), Published by the Government of the Republic of Botswana.

Mogotsi, K., Kgosikoma, O. E., & Lubinda, K. F. (2016). Wildlife-livestock interface, veterinary cordon fence damage, lack of protection zones, livestock theft and owner apathy: Complex socio-ecological dynamics in Foot and Mouth disease control in Southern Africa. *Pastoralism, 6*(1), 21.

National Research Council. (2005). *Endangered and threatened species of the Platte River*. National Academies Press.

OECD, Organisation for Economic Co-operation and Development. (2012). *Livestock diseases: Prevention, control and compensation schemes*. OECD Publishing. https://doi.org/10.1787/9789264178762-en

Osofsky, S. A. (2019). The global burden of (how we manage) animal disease: Learning lessons from Southern Africa. *Journal of Wildlife Diseases, 55*(4), 755–757.

Osofsky, S. A. (2020). The global burden of (how we manage) animal disease: learning lessons from Southern Africa. *Journal of Avian Medicine and Surgery, 34*(1), 103–105.

Owens, M., & Owens, D. (1984). *Cry of the Kalahari* (pp. 293–308). Houghton Mifflin Company.

Perkins, J. S. (1996). Botswana: Fencing out the equity issue. Cattleposts and cattle ranching in the Kalahari Desert. *Journal of Arid Environments, 33*, 503–517.

Perkins, J. S. (2020). Take me to the river along the African drought corridor: Adapting to climate change. *Botswana Journal of Agriculture and Applied Sciences, 14*(1), 60–71.

Perkins, J. S., & Ringrose, S. (1996). *Development co-operation objectives and the beef protocol: The case of Botswana: A study of livestock/wildlife/tourism/degradation linkages* (p. 65). Report to Metroeconomica Ltd/EU.

Peters, R., Ripple, W., Wolf, C., Moskwik, M., Carreo, N., Arroyo, G., Ceballos, G., Cordova, A., Dirzo, R., Ehrlich, P., & Flesch, A. (2018). Nature divided, scientists united: US-Mexico Border Wall threatens biodiversity and binational conservation. *BioScience, 68*(10), 740–743.

Pirie, T. J., Thomas, R. L., & Fellowes, M. D. (2017). Game fence presence and permeability influences the local movement and distribution of South African mammals. *African Zoology, 52*, 217–227. https://doi.org/10.1080/15627020.2017.1410074

Reisen, W. (2010). Landscape epidemiology of vector-borne diseases. *Annual Review of Entomology, 55*, 461–483.

Ross, K. (2003). *Okavango: Jewel of the Kalahari*. Struik Publishers.

Sanderson, E. W., Jaiteh, M., Levy, M. A., & Redford, K. H. (2002). The human footprint and the last of the wild. *BioScience, 52*, 891–904.

Scholes, R. J., & Biggs, R. (Eds.). (2004). *Ecosystem services in Southern Africa: A regional assessment*. Council for Scientific and Industrial Research, 76 pp.

Scoones, I., & Wolmer, W. (2008). *Foot-and-mouth disease and market access: Challenges for the beef industry in Southern Africa*. Institute of Development Studies.

Smith, D., King, R., & Allen, B. L. (2020). Impacts of exclusion fencing on target and non-target fauna: A global review. *Biological Reviews, 95*(6), 1590–1606.

Stone, M. T., & Nyaupane, G. P. (2018). Protected areas, wildlife-based community tourism and community livelihoods dynamics: Spiraling up and down of community capitals. *Journal of Sustainable Tourism, 26*(2), 307–324.
Taylor, R. D., & Martin, R. B. (1987). Effects of veterinary fences on wildlife conservation in Zimbabwe. *Environmental Management, 11*, 327–334.
Williamson, D., & Williamson, J. (1984). Botswana's fences and the depletion of Kalahari wildlife. *Oryx, 18*(4), 218–222.

Chapter 4
Community Benefits and Challenges of Wildlife Tourism in the Kgalagadi Transfrontier Park in South Africa

Tembi M. Tichaawa and Refiloe J. Lekgau

4.1 Introduction

Wildlife tourism is an important part of the tourism industry in Sub-Saharan Africa (Mangachena & Pickering, 2021). Most often occurring in protected areas such as national parks, game reserves and Transfrontier parks (or transfrontier conservation areas), wildlife tourism has grown to be one of the leading industries in Sub-Saharan Africa (Holland et al., 2022; Stone & Nyaupane, 2017). This form of tourism is conceptualised and geared to assist in job creation, local economic development, community empowerment, and natural resource conservation in the region (Warnholtz et al., 2022; Stone et al., 2022; Holland et al., 2022; Nyaupane et al., 2022). The rationale behind the use of wildlife tourism to attain these socio-economic and environmental goals stems from the fact that wildlife tourism occurs in communal spaces which are often characterized by their rural landscape and limited livelihood-generating activities (Thondhlana et al., 2015; Dikang & Muchapondwa, 2017; Stone et al., 2020). Furthermore, non-consumptive wildlife tourism encourages community involvement and appreciation for the conservation of natural resources (Stone & Nyaupane, 2017).

As such, the growing popularity and significance of wildlife tourism to Sub-Saharan African host communities has led to a wide exploration of the realities and issues around its socioeconomic impacts, the nature of community involvement participation, and debates around conservation and livelihoods (see Hulke et al., 2022; Moswete al., 2020; Thondhlana et al., 2015; Zhang et al., 2020; Spenceley et al., 2019; Holland et al., 2022). To a large extent, these studies have narrowed their

T. M. Tichaawa (✉) · R. J. Lekgau
School of Tourism & Hospitality, College of Business and Economics, University of Johannesburg, Johannesburg, South Africa
e-mail: tembit@uj.ac.za; rlekgau@uj.ac.za

© The Author(s), under exclusive license to Springer Nature Switzerland AG 2024

L. S. Stone, M. T. Stone (eds.), *Wildlife Tourism Dynamics in Southern Africa*, Geographies of Tourism and Global Change,
https://doi.org/10.1007/978-3-031-57252-4_4

focus to national parks, conservancy areas, and game reserves. To truly understand the nature of wildlife tourism and host community relationships and dynamics, further examination is required of other types of protected areas, such as Transfrontier parks.

In Southern Africa, Transfrontier parks were first established in the early 1990s (Chiutsi & Saarinen, 2017; Barrett, 2013). Also known as peace parks, Transfrontier parks are written to usher in a new conservation paradigm premised on supporting biodiversity, sustainable development, and regional peace and cooperation (Chiutsi & Saarinen, 2019). This was to counter the current pressing (and related) challenges of large-scale biodiversity loss and increasing poverty levels (Dikang & Muchapondwa, 2017). Several scholars have explained that transfrontier parks represent spaces of opportunity to attain a variety of local, national, and international socio-economic and environmental goals (Barrett, 2013; Chiutsi & Saarinen, 2017, 2019; Moswete & Thapa, 2015, 2018). Indeed, Bhatasara et al. (2013) attribute the rise in Transfrontier parks in the region to the global recognition that conservation activities cannot be restricted within geographical locations and that current environmental concerns require international cooperation to manage.

The current chapter focuses on the Kgalagadi Transfrontier Park (KTP), the first in Southern Africa. Shared by Botswana and South Africa, the KTP was formed to maintain ecological diversity and spur tourism and economic development to surrounding communities (Thondlhana et al., 2015). However, one of the major criticisms of Transfrontier parks such as the KTP lies with the actualization of the anticipated benefits, particularly by communities. For instance, Chiutsi and Saarinen (2017) observe that host communities are often marginalized and forgotten in tourism and conservation activities in Transfrontier parks. As such, there is a need for further investigations into wildlife tourism in Transfrontier parks and host communities. This chapter considers the South African side of the KTP (focusing on the Askham locality comprising of the Khomani San and Mier communities) and explores the benefits and challenges of wildlife tourism on these communities. The chapter adopted a qualitative approach and data was collected from key stakeholders in wildlife tourism in the Kalahari region, including park managers, representatives of tourism departments, community-based organisations, tourism businesses and community leaders.

4.2 The KTP in Context

The formal establishment of the KTP was followed by developmental linkages to the surrounding communities, with wildlife tourism identified as a vehicle for addressing such socio-economic and environmental goals (Moswete et al., 2020). this is because the establishment of the park has resulted in the displacement of indigenous populations that resided within the two national parks merged (i.e. the Kalahari Gemsbok National Park in South Africa and Gemsbok National Park and the Mabuasehube game reserve in Botswana) (Peace Parks Foundation, 2019; Thondhlana et al., 2015). In 1937, the then Bechunaland (now Botswana) displaced over 80 farmers when establishing the Gemsbok National Park (Peace Park Foundations, 2019). In later years in South Africa the San community, being the oldest indigenous community in the country was forcibly removed from Kalahari Gembok National Park in the 1970s (Schoon, 2013). In 1971, the Mabuasehube Game Reserve was added to the Gemsbok National Park and while these three conservation areas were separate, the park managers at the time had agreed to collaborate informally to manage the entire Kalahari ecosystem (Schoon, 2013; Peace Parks Foundation, 2019). In 1999, a joint management committee was formed by conservation authorities in Botswana (Department of Wildlife and National Park) and South Africa (SANParks) to formalize the de facto existence of the conservation area which occupies an area of 37,991 km^2 (Moswete & Thapa, 2015). By 2000, the KTP was open for tourists and the management committee established developmental objectives (Moswete & Thapa, 2015). The overarching objective of the KTP is to maintain the ecological integrity of the Kalahari region and subsequent objectives are to increase the conservation profile of the KTP internationally to enhance the tourism potential of both countries and to realise the economic potential of communities (Moswete & Thapa, 2016).

The KTP's objective to realise the economic potential of host communities stems from the exacerbated vulnerability faced by communities in their displacement from the national park and conservation policies which limited livelihood activities (such as hunting). In recognizing this, both countries set out to leverage the park to alleviate the socio-economic conditions prevalent in the adjacent communities through policies and institutional structures (Lekgau & Tichaawa, 2019). Interestingly, the establishment of the KTP coincided with South Africa's transition into a democratic society which resulted in emerging policies that recognized the restoration of land rights (Dikang & Muchapondwa, 2017). Indeed, Dikang and Muchapondwa (2017) explained that the Khomani San and Mier community had, under the land restitution programme, been awarded land inside and outside of the KTP. Further, the two countries also re-established their conservation and community development policies once their independence was gained (at different times). While the South African and Botswana conservation policies that govern conservation and tourism activities in the KTP were unpacked in a previous study (see Lekgau & Tichaawa, 2019), these strategic documents place the community as an important stakeholder in wildlife conservation and emphasize the role of tourism towards local development. Thondhlana et al. (2015) further clarify that policies advocating for community inclusion and participation in KTP are geared toward non-consumptive uses of the natural resources inside the park, the pursuit of livelihood activities in resettlement farms outside the park, and participation in wildlife tourism development and activities.

Wildlife tourism in the KTP features a diversity of wildlife, the natural Kalahari landscape, and community-based tourism ventures. The major drawcard includes both large mammals (such as cheetahs, leopards, brown hyena, silver foxes, black maned lions, gemsbok, springbok and wildebeest) and small mammals such as (Cape fox, aardwolf, black-footed cat and various bird species) (Botswana Tourism Organisation [BTO], 2015). There are a number of tourism infrastructures built, most of which lies on the South African side of the Park (Moswete & Thapa, 2018).

There are numerous activities offered to visitors, including game drives and self-drives. However, owing to the primary mandate being conservation, there is a limit to tourism development inside the Park (Tichaawa & Lekgau, 2020a). Additionally, a previous study revealed wildlife tourism is complemented by cultural tourism activities in the adjacent host communities (Tichaawa & Lekgau, 2020b).

4.3 Methodology

A qualitative research design grounded on the constructivist research paradigm was applied to this study. The constructivist paradigm views reality as consisting of a multiplicity of intangible cognitive constructs emerging from human experiences (Mittwede, 2012). The use of this paradigm enables the generation of insightful and deep experiences of host communities in wildlife tourism spaces. The study was divided into two phases, the first phase being the historical documentary analysis to understand the context in which the park was developed and how the community was integrated into this development. This analysis enabled the design of the second phase of the study which entailed key informant interviews.

The key informants included wildlife tourism stakeholders in the KTP, including park managers, representatives of the tourism department, representatives of conservation agencies involved in the KTP, tourism business owners, and representatives of community-led tourism organisations and community leaders. A mixture of purposive and snowball sampling was used in the collection of primary data. In the interviews, participants were identified according to their involvement in tourism and conservation activities in the park and later asked to recommend other individuals in the Khomani San and Mier communities that would be able to address the research focus. The interviews, which were conducted face-to-face, were exploratory in nature and consisted of open-ended questions, that aimed to explore the relationship between the community and the KTP. In total, 21 interviews were conducted in 2019. Having obtained prior permission, the interview discussions were recorded and transcribed verbatim. Thereafter, the data was loaded onto Atlas. ti software for the generation of codes and grouping of codes to develop the key themes, which are discussed in the following section.

4.4 Results and Discussion

4.4.1 Wildlife Tourism Development in the Askham Community

Wildlife tourism in the KTP has provided several benefits for host communities. The study participants underscore the KTP has been important for the development of wildlife tourism in the area. The interview discussions revealed that the attraction of wildlife tourists to the KTP led to developments in some accommodation facilities as well as other activities that complement the cultural and wildlife features of the area. This is demonstrated by the following quotes from accommodation and wildlife sanctuary owners, respectively:

> For me, the park is still one of the most special spaces there is. There are just bigger spaces that are untouched. That it is a protected area, so for me the park is one of the main [attractions] here, and it's also one of the main reasons that we established this lodge. For me to establish this lodge without the park would be useless.

> Yes, the whole point in starting this sanctuary is to complement the park. Because in the park you can't walk around so I thought this place needs a place where people can come and walk around and I teach them the tracks and plants and stuff like that and I had a very good bushman mentor.

The quotes above illustrate the park has spurred the development of tourism offerings in the region. In addition to the accommodation establishments one community representative contributed '*…we have a camping site, Erin Game Ranch, and most of the traditional walking safaris and cultural safaris in the area can be done by the community [for tourists]*'. Furthermore, the development of wildlife tourism in the region was followed by the establishment of a tourist route to the KTP. A representative of the route provided:

> The Red Dune Route is all the guest houses, for the tourists, from Upington up to the Park. So, they have other activities that the Park does not usually give to the tourists. For example, they can go into the [sand] dunes and have sundowners at the dunes. Or they can do sand dune surfing, and the other one is they can do 4×4-ing in the dunes. And those are the activities that people cannot do in the Park. It's a long way to the Park for the people in South Africa, and even overseas. Just to come to the Park, it's long-distance driving, so they normally sleep 1 or 2 nights outside, then they sleep inside [the Park], then outside again, before they head off to Upington, or via the airplane. So that's other activities that the community is giving the tourists, to experience the Kalahari. (R9)

It is clear from the above quotes that the KTP has contributed greatly to the development of wildlife tourism in the Askham community. These findings support the assertions of several scholars who have concurred the advantage of tourism from protected areas is its dependence on relatively undisturbed and pristine environments (see for example Stone et al., 2022; Black & Cobbinah, 2018; Stone & Nyaupane, 2017). Indeed, Black and Cobbinah (2018) underline natural and cultural spaces attract significant tourist attention. Additionally, these findings suggest that this form of tourism has experienced a boost in economic activities emerging from wildlife tourism, which further extends to include livelihood activities, which are underlined in the following subsection of the chapter.

4.4.2 Livelihood Activities in Askam

One prominent argument for wildlife tourism development in developing countries is that it is not selective on the level of development, but rather able to suit rural regions in countries where there are limited economic activities for communities to sustain their livelihoods. Indeed, studies such as Pour et al. (2018) and Soe and Yeo-Change (2019) find that rural livelihoods are dependent on natural resources found within the vicinity of communities in which the only livelihood activities are hunting, fishing, and farming. Unfortunately, findings from the study show that there are limited economic activities in the Askham community as consumptive hunting is prohibited inside the park and the community members lament the lack of land to farm beyond subsistence levels:

> This field, it's too small for the community. Now you can't even rear more than 200 sheep. Now, this thing, because I'm [the] one who wants this farm to go forward, I don't want to be a small farmer, I want to go big, but I can't because of [the] land.

> The other problem that we have here is this place for us is too small to farm, you understand. So, most of us look up to Kgalagadi [Transfrontier Park], not that we just want to but because Upington is too far. It is about 200km or something. So [to get there] we must get someone to give us a lift because most of us don't have vehicles. And now sometimes you get a lift, and they will tell you 'I can help you from here, but from there I can't help you'. Now you see with that lift, what was the point? Because now I must struggle again for a lift from that side. And most of the businesses in Upington now are full, we look up to the Kgalagadi.

The remoteness of Askham has limited the ability of the community to find employment opportunities outside of the community. As such, the KTP constitutes a major role player in the socio-economic development of the adjacent communities. These findings support the positioning of wildlife tourism as an alternative livelihood strategy in such communities (Holland et al., 2022; Nyaupane et al., 2022; Stone et al., 2020). Some community members had gone further to add that the only income generation opportunities are tied to the Park:

> Our salaries each year is about 2.5 million, going to the small village of Welkom. So I mean take that out, take the park out, you sit with how many families that don't have an income?

> There is going to be another lodge [with] about twenty-eight beds. And [with] this lodge, the community will supply the staffing and we will also supply one supervisor who will sleep there every night. The staff will be ferried in every day and ferried out every day. And obviously, there is a profit share on the lodge. Exactly, how much? I am not sure exactly because that is something the park manager and the regional manager deal with. But they get the majority of the shares, and we get the rest.

The findings presented form a compelling argument for the contribution of the Park towards fostering economic opportunities for the surrounding Askham community. In fact, the quote above focuses on revenue sharing with the community as co-owners of the KTP. Revenue sharing (or profit sharing) is a form of benefit sharing that refers to revenues gained from tourism (through partnerships, concessions, and other revenue-earning opportunities (Spenceley et al., 2019). Benefit sharing is highly contested within protected areas and tourism discourse (see for instance Thondhlana et al., 2015; Spenceley et al., 2019; de Koning & De Beer, 2014). These studies point out that communities adjacent protected areas often receive little, or no profits generated and those that do receive some profits rely on the success of the tourism industry in the protected area. The quote above thereby suggests a similar situation, where the Park retains a majority share of the tourism profits, to be used for conservation activities. Additionally, other participants added that other programmes have contributed to livelihood activities. For instance, one community member mentioned:

> Yes, they also have the EWP [Extended Public Works Programmes] which is integrated into SANParks. So they go out to the communities they gave contract work inside the Park, so they are not just focusing on the SANParks work, they also create jobs for maybe tree cutting, roads and those kinds of works, in conjunction with the government, so they are doing that.

While the findings praise the contributions of the Park towards creating some livelihood opportunities for the community, it must be recognized that this can only occur at a limited scale. The KTP can only grow up to a limited scale to ensure that the ecological integrity of the protected area is not disrupted by tourism activity (Tichaawa & Lekgau, 2020a). This places a limit to tourism development in the region. Further, the centrality of wildlife tourism in the KTP to employment in Askham places the community at great risk. This challenging situation is a common feature in South Africa's protected area landscape as De Koning and De Beer (2014) argued that while nature-based tourism should not replace the diverse livelihoods in rural communities, surrounding lands of protected areas are unsuitable for farming (due to the quality of the land) which result in protected areas (and tourism) being considered central to economic development.

4.4.3 Socio-cultural Contributions of Wildlife Tourism to Askham Communities

Another interesting theme that emerged from the study was the positive socio-cultural contributions of wildlife tourism to the community. In terms of the social impact, the study found that the few wildlife tourism establishments (such as offering accommodation and restaurant) play an important role in supporting social groups within the community. This is illustrated by the following quote taken from an owner of a tourism establishment:

> We do a lot for the little school. We buy books, we buy sweaters and we buy what we can. We don't always do enough but we need somebody to initiate. Because running a business like this, it doesn't always give you time to go find places to support. We had a lady now about 5 weeks ago, that [was] doing some nice work. She was actually pairing up the school principals with business owners. She chose me, so what we [are] going to do is a workshop with the school to try and help them to find better funding for certain things. So if they need something, they can come to the business owners, and we want to sit down all business owners and say 'Listen here, in our small community, this school, we need this and this, we need this paint it, we need to do this' So that we get the community to actually find some money to say 'okay cool, we are going to do this for them, and this'.

Despite the challenges of running tourism establishments in the region, the study found that some tourism business owners have made some efforts to support schools and local churches on several initiatives. This demonstrates a high level of social capital in the community, in that there are strong social structures and networks embedded into the community that allows them to attain positive outcomes, such as social change (Zhang et al., 2020). There could be a sense of responsibility from business owners, owing the socio-economic conditions in the community as the limited economic activities available as well as limited businesses (of any economic sectors) in the area. Additionally, the current study further revealed there to be cultural contributions made by wildlife tourism to the community, which had assisted in the preservation of cultural values. In speaking about the continuation of a long tradition, one community member had this to say:

> So in terms of hunting, the community does have traditional rights, its actually the only community in Africa that has got the right back to hunt traditionally within the park. We're currently working out, understandable protocols on our own and then give it to SANParks, everyone has the right, but only hunting traditionally not commercially. So its not with a rifle, its with a bow and arrow, on foot and not with a vehicle. So here it is signed sealed and delivered, the community have that right. The reason behind hunting not being commercial is the emotional process, cultural process, it's a lot of information that we are not willing to sell to, share with SANParks, and by this stage, they understand fully. Previously they didn't understand, they 'let them come hunt, we want to see them come hunt' but that is actually not the case, the hunting process is a fully traditional process which we don't want tourists on.

Further, the recognition of the KTP as a co-managed protected area must take into account its original inhabitants (i.e. the Khomani San). While forming a small share of the global population, a major share of the world's healthy ecosystem is home to indigenous communities such as the Khomani San (Warnholtz et al., 2022). Warnholtz et al. (2022) contend that there is a strong rooted relationship between the natural ecological system and their sociocultural system. As such, the findings suggest that the KTP management has allowed traditional activities such as hunting and the connection to the veld school to ensure the continuity of the Khomani San culture. This has further been recognised at policy levels, particularly with the KTP (Thondhlana et al., 2015). While wildlife tourism does not directly contribute to the preservation of this traditional activity, the wildlife tourism experience in the Kalahari is one infused with cultural features of the Khomani San and Mier communities (Tichaawa & Lekgau, 2020b), and the attraction of this culture by wildlife tourists provided an added incentive towards cultural conservation. However, despite the interlinkage of wildlife tourism and cultural tourism within the Askham community, the above quote further highlights the community's intention to preserve some cultural practices from tourism. This could be to avoid any form of commodification that could result from tourism's interaction with local culture.

4.4.4 Reach of Wildlife Tourism Benefits

Emerging from the findings is the variation of wildlife tourism benefits to the wider community. The community representatives noted that the communities adjacent to the KTP, on the South African side, benefit differently from wildlife tourism initiatives inside the protected area. Two community representatives summed up:

> Because last year [2018] they built another lodge in Kgalagadi, where they were spilt away from the other ones, the Mata Mata one and Nossob. They also built a lodge there, the Khomani San because they also have a piece of the Kgalagadi for them. But they say it's only the Khomani San who can go that side, if you are not er registered as a Khomani San then you can't go work there. But most, they go 50/50 there in the Kgalagadi, then they tell other story, because there are lot of stories by us here. Kgalagadi [Transfrontier Park] will take their people and Khomani San will take their people. That's how they will do it.

> For the community, the park is not doing as great as the community wants the park to do. That's why, the Welkom community said that the park isn't bringing enough activities, or they are not as much involved in the park as they want to be. So with, with regards to tourism and conservation, there is nothing that I can say the park is ever doing with the community, with our community here.

This represents one of the negative impacts of wildlife tourism, in that the benefits of the sector as well as efforts to involve communities are not being felt by the wider community members (Stone & Nyaupane, 2018). This emerges as a result of the recognized ownership of the protected area, where in the case of the KTP, the Khomani San community had larger interests in the Park and therefore a significant portion of the benefits (relating to profits shared, jobs created, training opportunities) are prioritized to this community. This leads to different levels of affluence in the wider communities. Further, the Khomani San's involvement in wildlife tourism and conservation in the area and the KTP is directed through the community property association (CPA), and this community organisation can effectively use their available resources (i.e, social networks and access to the KTP) to better the poor socio-economic conditions of the Khomani San. One representative of this community explained:

> In terms of capacity building, there are guide training and also NCTA [Northern Cape Tourism Authority] which is responsible for marketing in the provinces. So the NCTA would take our small businesses to Indaba for economic development. So its two folds, here they fund it from economic development and NCTA also transfers marketing skills and access to other markets, like the European markets

The efforts of the Khomani San to spur development for their community has resulted in a clear disparity with other surrounding communities in Askham. Other community members had argued for greater inclusion of the wider community in such initiatives, with one community member stating:

> Because we also have a lot of tourist attractions here in Askham. And the young people here in Askham could also be trained in tourism, like the young people at Andrievale because there are youth camps that are taking place at the Khomani San, and they also involve the Askham youth, but not in a way that we would like it to take place.

4.4.5 Challenges with KTP Management

While recognizing the benefits of wildlife tourism to the community, the study participants further identified challenges to wildlife tourism development which subsequently affects the sector's contribution to the community. Some of the participants highlighted that the Park operated in isolation to the host community. For instance, one business owner lamented "*The problem I would say is that the park is [that] it is busy with its own affairs. Also, they don't really interact too much with the people here and the tragedy is that they did before*". Unfortunately, the separation of protected areas and surrounding host communities is common within Sub-Saharan African communities (see works such as Bhatasara et al., 2013; Chiutsi & Saarinen, 2019; Schoon, 2013). Indeed, the founding conservation ideology was solely concerned with preserving the ecosystem and neglected human needs and societal issues (Stone et al., 2022). Additionally, other participants expressed concern over the effect that change in park management has on tourism development in the KTP and the subsequent consequences on wildlife tourism benefits for the community. The following quote encapsulates such views:

> The Khomani San and the Mier people, we have a lodge inside the KTP, that is what we call the !Xaus lodge. And also, there is another lodge that is going to be built, within the Park in the near future that is going to be called the Nxaba lodge, and from that, we will be able to benefit from tourism. But as it is now we are not really benefitting from the !Xaus lodge, because every time there comes a new management team, they have to start at the beginning, taking time in managing and rewriting the managing structures and so forth. But the financial benefit, also empowering people and capacity building at !Xaus lodge, is a little bit absent.

The findings are consistent with previous works on protected area management and communities (Zhang et al., 2020; Stone et al., 2022; Black & Cobbinah, 2018). Interestingly, this study denotes a change in management affects the progress of tourism development in the KTP, which affects the livelihoods (and other benefits) tied to wildlife tourism. This is a challenge for this community due to the historically contentious relationship between two stakeholder groups over natural resource use, ownership, and participation in wildlife conservation and tourism. While there have been great strides to positively change this relationship (Lekgau & Tichaawa, 2019), the change in park management and processes within the KTP could affect these strides. In fact, Black and Cobbinah (2018) concluded that, often, the costs of conservation and tourism were greater than the benefits, and over time could lead to negative attitudes from community members. As such, the delays in the change of management and the limited capacity-building initiatives represent areas that need to be addressed and prioritized to maintain the working relationship between KTP's wildlife tourism and conservation and the Askham community. It is worth adding that this situation is further exacerbated by the dependence of the community on the KTP for their livelihoods.

Moreover, some of the participants mentioned delayed decision-making as another challenge with the KTP. Two tourism business owners explained:

> They [park managers] must just ask their bosses. It's not a decision that they make now. They just tell us we must now wait because the park manager there, sometimes they can make the decisions if it is not so big, then he makes his own decisions and he say 'now I will come tomorrow, or I will do it in next week', but [only] if its immediate now, you need it

> now. They tell you if it is the small problems we help immediately but if it is the big ones, we just arrange with their bosses and then he comes back and he will tell us.

> There are a lot of processes, that you must run and wait, it's not that you just get up. If you ask for it, then you say no you can go. You must also wait for it.

Indeed, transfrontier parks have long governance structures which have increased with the recognition of local community structures. This suggests that big decisions will require more time as approvals will have to be obtained at various levels with the KTP governance system; this could slow down some development initiatives.

4.5 Conclusion

The chapter explored the benefits and challenges of wildlife tourism to host communities, using the KTP and the Askham community. The findings demonstrate the ability of wildlife tourism in protected areas to spur economic development in surrounding communities. However, while the park had resulted in some tourism development (through the establishment of accommodation facilities, activities, and the formulation of the route) which had a positive effect on the livelihood options, wildlife tourism, and the KTP represent the main employers in the region which leads to an unsustainable dependence. The tourism industry is inherently vulnerable and the 2020 health crisis which led to the halt of all tourism activity undoubtedly destabilised an already vulnerable community. Moreover, the findings highlight that the governance and change management process hinders tourism in the KTP and subsequently the extent to which the community can capitalise on wildlife tourism in the KTP.

This chapter contributes to the few available studies on wildlife tourism in the Kgalagadi Transfrontier Park. Indeed, most of the previous studies have focused on the relationship between the Park and host communities, focusing on institutions and governance structures (Thondhlana et al., 2015; Moswete et al., 2012, 2020), community-based organisations (and/or trusts) (Moswete & Thapa, 2015, 2018), the livelihood interlinkages with natural resources (Thondhlana et al., 2012; Tichaawa & Lekgau, 2020a) and the nature of community participation in wildlife tourism (Lekgau & Tichaawa, 2019, 2021). This study draws attention to wildlife tourism development in Askham (as a policy imperative to assist in economic development) and the benefits and challenges of this form of tourism for the local communities. Further, the interrelationship between conservation, tourism, and community livelihoods play an important role in protected area sustainability (Zhang et al., 2020), and this chapter contributes to understanding this relationship further by considering the intentions of the KTP in its establishment phase and the present contribution of wildlife tourism to the community.

References

Barrett, G. (2013). Markets of exceptionalism: Peace parks in Southern Africa. *Journal of Contemporary African Studies, 31*(3), 457–480. https://doi.org/10.1080/02589001.2013.808876

Bhatasara, S., Nyamwanza, A. M., & Kujinga, K. (2013). Transfrontier parks and development in Southern Africa: The case of the Great Limpopo Transfrontier Park. *Development Southern Africa, 30*(4–5), 629–639. https://doi.org/10.1080/0376835X.2013.837377

Black, R., & Cobbinah, P. B. (2018). Local attitudes towards tourism and conservation in rural Botswana and Rwanda. *Journal of Ecotourism, 17*(1), 79–105. https://doi.org/10.1080/14724049.2016.1258074

Botswana Tourism Organisation (BTO). (2015). *Travel companion: Kgalagadi/Central*. BTO.

Chiutsi, S., & Saarinen, J. (2017). Local participation in transfrontier tourism: Case of Sengwe community in Great Limpopo Transfrontier conservation area, Zimbabwe. *Development Southern Africa, 34*(3), 260–275. https://doi.org/10.1080/0376835X.2016.1259987

Chiutsi, S., & Saarinen, J. (2019). The limits of inclusivity and sustainability in Transfrontier peace parks: Case of Sengwe community in Great Limpopo Transfrontier conservation area, Zimbabwe. *Critical African Studies, 11*(3), 348–360. https://doi.org/10.1080/21681392.2019.1670703

de Koning, M., & De Beer, E. (2014). Land restitution and income generation in protected areas in Mpumalanga, South Africa. *Development in Practice, 24*(1), 118–128. https://doi.org/10.1080/09614524.2014.866214

Dikgang, J., & Muchapondwa, E. (2017). Local communities' valuation of environmental amenities around the Kgalagadi Transfrontier Park in Southern Africa. *Journal of Environmental Economics and Policy, 6*(2), 168–182. https://doi.org/10.1080/21606544.2016.1240631

Holland, K. K., Larson, L. R., Powell, R. B., Holland, W. H., Allen, L., Nabaala, M., Tome, S., Seno, S., & Nampushi, J. (2022). Impacts of tourism on support for conservation, local livelihoods, and community resilience around Maasai Mara National Reserve, Kenya. *Journal of Sustainable Tourism, 30*(11), 2526–2548. https://doi.org/10.1080/09669582.2021.1932927

Hulke, C., Kalvelage, L., Kairu, J., Revilla Diez, J., & Rutina, L. (2022). Navigating through the storm: Conservancies as local institutions for regional resilience in Zambezi, Namibia. *Cambridge Journal of Regions, Economy and Society, 15*(2), 305–322. https://doi.org/10.1093/cjres/rsac001

Lekgau, R. J., & Tichaawa, T. M. (2019). Effects of institutional arrangements and policies on community participation in wildlife tourism in Africa. *Geo Journal of Tourism and Geosites, 27*(4), 1280–1295. https://doi.org/10.30892/gtg.27414-433

Lekgau, R. J., & Tichaawa, T. M. (2021). Community participation in wildlife tourism in the Kgalagadi Transfrontier Park. *Tourism Review International, 25*(2–3), 139–155. https://doi.org/10.3727/154427220X16059054538746

Mangachena, J. R., & Pickering, C. M. (2021). Why are some animals popular with wildlife tourists: Insights from South Africa. *Journal of Ecotourism, 1–17*. https://doi.org/10.1080/14724049.2021.2019261

Mittwede, S. K. (2012). Research paradigms and their use and importance in theological inquiry and education. *Journal of Education and Christian Belief, 16*(1), 23–40. https://doi.org/10.1177/205699711201600104

Moswete, N., & Thapa, B. (2015). Factors that influence support for community-based ecotourism in the rural communities adjacent to the Kgalagadi Transfrontier Park, Botswana. *Journal of Ecotourism, 14*(2–3), 243–263. https://doi.org/10.1080/14724049.2015.1051537

Moswete, N., & Thapa, B. (2016). Factors that influence the support for community-based ecotourism in the rural communities adjacent to the Kgalagadi Transfrontier Park, Botswana. *Journal of Ecotourism, 12*(2–3), 243–263. https://doi.org/10.1080/14724049.2015.1051537

Moswete, N., & Thapa, B. (2018). Local communities, CBOs/trusts, and people–park relationships: A case study of the Kgalagadi Transfrontier Park, Botswana. *The George Wright Forum, 35*(1), 96–108. https://doi.org/10.3390/su12041524
Moswete, N. N., Thapa, B., & Child, B. (2012). Attitudes and opinions of local and national public sector stakeholders towards Kgalagadi Transfrontier Park, Botswana. *International Journal of Sustainable Development & World Ecology, 19*(1), 67–80. https://doi.org/10.1080/13504509.2011.592551
Moswete, N., Thapa, B., Darley, K., & W. (2020). Local communities' attitudes and support towards the Kgalagadi Transfrontier Park in Southwest Botswana. *Sustainability, 12*(4), 1524. https://doi.org/10.3390/su12041524
Nyaupane, G. P., Poudel, S., & York, A. (2022). Governance of protected areas: An institutional analysis of conservation, community livelihood, and tourism outcomes. *Journal of Sustainable Tourism, 30*(11), 2686–2705. https://doi.org/10.1080/09669582.2020.1858089
Peace Parks Foundation (2019). *Kgalagadi*. Available from: https://www.peaceparks.org/tfcas/kgalagadi/ (Accessed 18 August 2019).
Pour, M. D., Barati, A. A., Azadi, H., & Scheffran, J. (2018). Revealing the role of livelihood assets in livelihood strategies: Towards enhancing conservation and livelihood development in Hara Biosphere Reserve in Iran. *Ecological Indicators, 94*, 336–347. https://doi.org/10.1016/j.ecolind.2018.05.074
Schoon, M. (2013). Governance in transboundary conservation: How institutional structure and path dependence matter. *Conservation and Society, 11*(4), 420–428. https://doi.org/10.4103/0972-4923.125758
Soe, K. T., & Yeo-Chang, Y. (2019). Perceptions of forest-dependent communities towards participation in forest conservation: A case study in Bago Yoma, South Central Myanmar. *Forest Policy and Economics, 100*, 129–141. https://doi.org/10.1016/j.forpol.2018.11.009
Spenceley, A., Snyman, S., & Rylance, A. (2019). Revenue sharing from tourism in terrestrial African protected areas. *Journal of Sustainable Tourism, 27*(6), 720–734. https://doi.org/10.1080/09669582.2017.1401632
Stone, M. T., & Nyaupane, G. P. (2017). Ecotourism influence on community needs and the functions of protected areas: A systems thinking approach. *Journal of Ecotourism, 16*(3), 222–246. https://doi.org/10.1080/14724049.2016.1221959
Stone, M. T., & Nyaupane, G. P. (2018). Protected areas, wildlife-based community tourism and community livelihoods dynamics: Spiraling up and down of community capitals. *Journal of Sustainable Tourism, 26*(2), 307–324. https://doi.org/10.1080/09669582.2017.1349774
Stone, M. T., Lenao, M., & Moswete, N. (2020). *Natural resources, tourism and community livelihoods in Southern Africa: Challenges of sustainable development*. Routledge.
Stone, M. T., Stone, L. S., & Nyaupane, G. P. (2022). Theorizing and contextualizing protected areas, tourism and community livelihoods linkages. *Journal of Sustainable Tourism, 30*(11), 2495–2509. https://doi.org/10.1080/09669582.2021.2003371
Thondhlana, G., Vedeld, P., & Shackleton, S. (2012). Natural resource use, income and dependence among San and Mier communities bordering Kgalagadi Transfrontier Park, southern Kalahari, South Africa. *International Journal of Sustainable Development & World Ecology, 19*(5), 460–470. https://doi.org/10.1080/13504509.2012.708908
Thondhlana, G., Shackleton, S., & Blignaut, J. (2015). Local institutions, actors, and natural resource governance in Kgalagadi Transfrontier Park and surrounds, South Africa. *Land Use Policy, 47*, 121–129. https://doi.org/10.1016/j.landusepol.2015.03.013
Tichaawa, T. M., & Lekgau, R. J. (2020a). Leveraging wildlife tourism for employment generation and sustainable livelihoods: The case of the Kgalagadi Transfrontier Park, Southern Africa. *Bulletin of Geography. Socio-economic Series, 49*, 93–108. https://doi.org/10.2478/bog-2020-0026
Tichaawa, T., & Lekgau, R. J. (2020b). Nature and characteristics of wildlife tourism in the Kgalagadi Transfrontier Park. *Acta Universitatis Danubius. Œconomica, 16*(5) https://dj.univ-danubius.ro/index.php/AUDOE/article/view/303

Warnholtz, G., Ormerod, N., & Cooper, C. (2022). The use of tourism as a social intervention in indigenous communities to support the conservation of natural protected areas in Mexico. *Journal of Sustainable Tourism, 30*(11), 2649–2664. https://doi.org/10.1080/09669582.2020.1860069

Zhang, Y., Xiao, X., Zheng, C., Xue, L., Guo, Y., & Wu, Q. (2020). Is tourism participation in protected areas the best livelihood strategy from the perspective of community development and environmental protection? *Journal of Sustainable Tourism, 28*(4), 587–605. https://doi.org/10.1080/09669582.2019.1691566

Chapter 5
Nature-based Tourism in Zimbabwe: Sustainability Issues

Ngoni C. Shereni and Michael Chambwe

5.1 Introduction

There is a strong relationship between tourism and natural areas conservation as evidenced by the increasing demand to engage with wildlife in conservation areas (Wolf et al., 2019). This gives rise to nature-based tourism, a growing trend, which is characterised by visitation to natural areas to enjoy wildlife and natural attractions which in most cases contributes to biodiversity conservation and benefits host populations (Acquah et al., 2016; Spenceley, 2003). The engagement with wildlife in nature-based tourism destinations may include consumptive and non-consumptive activities (Thapa et al., 2022; Wolf et al., 2019). Globally, nature-based tourism is a significant sector of the tourism industry attracting visitors to mostly protected areas (Winter et al., 2020). It is estimated that in 2018 nature-based tourism contributed US$ 120 billion to global GDP, accounting for 4.4% of direct tourism GDP and providing 9.1 million jobs (WTTC, 2019).

Tourism in the Global South plays a crucial role in facilitating development, creating employment, contributing to poverty reduction, and promoting sustainable economic growth and social inclusiveness (Mushawemhuka et al., 2018; UNWTO, 2015). Over half of international tourist arrivals in Africa are for leisure purposes and a significant number of these arrivals end up engaging in nature-based tourism (UNWTO, 2015). Nature-based tourism is a priority sub-sector of the tourism industry in most African countries endowed with varied wildlife species (Hambira et al., 2022; Spenceley, 2006). East and Southern Africa are regarded as leading

N. C. Shereni (✉)
School of Tourism & Hospitality, University of Johannesburg, Johannesburg, South Africa

M. Chambwe
Tourism Research in Economics, Environs and Society (TREES), North West University, Potchefstroom, South Africa

© The Author(s), under exclusive license to Springer Nature Switzerland AG 2024

L. S. Stone, M. T. Stone (eds.), *Wildlife Tourism Dynamics in Southern Africa*, Geographies of Tourism and Global Change,
https://doi.org/10.1007/978-3-031-57252-4_5

destinations for nature-based tourism in Africa and the rest of the world (Price, 2017). Consequently, a number of countries in southern Africa have identified nature-based tourism as a vehicle for spearheading sustainable rural development (Tervo-Kankare et al., 2018). Countries in Southern Africa like Zimbabwe, South Africa, Namibia, Botswana, Zambia and Tanzania have come up with various institutional frameworks to derive benefits from their viable nature-based tourism products (Gronau et al., 2017, Stone, 2006).

Undoubtedly, nature-based tourism contributes to the sustainability discourse by providing benefits to individuals, the community and the society (Segobye et al., 2022; Winter et al., 2020). Historically, various collaborative governance strategies have been conceptualised in communities settled around protected areas to ensure that they derive benefits from nature-based tourism (DeGeorges & Reilly, 2009). These strategies are commonly known as Community-Based Natural Resource Management practices (CBNRM). In sub-Saharan Africa, nature-based tourism and CBNRM programmes have been integrated to drive conservation and economic development goals (Gronau et al., 2017). The Communal Areas Management Programme For Indigenous Resources (CAMPFIRE) is considered a pioneer programme in Southern Africa and has resulted in communities collaborating with various stakeholders to manage resources within their localities (DeGeorges & Reilly, 2009; Shereni & Saarinen, 2021). This programme has been replicated in other different contexts with various levels of success within Southern Africa (Dube, 2019; Hoole & Berkes, 2010; Musavengane & Simatele, 2016).

Due to the changing dynamics in nature-based tourism and changes in the socio-economic conditions in Zimbabwe in the last two decades, CAMPFIRE has been facing operational constraints (Shereni & Saarinen, 2021). Local communities are no longer benefitting as before and the programme has become irrelevant to the communities (Dube, 2019). There is now a greater need for the re-evaluation of the programme so that it remains relevant both as a tool for sustainable rural development and for the achievement of conservation goals. In this chapter, we aim to interrogate CAMPFIRE as a conservation programme, overview its historical development as well as highlight its successes over time. The specific objectives of this chapter are to; provide a historical development of the CAMPFIRE programme; evaluate CAMPFIRE development in Zimbabwe; Situate CAMPFIRE in current wildlife conservation dynamics and highlight lessons learnt over time and make recommendations on how to improve nature-based tourism and community participation in Zimbabwe.

5.2 CAMPFIRE Development in Zimbabwe

Just like in many African countries, the mandate of preserving wildlife and protected areas in Zimbabwe has historically been the prerogative of the country's Wildlife Conservation Authority, the Department of National Parks and Wildlife Management (DNPWM) (Logan & Moseley, 2002). The initial approaches adopted (pre-CAMPFIRE) for conservation meant that only the state would regulate the legal exploitation of wildlife; this resulted in the exclusion of indigenous communities as they were dually denied access to wildlife resources and the land, all of which they had access to prior to the circumstances of colonial pillage (Cumming, 1989). The confinement of indigenous communities to communal lands resulted in increased land pressure for subsistence agriculture and settlements. Limited access to land and wildlife changed the local perspectives on the usefulness of wildlife in their communities, which when not hunted illegally, became a nuisance and liability to their livelihoods (Frost & Bond, 2008; Hasler, 1999).

In line with the submissions of Child (1985), the Wildlife Conservation Act of 1960 set the tone for new wildlife management strategies that were based on economic incentivisation which then gave an impetus to the passing of the Parks and Wildlife Act of 1975. This Act furnished landowners with the right to usage and management of wildlife within their boundaries for economic benefit. While this arguably provided a strong rationale for wildlife conservation to the landowners (mostly commercial farmers and ranch owners), this policy remained unattractive to indigenous inhabitants of communal lands (Child, 1985; Martin et al., 1977).

According to Martin et al. (1977), the Wildlife Industries New Development for All (WINDFALL) project was the first genuine attempt to incorporate indigenous communities in wildlife management and conservation as it allowed for the provision of economic benefits from the wildlife to communities. The project's objectives were to reduce human-wildlife conflicts by changing the locals' attitude towards wildlife conservation through sharing of revenue from wildlife activities. This project, however, did not register much success due to a myriad of reasons. It was not based on wildlife found on communal land but on those found on state land, the locals were excluded from decision making, it was based on state land, and revenues generated from wildlife seldom found their way to communities (Martin, 1986; Martin et al., 1977; Thomas, 1991). As such, WINDFALL could not sustain the link between economic benefits and wildlife conservation, which the project proponents had discovered to be critical.

The failure of WINDFALL as a strategy to conserve wildlife with the support and buy-in of the communities, coupled with the government's strong resolve for localised planning and management of indigenous resources, propelled the DNPWM (with the help and funding of USAID, the World Wildlife Fund, the University of Zimbabwe Centre for Applied Social Science, and a non-governmental organisation, Zimbabwe Trust) to invent the CAMPFIRE programme that gave rural communities full ownership, control, and management of their wildlife (Logan & Moseley, 2002; Muboko & Murindagomo, 2014). The programme was premised on the idea that environmental conservation would have higher chances of success when local communities living around wildlife sanctuaries were given access to exploit such natural resources, albeit sustainably, for their benefit (Child, 1985; Cumming, 1989; Mberengwa, 2002). Since communal lands were mostly semi-arid and unfit for commercial agriculture yet they had free roaming wildlife, the programme sought to encourage tourism within these areas by funding wildlife ecology research, and wildlife enterprises that allowed for trophy hunting (Mberengwa, 2002; Muboko & Murindagomo, 2014; Murombedzi, 1997).

With the CAMPFIRE programme (the pioneering community-based wildlife conservation initiative to view wildlife as a sustainable, lucrative resource), communities were empowered to benefit materially from the economic proceeds of wildlife enterprises as profits would be distributed among individual households and respective communities would dictate (Muboko & Murindagomo, 2014). Funds received from these enterprises would, on the one hand, fund the conservation of wildlife and research facilities, and on the other, facilitate the development of communities through the construction and refurbishment of various infrastructure such as schools, clinics, and boreholes (Cumming, 1989; Frost & Bond, 2008; Mutandwa & Gadzirayi, 2007).

5.3 CAMPFIRE Success Story

Most studies on CAMPFIRE in the last two decades focused more on the weaknesses and failures of CAMPFIRE. Recent literature seems to suggest that CAMPFIRE failed to deliver on its key objectives of conservation and poverty alleviation. However, there is no doubt that many ecological and socio-economic benefits have been reaped over the years in the conservation of natural resources and in uplifting the well-being of communities through CAMPFIRE (Chigonda, 2018). In its formative years, CAMPFIRE was hailed for replacing state-centred conservation models which dismally failed to integrate communities with wildlife resources (Balint & Mashinya, 2008). Before CAMPFIRE was initiated, a legally recognised institutional framework to involve local communities in the management and benefits of natural resources within their communities didn't exist.

It is reported that safari hunting, which is the major revenue earner for CAMPFIRE, saw an increase in revenue from USD350 000 in 1989 to USD two million in 2001 earning producer communities an estimated USD20.3 million (Child et al., 2003; Roe & Nelson, 2009). Revenue earned in CAMPFIRE areas made it easy to fund conservation efforts and provide direct benefits to communities. CAMPFIRE ensured the sustainable use of resources by communities and raised income through activities such as trophy hunting, selling live animals, tourism, selling game meat and community-based beekeeping (Stone, 2006). The benefits brought about through CAMPFIRE made communities see wildlife as a resource they can manage at a local level and benefit from. Consequently, this led to reduced human-wildlife conflict as communities made concerted efforts to protect and conserve wildlife, especially from the menace of poachers (Gandiwa et al., 2013). In a sense, CAMPFIRE motivated conservation efforts by the local communities as they could see direct benefits accruing at both the household and community level.

CAMPFIRE is credited for increases in local income which also led to improvements in the standard of living (Mutandwa & Gadzirayi, 2007). The programme provided people in rural areas who ordinarily have limited economic means with an opportunity to monetise their wildlife resources for their benefit (Tichaawa & Mhlanga, 2015). Some of the benefits accrued at the community level from revenue earned through CAMPFIRE, including the sinking of boreholes, the construction of schools and clinics, the purchasing of grinding mills and provision of game meat to communities in cases where the trophy carcass is edible (Chigonda, 2018; Zunza, 2012). CAMPFIRE presents employment opportunities to locals as game rangers, guides, cooks and skinners among others (Zunza, 2012).

A look at specific areas where CAMPFIRE was initiated shows varied levels of success. In Mahenye, which borders Gonarezhou National Park, CAMPFIRE is rated for infrastructural development in the community, creating employment opportunities, providing cash dividends to households, electrification of homes, provision of piped water for the community, construction of schools, grinding mills and a donation of three lodges from a key stakeholder (Gandiwa et al., 2013; Zunza, 2012). In communities adjacent to Hwange National Park, CAMPFIRE is credited for the construction of a community hall, funding income generating projects for locals, the provision of game meat for the community and drilling boreholes among other benefits (Shereni & Saarinen, 2021).

5.4 CAMPFIRE in Current Wildlife Conservation Dynamics

The three principal tenets upon which CAMPFIRE gained regional and international acclaim have been its ability to: (i) convince communities to perceive wildlife as both an agricultural resource and a form of agriculture; (ii) defuse land-use tensions between wildlife conservation and arable agriculture; and (iii) create harmony between game management and communal subsistence agriculture in a complementary fashion that brings revenue to communities (Vincent et al., 2022). Nonetheless, the need for efficiency and effectiveness in community nature-based conservation is continually under threat by rising elements such as expanding human land use, climate change, invasive species, unmet conservation goals, and social, political and economic limitations (Biggs et al., 2019).

The most contentious issue in conservation policy is undoubtedly how national parks, wildlife sanctuaries, and other protected areas affect the people who live nearby (Gandiwa et al., 2014; Muboko & Murindagomo, 2014). Regrettably, forced eviction, terror, torture, and denial of access to resources from protected regions are still commonplace in many communities (Biggs et al., 2019; Chigonda, 2018; Ntuli & Muchapondwa, 2018). Protected areas are increasingly burdened by rising levels of poverty in developing economies, where for their survival, they must be a land use option, just like others, that is compatible with sustainable development (Nthiga et al., 2015).

Conflicts between people and wildlife are another issue that local populations around protected areas continue to grapple with. Humans are especially prone to conflict with elephants, large carnivores, ungulates, primates, and other large mammals because of these animals' extensive home-range requirements, which force them to compete with people for scarce space and resources (Hoffmann, 2022). This significantly hurts people and destroys their crops and livestock, which worsens the communities' poverty profiles (Mutanga et al., 2017). Along with increasing competition for grazing and expenses of missed opportunities, animals also spread diseases like foot and mouth, anthrax, and rabies to both humans and cattle (Chigonda, 2018; Ntuli & Muchapondwa, 2018). Such human-wildlife conflicts often lead to poaching and retaliatory killings which threaten biodiversity conservation (Hoffmann, 2022). Although recent studies have revealed that fencing has ecological costs due to the obstruction of migration pathways and restriction of biodiversity range use, which may result in overstock, inbreeding, and isolation, fencing may be useful in minimizing human-wildlife conflicts (Tchakatumba et al., 2019; Vincent et al., 2022).

The distribution of expenses between those who gain from conservation and those who must bear the costs of damage may be accomplished using damage compensation (Mutanga et al., 2017; Ntuli & Muchapondwa, 2018). Ex-post compensation, where the harm is reimbursed after it has occurred, and compensation in advance, where payments are based on an estimate of the projected loss and are received independently from when the loss occurs, are the two main types of compensation arrangements (Ntuli & Muchapondwa, 2018). In situations of extreme poverty, Chigonda (2018) contends that compensation is generally inexpensive to administer and is widely accepted by the local populace.

The path to socially just conservation is still rocky and littered with challenges. If significant outcomes are to be realised on the ground, additional effort is required. Conservationists need to understand that isolated protected areas cannot endure in the face of extreme poverty without being seen to be making a difference in the lives of the inhabitants. This is because local communities close to national parks and protected areas are increasingly seen as playing a significant role in achieving conservation and sustainability goals because of their ongoing interactions with the environment through resource use, recreational opportunities, and familiarity with the area.

5.5 Lessons Learnt and Recommendations

The failures and success of CAMPFIRE brought to the fore the realisation that if any CBNRM programme is to achieve the desired goals authority should be devolved to local communities (Balint & Mashinya, 2008). The current model where the Rural District Councils have more authority to run the programme over the local communities has proven to be faulty (Shereni & Saarinen, 2021). The RDCs' reluctance to devolve authority to the local communities and lack of transparency in revenue sharing has resulted in numerous conflicts and distrust (Muboko & Murindagomo, 2014; Musavengane & Simatele, 2016). Stakeholder trust is crucial in the successful implementation of a CBNRM programme. Therefore, there is a need to remodel CAMPFIRE so that full authority is devolved to the producer communities and they take full control of the programme (Ntuli & Muchapondwa, 2017). In addition, there is need to improve transparency and build consensus among the key stakeholders involved in CAMPFIRE, especially the local communities (Chiutsi & Saarinen, 2017).

Without capacitating the elected local committees, the success of CAMPFIRE cannot be guaranteed. The local committees have a huge responsibility of managing the project and their responsibilities span to include *inter alia* making decisions on behalf of the community, engaging with various stakeholders, managing revenues as well as overseeing community development initiatives (Balint & Mashinya, 2008). Such responsibilities require a certain set of skills in project management. It has been noted that the composition of local committees largely lacks the capacity to manage projects of such magnitude (Shereni & Saarinen, 2021). There is, therefore a need to ensure that community representatives are capacitated through training programmes to make them active and equal partners in the running of the programme (Saarinen & Lenao, 2014).

The influence of traditional and political elites is indeed a major threat to the success of CAMPFIRE as a programme (Balint & Mashinya, 2008). This is especially notable when people without a direct interest in the community are put at the forefront of a project like CAMPFIRE, they end up making decisions that benefit themselves (Dube, 2019). This calls for a stronger institutional framework that sanctions the rights and gives full authority to local communities. One of the misgivings of CAMPFIRE is the lack of a legal framework to enforce administrative guidelines especially when it comes to revenue sharing (Muboko & Murindagomo, 2014). Musavengane and Simatele (2016) argued that specific legislations and policies are important in supporting community participation in CBNRM. Without strong institutions or supportive legislation and policies, CBNRM projects become susceptible to capture by a few privileged individuals who then subordinate the interests of the community members.

The end of donor funding towards CAMPFIRE from the year 2000 exposed the operational deficiencies of the programme (Gandiwa et al., 2013). The government has also failed to fund conservation efforts in CAMPFIRE areas without assistance from Non-Governmental Organisations (Chiutsi & Saarinen, 2017). This ignited debates on the ability of CAMPFIRE to sustain itself without external support. In as much as external support is important for the success of a CBNRM programme, it is integral to ensure that the project sustains itself. In this case, it will be ideal to invest in CBT projects that can be properly run by communities and generate benefits for them. In addition, the sustainability of CAMPFIRE can be ensured by establishing business partnerships between communities and private business entities (Stone, 2006).

Increased human-wildlife conflicts have been recorded in areas where the wildlife population increased due to conservation efforts (Gandiwa et al., 2013). This defeats the essence of the CAMPFIRE programme which seeks to link the costs and benefits derived from natural resources. One of the fundamental principles of CAMPFIRE is to encourage people to conserve natural resources like wildlife and in return get compensated for their troubles from the wildlife (Dube, 2019). However, this assumption that for as long as communities get compensated they will tolerate destruction caused by wildlife is not only flawed but is over simplistic (Balint & Mashinya, 2008). Compensation needs to exceed the cost of damage and tangible benefits should accrue to households otherwise the CAMPFIRE programme will suffer from negative perceptions (Muboko & Murindagomo, 2014). More so, there is need to have an appreciation of the heterogeneous nature of different communities as many might have different priorities that conflict with the goals of CAMPFIRE (Balint & Mashinya, 2008).

The land reform programme initiated in the year 2000 changed the land use patterns in previous wildlife areas (Mutandwa & Gadzirayi, 2007). Some conservancies that used to be at the centre of CAMPFIRE are now occupied by resettled farmers (Shereni & Saarinen, 2021). The implication of this is that land use in CAMPFIRE areas is now a mixture of wildlife production and subsistence agriculture. This brings about competing interests and divergent goals as a result of different priorities emanating from the different land use patterns (Musavengane & Kloppers, 2020). CAMPFIRE in those areas needs to be remodelled to suit the need of the resettled farmers as well as communities that have historically depended on wildlife for survival.

Recent studies have shown that local communities hold negative perceptions towards CAMPFIRE (Dube, 2019). This stems from diminishing benefits emanating from the programme as well as CAMPFIRE's failure to live up to its potential of alleviating poverty within communities. In most cases for communities settled adjacent to protected areas, the costs outweigh the benefits of CAMPFIRE (Matseketsa et al., 2018). Tichaawa and Mhlanga (2015) argue that to change this perception the programme needs to impact the economic circumstances of the villagers and also ensure that they are compensated fairly and equitably for wildlife damage. There is also a need to ensure active participation by community members in CAMPFIRE projects and the drawing of clear guidelines that define the responsibilities of all the stakeholders involved (Chiutsi & Saarinen, 2017).

5.6 Conclusion

This chapter has traced the CAMPFIRE programme's historical development, achievements, constraints, and challenges, situated it in the current wildlife conservation dynamics and has brought forward recommendations for future practice. Imperatively, the following conclusions can be drawn. The development of CAMPIRE demonstrates a concerted effort by Zimbabwe's DNPWM, over time to come up with conservation policies that were relevant to the circumstances of the times—in this case, the support and buy-in of local communities proved paramount. As the needs and situation of communities continually change, the historical review of CAMPFIRE development in Zimbabwe highlights, above all, the need for perpetual interventions that remain relevant to the plight of communities and that of conservation.

This chapter has also shown that the reasons the CAMPFIRE programme has been largely successful were due to its cost-benefit sharing scheme where all parties that contribute to conservation efforts are rewarded. Without material and economic benefits accruing to local communities, yearned wildlife management remains elusive. This chapter demonstrates that the success of biodiversity conservation is contingent upon local communities' attitudes towards wildlife which is positively influenced by the sustenance of material and economic benefits.

The study also highlights the challenges with which communities under the CAMPFIRE regime continue to grapple with, such as human-wildlife conflict, uneven distribution of benefits, expanding human land use, climate change, and increased rural poverty. Summarily, the study recommends the remodelling of CAMPFIRE approaches to suit the current land use patterns, especially in areas where there has been land reform; reimagined approaches that provide full authority and control to producer economies; improvement of community representation, transparency, and communication; as well as fostering of private partnerships to circumvent the funding constraints.

Accordingly, by proving recommendations to improve CBNRM and CAMPFIRE the chapter makes practical contributions. In the same breath, this chapter contributes to the literature discussion on the continued relevance of CBNRM and CAMPFIRE in the current community dynamics.

References

Acquah, E., Dearden, P., & Rollins, R. (2016). Nature-based tourism in Mole National Park, Ghana. *African Geographical Review, 35*(1), 53–69. https://doi.org/10.1080/19376812.2015.1088389

Balint, P. J., & Mashinya, J. (2008). CAMPFIRE through the lens of the "commons" literature: Nyaminyami rural district in post-2000 Zimbabwe. *Journal of Southern African Studies, 34*(1), 127–143. https://doi.org/10.1080/03057070701832924

Biggs, D., Ban, N. C., Castilla, J. C., Gelcich, S., Mills, M., Gandiwa, E., Etienne, M., Knight, A. T., Marquet, P. A., & Possingham, H. P. (2019). Insights on fostering the emergence of robust conservation actions from Zimbabwe's CAMPFIRE program. *Global Ecology and Conservation, 17*, e00538. https://doi.org/10.1016/j.gecco.2019.e00538

Chigonda, T. (2018). More than just story telling: A review of biodiversity conservation and utilisation from precolonial to postcolonial Zimbabwe. *Scientifica*, 1–11. https://doi.org/10.1155/2018/6214318

Child, G. (1985). *Managing wildlife for people in Zimbabwe. National parks, conservation, and development. The role of protected areas in sustaining society*. Proceedings of the World Congress on National Parks, Bali, Indonesia, 11–12 October 1982.

Child, B., Jones, B., Mazambani, D., Mlalazi, A., & Moinuddin, H. (2003). Final evaluation report: Zimbabwe natural resource management programme - USAID/Zimbabwe strategic objective No.1. USAID, Washington, DC.

Chiutsi, S., & Saarinen, J. (2017). Local participation in transfrontier tourism: Case of Sengwe community in Great Limpopo transfrontier conservation area, Zimbabwe. *Development Southern Africa, 34*(3), 260–275. https://doi.org/10.1080/0376835X.2016.1259987

Cumming, D. (1989). Commercial and safari hunting in Zimbabwe. In R. J. Hudson, K. R. Drew, & B. L.M.· (Eds.), *Wildlife production systems: Economic utilisation of wild ungulates* (pp. 147–169). Cambridge University Press.

DeGeorges, P. A., & Reilly, B. K. (2009). The realities of community based natural resource management and biodiversity conservation in Sub-Saharan Africa. *Sustainability, 1*, 734–788. https://doi.org/10.3390/su1030734

Dube, N. (2019). Voices from the village on trophy hunting in Hwange district, Zimbabwe. *Ecological Economics, 159*, 335–343. https://doi.org/10.1016/j.ecolecon.2019.02.006

Frost, P. G. H., & Bond, I. (2008). The CAMPFIRE programme in Zimbabwe: Payments for wildlife services. *Ecological Economics, 65*(4), 776–787. https://doi.org/10.1016/j.ecolecon.2007.09.018

Gandiwa, E., Heitkönig, I. M. A. A., Lokhorst, A. M., Prins, H. H. T. T., & Leeuwis, C. (2013). CAMPFIRE and human-wildlife conflicts in local communities bordering northern Gonarezhou National Park, Zimbabwe. *Ecology and Society, 18*(4). https://doi.org/10.5751/ES-05817-180407

Gandiwa, E., Zisadza-Gandiwa, P., Muboko, N., Libombo, E., Mashapa, C., & Gwazani, R. (2014). Local people's knowledge and perceptions of wildlife conservation in Southeastern Zimbabwe. *Journal of Environmental Protection, 5*(6), 457–481. https://doi.org/10.4236/jep.2014.56050

Gronau, S., Winter, E., & Grote, U. (2017). Modelling nature-based tourism impacts on rural development and conservation in Sikunga Conservancy, Namibia. *Development Southern Africa, 34*(3), 276–294. https://doi.org/10.1080/0376835X.2016.1269638

Hambira, W. L., Stone, L. S., & Pagiwa, V. (2022). Botswana nature-based tourism and COVID-19: Transformational implications for the future. *Development Southern Africa, 39*(1), 51–67. https://doi.org/10.1080/0376835X.2021.1955661

Hasler, R. (1999). *An overview of the social, ecological and economic achievements and challenges of Zimbabwe's CAMPFIRE programme*. International Institute for Environment and Development. Retrieved 20 March 2023 from https://www.iied.org/7796iied

Hoffmann, S. (2022). Challenges and opportunities of area-based conservation in reaching biodiversity and sustainability goals. *Biodiversity and Conservation, 31*(2), 325–352. https://doi.org/10.1007/s10531-021-02340-2

Hoole, A., & Berkes, F. (2010). Breaking down fences: Recoupling social-ecological systems for biodiversity conservation in Namibia. *Geoforum, 41*(2), 304–317. https://doi.org/10.1016/j.geoforum.2009.10.009

Logan, B. I., & Moseley, W. G. (2002). The political ecology of poverty alleviation in Zimbabwe's communal areas management programme for indigenous resources (CAMPFIRE). *Geoforum, 33*(1), 1–14. https://doi.org/10.1016/S0016-7185(01)00027-6

Martin, R. (1986). *Communal areas management programme for indigenous resources*. International Development Research Centre. Retrieved 20 March 2023 from https://idl-bnc-idrc.dspacedirect.org/handle/10625/2581

Martin, R., Drury, M., & Edwards, T. (1977). *Operation WINDFALL*. Department of National Parks and Wildlife Management.

Matseketsa, G., Chibememe, G., Muboko, N., Gandiwa, E., & Takarinda, K. (2018). Towards an understanding of conservation-based costs, benefits, and attitudes of local people living adjacent to save valley conservancy, Zimbabwe. *Scientifica, 2018*, 1–9. https://doi.org/10.1155/2018/6741439

Mberengwa, I. (2002). *The Communal Areas Management Program for Indigenous Resources (CAMPFIRE) and rural development in Zimbabwe's marginal areas: A study in sustainability* [UMI Dissertation Services]. WorldCat.org. Ann Arbor, MI.

Muboko, N., & Murindagomo, F. (2014). Wildlife control, access and utilisation: Lessons from legislation, policy evolution and implementation in Zimbabwe. *Journal for Nature Conservation, 22*(3), 206–211. https://doi.org/10.1016/j.jnc.2013.12.002

Murombedzi, J. (1997). Management of the land and resources of the Masoka community of Dande communal lands, Zimbabwe. *Society & Natural Resources, 10*(4), 405–408. https://doi.org/10.1080/08941929709381038

Musavengane, R., & Kloppers, R. (2020). Social capital: An investment towards community resilience in the collaborative natural resources management of community-based tourism schemes. *Tourism Management Perspectives, 34*, 100654. https://doi.org/10.1016/j.tmp.2020.100654

Musavengane, R., & Simatele, D. M. (2016). Community-based natural resource management: The role of social capital in collaborative environmental management of tribal resources in KwaZulu-Natal, South Africa. *Development Southern Africa, 33*(6), 806–821. https://doi.org/10.1080/0376835X.2016.1231054

Mushawemhuka, W., Rogerson, J. M., & Saarinen, J. (2018). Nature-based tourism operators' perceptions and adaptation to climate change in Hwange National Park, Zimbabwe, Bulletin of Geography. *Socio-Economic Series, 42*(42), 115–127. https://doi.org/10.2478/bog-2018-0034

Mutandwa, E., & Gadzirayi, C. T. (2007). Impact of community-based approaches to wildlife management: Case study of the CAMPFIRE programme in Zimbabwe. *International Journal of Sustainable Development and World Ecology, 14*(4), 336–344. https://doi.org/10.1080/13504500709469734

Mutanga, C. N., Muboko, N., & Gandiwa, E. (2017). Protected area staff and local community viewpoints: A qualitative assessment of conservation relationships in Zimbabwe. *PLoS One, 12*(5), e0177153. https://doi.org/10.1371/journal.pone.0177153

Nthiga, R. W., Van der Duim, R., Visseren-Hamakers, I. J., & Lamers, M. (2015). Tourism-conservation enterprises for community livelihoods and biodiversity conservation in Kenya. *Development Southern Africa, 32*(3), 407–423.

Ntuli, H., & Muchapondwa, E. (2017). A bioeconomic analysis of community wildlife conservation in Zimbabwe. *Journal for Nature Conservation, 37*, 106–121. https://doi.org/10.1016/j.jnc.2017.04.003

Ntuli, H., & Muchapondwa, E. (2018). The role of institutions in community wildlife conservation in Zimbabwe. *International Journal of the Commons, 12*(1), 134–169. https://doi.org/10.18352/ijc.803

Price, R.A. (2017). *The Contribution of Wildlife to the Economies of Sub Saharan Africa.* K4D Helpdesk Report. Brighton, UK: Institute of Development Studies.

Roe, D., & Nelson, F. (2009). The origins and evolution of community-based natural resource management in Africa. In D. Roe, F. Nelson, & C. Sandbrook (Eds.), *Community management of natural resources in Africa: Impacts, experiences and future directions.* International Institute for Environment and Development.

Saarinen, J., & Lenao, M. (2014). Integrating tourism to rural development and planning in the developing world. *Development Southern Africa, 31*(3), 363–372. https://doi.org/10.1080/0376835X.2014.888334

Segobye, A. K., Mpolokang, M., Shereni, N. C., Mago, S., & Seleka, M. (2022). Community based tourism as a pathway towards sustainable livelihoods and well-being in Southern Africa. In J. Saarinen, B. Lubbe, & N. N. Moswete (Eds.), *Southern African perspectives on sustainable tourism management. Geographies of tourism and global change.* Springer. https://doi.org/10.1007/978-3-030-99435-8_9

Shereni, N. C., & Saarinen, J. (2021). Community perceptions on the benefits and challenges of community-based natural resources management in Zimbabwe. *Development Southern Africa, 28*(6), 879–895. https://doi.org/10.1080/0376835X.2020.1796599

Spenceley, A. (2003). Managing sustainable nature-based tourism in Southern Africa: A practical assessment tool (Doctoral dissertation). University of Greenwich. https://doi.org/10.13140/RG.2.1.4495.0889

Spenceley, A. (2006). Tourism in the Great Limpopo Transfrontier Park Tourism in the Great Limpopo Transfrontier Park. *Development Southern Africa, 23*(5), 649–667. https://doi.org/10.1080/03768350601021897

Stone, M. T. (2006). *Community based natural resources management (CBNRM) and tourism: The Nata bird sanctuary project, Central District Botswana.* University of Witwatersrand, Johannesburg.

Tchakatumba, P. K., Gandiwa, E., Mwakiwa, E., Clegg, B., & Nyasha, S. (2019). Does the CAMPFIRE programme ensure economic benefits from wildlife to households in Zimbabwe? *Ecosystems and People, 15*(1), 119–135. https://doi.org/10.1080/26395916.2019.1599070

Tervo-Kankare, K., Saarinen, J., Kimaro, M. E., & Moswete, N. N. (2018). Nature-based tourism operators' responses to changing environment and climate in Uis, Namibia. *African Geographical Review, 37*(3), 273–282. https://doi.org/10.1080/19376812.2017.1286246

Thapa, K., King, D., Banhalmi-Zakar, Z., & Diedrich, A. (2022). Nature-based tourism in protected areas: A systematic review of socio-economic benefits and costs to local people. *International Journal of Sustainable Development and World Ecology, 29*(7), 625–640. https://doi.org/10.1080/13504509.2022.2073616

Thomas, S. J. (1991). T*he legacy of dualism and decision-making: The prospects for local institutional development in "CAMPFIRE"*, Second Annual Conference of the International Association for the Study of Common Property (IASCP), Winnipeg, Canada.

Tichaawa, T. M., & Mhlanga, O. (2015). Community perceptions of a community-based tourism project: A case study of the CAMPFIRE programme in Zimbabwe. *African Journal for Physical, Health Education, Recreation and Dance, 21*(Suppl. 1), 55–67. http://reference.sabinet.co.za/document/EJC182968

UNWTO. (2015). *Towards measuring the economic value of wildlife watching tourism in Africa.* Madrid, Spain.

Vincent, J., De Wit, A. H., & Nigel, L. W. (2022). Evaluating the relevance of the CAMPFIRE principles to wildlife utilization in Northern Zimbabwe. *African Journal of Wildlife Research, 52*(1). https://doi.org/10.3957/056.052.0056

Winter, P. L., Selin, S., Cerveny, L., & Bricker, K. (2020). Outdoor recreation, nature-based tourism, and sustainability. *Sustainability, 12*(18), 1–12.

Wolf, I. D., Croft, D. B., & Green, R. J. (2019). Nature conservation and nature-based tourism: A Paradox ? *Environments, 6*(104), 1–22.

WTTC. (2019). *Global wildlife tourism generates five times more revenue than illegal wildlife trade annually.* World Travel & Tourism Council. Retrieved 3 May 2023, from https://wttc.org/news-article/global-wildlife-tourism-generates-five-times-more-revenue-than-illegal-wildlife-trade-annually

Zunza, E. (2012). *Local level benefits of CBNRM: The case of Mahenye Ward* (Masters Dissertation). University of Zimbabwe.

Part III
The Dynamics of Tourism-Based Wildlife Economies: Implications for Sustainability

Chapter 6
The State of Wildlife Tourism in the Southern African Region: The Case of the Southern African Development Community (SADC)

Tariro Kamuti

6.1 Introduction

This chapter aims to uses extant literature to synthesize the state of wildlife tourism in Southern Africa, using the case of countries found in the Southern African Development Community (SADC). The lack of a systematic assessment of wildlife tourism performance in the region has motivated the execution of this chapter. The systematic assessment covers the wildlife experience, which is the core of wildlife tourism, and some issues surrounding wildlife tourism in the region. Wildlife tourism is a travel activity that involves visiting natural environments and observing, interacting with, or learning about various forms of wildlife in their natural habitats (Samal & Dash, 2023). This type of tourism has gained popularity over the years as people became more interested in conservation, education and documentary, research, and thrill-seeking purposes (Mariki, 2013; Samal & Dash, 2023). Wildlife tourism can take many different forms, including safari expeditions, bird-watching trips, whale watching, jungle hikes, and nature photography tours. By providing experiences in nature for people, wildlife tourism seeks to encourage environmental conservation and raise awareness about the importance of protecting wildlife and their habitat (Cong, et al., 2014).

Wildlife tourism is a significant contributor to the economy of Southern Africa. It is characterised by the interaction of humans with various species of wild animals, their habitats, and related natural resources (Buckley & Chauvenet, 2022). This region is home to a vast range of animal species, including elephants, rhinos, lions, leopards, buffalos, and many other wildlife species, making it a popular destination

T. Kamuti (✉)
Centre for Gender and Africa Studies, University of the Free State, Bloemfontein, South Africa
e-mail: KamutiT1@ufs.ac.za

© The Author(s), under exclusive license to Springer Nature Switzerland AG 2024

L. S. Stone, M. T. Stone (eds.), *Wildlife Tourism Dynamics in Southern Africa*, Geographies of Tourism and Global Change,
https://doi.org/10.1007/978-3-031-57252-4_6

for safari and wildlife enthusiasts (Cong, et al., 2014). In this chapter, Southern Africa refers to the southern part of the African continent, which is made up of the member states of the Southern African Development Community (SADC). These countries include Angola, Botswana, Comoros, Democratic Republic of Congo, Eswatini, Lesotho, Madagascar, Malawi, Mauritius, Mozambique, Namibia, Seychelles, South Africa, Tanzania, Zambia, and Zimbabwe.[1] Wildlife is taken to mean all living non-human species (flora and fauna) found on all terrestrial, freshwater, and marine ecosystems that includes "all living organisms … in their natural habitat" (Kunedzimwe et al., 2021, p. 229). This is a holistic view that recognises the totality of ecosystems that constitute the various landscapes which are also panoramic and provide the natural environments, that eventually give a terrific tourist experience.

In this chapter, wildlife tourism is taken as part of nature-based tourism. Nature-based tourism defines "tourism that use[s] natural resources in a wild or undeveloped form, [or] travel for the purpose of enjoying undeveloped natural areas or wildlife" (IUCN ESARO, 2020, p. xvi). Nature-based tourism is important to fulfil the roles of uplifting and sustaining people's livelihoods and contributing to conservation (Roe et al., 1997; Samal & Dash, 2023). The chapter proceeds with the conceptualisation of the wildlife experience, the core of wildlife tourism, where tourists travel to various places to interact with wildlife. This will be followed by a look at the places where the wildlife is found, broadly classified as state-protected areas, privately owned and community-based wildlife tourist attractions. Wildlife tourism in Southern Africa comes next by examining unique and major tourist attractions SADC countries with a nuanced discussion of the wildlife experiences that tourists enjoy. Wildlife tourism has issues that plague the sector, these will be addressed to provide an overall view of the state of the sector. Following that is a section providing a snapshot of challenges associated with wildlife tourism, the impact of the global COVID-19 pandemic on the sector and lastly the effects of natural disasters. This will be followed by concluding remarks to wrap up the chapter.

6.2 The Wildlife Experience

The core of wildlife tourism is the wildlife experience, where tourists have an encounter with wildlife. Tourists travel to attractions where wildlife is found for them to interact with it. This interaction is varied for instance, according to whether wildlife tourism is consumptive or non-consumptive. While wildlife tourism is a leisure activity, it is acknowledged that an encounter with wildlife in its natural setting has positive effects on someone's mental being (Buckley & Chauvenet, 2022). The desire to interact with nature draws tourists to places of attractions where wildlife is found, thus making wildlife tourism a worldwide phenomenon. The

[1] Member States https://www.sadc.int/member-states accessed: 26 March 2023.

wildlife experience is personal; hence it is varied depending on the perceptions of those encountering the wildlife in its natural settings (Patterson et al., 1998).

The movement of tourists to places of attraction is inherently associated with economic flows to tourist destinations. Countries in Southern Africa benefit from wildlife tourism by hosting both local and international tourists who spend money on travel, food, accommodation, park entry fees, tourism levies, and other goods and services during their wildlife tourism experiences. It is those revenues that contribute to each country's Gross Domestic Product (GDP), provide employment, stimulate local economic development, and help in financing the conservation of natural resources that provide the wildlife experience.

6.3 An Overview of Wildlife Distribution in Southern Africa

Historically, humans have co-existed with wildlife, though the nature of this co-existence has been changing (Cong et al., 2014). For instance, with civilisation and population growth, there have been rapid changes in land use, with increasing encroachment of human settlements and other human activities into areas that were previously occupied by wildlife, leading to habitat loss (Vieira-Alencar, et al., 2023). The governance of wildlife has also been instituted in a way that has allowed the designation of wildlife areas as protected and conservation areas under various kinds of management approaches.

This partly explains the development of wildlife tourism as people who are fascinated with nature and wildlife take time to travel to areas where they can experience wildlife in its natural state. However, in some instances, the designation of these protected and conservation areas has also involved the displacement of indigenous people by dispossessing them of their land to pave way for wildlife (Child, 2009a; Diallo, 2015). This has been a source of tension between state authorities and affected communities (Sirima & Backman, 2013; Musavengane & Leonard, 2022). The areas where wildlife is found are classified into three major groups (public protected areas, privately owned protected areas and those managed by local communities) based on their governance structures and processes.

Wildlife is mainly found in state or public protected areas. These areas have been designated and proclaimed by the state under laws within their jurisdiction as areas that are publicly managed under institutional structures and processes established from their colonial past (Child, 2009a). States have established various kinds of agencies at national, regional, or local levels to preside over the management of such protected and conservation areas. There are a variety of national designations of protected and conserved areas with the SADC countries. For example, in Zimbabwe, the national designations of protected and conserved areas include sanctuaries, national parks, national monuments, wildlife management areas, botanical reserves, nature reserves, recreational parks, botanical gardens, safari areas, state forests and protected forests (IUCN ESARO, 2020).

Wildlife is also found on privately owned land. These wildlife areas are also allowed under the law, but they have some level of autonomy (in terms of ownership and management) which is distinct from the conditions of governance associated with public protected areas. These wildlife areas consist of private nature reserves, wildlife ranches, game farms and conservancies. Privately owned protected areas play a critical role in biodiversity and thus nature conservation, but their thrust is profit-making. They package the wildlife experience in a way that takes advantage of the tourism value chain to attract local and international tourists. These packages include both consumptive and non-consumptive wildlife tourism services depending on what drives each private protected area. Consumptive wildlife tourism includes what should be sustainable utilisation of wildlife and trophy hunting is the most common (Bichel & Hart, 2023). However, some of the privately owned protected areas, for example in South Africa are found in areas where the same land has been the subject of land restitution or redistribution. This has posed questions about the equitable distribution of resources, and prioritisation of livelihoods for different stakeholders thereby causing tension and conflict in such situations (Kamuti, 2022b; Musavengane & Leonard, 2022).

Some protected and conservation areas are managed by local communities under various management arrangements that reflect and represent the interests of the communities. Proceeds from wildlife tourism generated within those areas are usually meant for the benefit of people within those local communities and these protected areas will be under what has popularly been known as Community Based Natural Resources Management (CBNRM) (Child, 2009a). CBNRM is a well-recognised strategy in SADC, where there is a CBNRM Forum that is made up of a diversity of actors ranging from non-governmental organisations, government agencies, and research institutions (SADC Secretariat, 2013). These actors are from Botswana, Malawi, Namibia, Mozambique, Zambia, and Zimbabwe working towards advancing CBNRM principles and facilitating their incorporation into each country's statutes to uplift the livelihoods of local communities (Soto, 2009; SADC Secretariat, 2013). For example, in Namibia, they have 86 communal conservancies covering around a fifth of the country's land area and sustaining over 230,000 people by bestowing to them the rights to use the land while benefitting from wildlife (Conservation Frontlines, 2023). The communal conservancies have served as empowerment of local people while serving the purpose of conservation of wildlife thus promoting sustainable development.

6.4 An Overview of Wildlife Tourism in Southern Africa

Southern Africa is known for its diverse wildlife and offers some great opportunities for wildlife tourism. Wildlife tourism is a potentially beneficial and sustainable industry as it can generate income for rural areas and support wildlife conservation initiatives, as well as create job opportunities for local communities (Ranke et al., 2023). In Southern Africa, national parks and wildlife reserves constitute a formidable tourist attraction that earns foreign currency to the host nations while helping to conserve wildlife populations from illegal activities and human encroachment of animal habitats (Ferreira, 2006; Mtahiko, 2004; Ranke et al., 2023). A major drawcard for wildlife tourism in Southern Africa is the 'Big Five', which is a group of five charismatic species of the elephant, lion, leopard, rhinoceros, and buffalo which are abundantly found in the countries' national parks and wildlife reserves. With the increase in nature-based tourism, there has been anticipation that people's standards of living would improve, especially those found in the vicinity of these protected areas (Ferreira, 2006).

South Africa offers some of the best wildlife tourism experiences in Southern Africa, with its popular national parks that include Kruger National Park, Addo Elephant National Park, and Hluhluwe-Imfolozi Park. The Kruger National Park is the African continent's initial game reserve (Ferreira, 2006). The private game reserves around Kruger National Park though they cover around 12% of the total area under protection, contribute significantly to wildlife tourism in terms of jobs, tax and revenue in relation to the gross domestic product (Chidakel et al., 2020). Botswana is well-known for its safari experiences in the Okavango Delta and the Central Kgalagadi Game Reserve, Gemsbok National Park and Chobe National Park being the few largest within the list of public protected areas in the country (Child, 2009b). Namibia is a popular destination for wildlife enthusiasts looking for an off-the-beaten-track experience. Etosha National Park is one of the most popular wildlife areas in Namibia. Zimbabwe offers some excellent wildlife tourism opportunities in wildlife areas including the Hwange National Park, Gonarezhou National Park and Mana Pools National Park. Zambia is famous for its walking safaris and offers excellent wildlife tourism experiences in South Luangwa National Park and Lower Zambezi National Park. Comoros is a beautiful archipelago nation located in the Indian Ocean, comprised of four major islands (Walsh, 2007). Although Comoros is not as popular as some of the other tourist destinations in the region, it offers unique opportunities for wildlife tourism. The islands are home to a diverse range of flora and fauna species, many of which are endemic to the region while others have been introduced from elsewhere like Madagascar to the islands for example, the guineafowl which was domestic but has become feral (Walsh, 2007). Visitors to Comoros can enjoy wildlife experiences such as whale watching, turtle nesting, mammal watching and bird watching.

The Democratic Republic of Congo has several national parks and game reserves that are popular for wildlife tourism. Virunga National Park is home to endangered mountain gorillas, chimpanzees, elephants, and hippos, among other wildlife. Visitors can go on guided gorilla treks or climb the active Nyiragongo volcano. Kahuzi-Biega National Park is known for its population of eastern lowland gorillas, as well as forest elephants, leopards, and other primates. Garamba National Park is a UNESCO World Heritage site and is home to elephants, giraffes, lions, and a variety of antelopes, among other species. Salonga National Park is the largest tropical rainforest reserve in Africa and is home to bonobos, forest elephants, and several bird species. These national parks and game reserves offer visitors an opportunity to witness some of Africa's most iconic animals in their natural habitats. However, ongoing conflicts and security concerns in the Democratic Republic of Congo, are a blight to wildlife tourism.

Several national parks and reserves that are popular for wildlife tourism offer unique and exciting opportunities to see a range of wildlife in Malawi. Liwonde National Park is home to various animals such as elephants, hippos, crocodiles, and lions and it is also a great place to go bird watching and see species such as the African Fish Eagle. Nyika National Park located in the north of Malawi, offers stunning landscapes and a chance to see unique wildlife such as the eland and roan antelope. Kasungu National Park is known for its large population of elephants, as well as other animals such as buffalo, impala, and kudu. Majete Wildlife Reserve, once an area destroyed by poaching, has been restored and is now home to various animals including the Big Five. Nkhotakota Wildlife Reserve, home to over 280 species of birds and various animals such as elephants, buffaloes, and leopards, also offers great opportunities for hiking and camping. Lake Malawi is also a huge attraction with a wide diversity of species of fish, some of which are endemic (Tweddle, 2000). Mozambique is home to several national parks, game reserves, and protected areas that offer great opportunities for wildlife tourism now with some protected areas having gone through restoration after a protracted civil war that destroyed conservation infrastructure, destroyed wildlife and stalled management processes (Child, 2009a; Soto, 2009; Diallo, 2015). Some of the most popular wildlife destinations in Mozambique offer a unique experience for wildlife enthusiasts. Gorongosa National Park is in the central part of Mozambique and is known for its diverse wildlife, including elephants, lions, hippos, and various species of antelopes, though its formation involved the displacement of people elsewhere in southern Africa (Diallo, 2015). Niassa Reserve is the largest conservation area in Mozambique, covering an area of over 42,400 square kilometres (Noe, 2015). It is home to a wide range of wildlife, including lions, leopards, elephants, and buffalos. Quirimbas National Park situated in the north-eastern part of Mozambique is a popular destination for marine wildlife enthusiasts. It is home to several species of whales, dolphins, turtles, and other sea creatures. Maputo Elephant Reserve located in the southern part of Mozambique, is home to a significant population of elephants, as well as other wildlife such as zebras, wildebeests, and several species of antelopes. Bazaruto Archipelago Marine Park is situated off the coast of Mozambique and is a popular destination for marine wildlife enthusiasts. It is home to diverse marine life, including dolphins, whales, sharks, and various species of fish.

Tanzania is famous for its abundant wildlife and has several national parks, game reserves as well as conservation areas that offer a memorable wildlife tourism experience (Mtahiko, 2004). The country has about 300,000 square kilometres of land under protection which is equivalent to about a third of its territory (Ranke et al., 2023). Serengeti National Park is the world-renowned park for the Great Migration of wildebeest and zebras. Ngorongoro Conservation Area is a UNESCO World Heritage Site and home to the stunning Ngorongoro Crater, which offers some of the best game viewing opportunities in Tanzania. Tarangire National Park is known for its large herds of elephants and baobab trees. Selous Game Reserve is the largest game reserve in Africa and a designated UNESCO World Heritage Site. Lake Manyara National Park is known for its tree-climbing lions and diverse birdlife. Ruaha National Park covering around 10,000 square kilometres is one of Tanzania's largest and lesser-visited parks, with impressive wildlife and striking landscapes (Mtahiko, 2004). Though Ruaha National Park is part of a huge ecological system of about 45,000 square kilometres including game reserves, it has an area of around 6022 that is designated as a Wilderness Zone, where there are limited visits (Mtahiko, 2004). Gombe Stream National Park is home to chimpanzees, baboons, and other primates, as well as a range of birds and butterflies. Mahale Mountains National Park is another chimpanzee habitat, which also offers stunning views of Lake Tanganyika. Katavi National Park, the third-largest national park in Tanzania, is known for its vast buffalo herds and abundance of hippos. Overall, Tanzania has 22 national parks with five major parks namely, Serengeti, Kilimanjaro, Tarangire, Lake Manyara and Arusha contributing the largest share of revenue, about 95% per year (Ranke et al., 2023).

Mauritius is not typically known for its wildlife tourism, as it is a small island nation with limited natural habitats. However, there are a few parks and reserves that are popular for their unique flora and fauna. Black River Gorges National Park is the largest national park in Mauritius and is home to many indigenous plants and animals, including the Mauritius kestrel and pink pigeon. Île aux Aigrettes—This is a small island reserve off the coast of Mauritius that is home to many rare and endangered species, including the Mauritius fody bird, olive white-eye bird, and Aldabra giant tortoise. La Vanille Nature Park is in the southern part of the island and is home to many giant tortoises, crocodiles, and other reptiles. Seychelles is known for its stunning beaches and coral reefs, but wildlife tourism is not a significant draw in the country as it does not have any big game animals. However, visitors can still experience the local fauna by visiting several nature reserves and parks. North Island Reserve, a private island, features several conservation projects, including sea turtle protection and reforestation. Visitors can see several endemic bird species, such as the Seychelles blue pigeon and the Seychelles warbler. Cousin Island Special Reserve is a protected national park and bird sanctuary that is home to several species of rare birds, including the Seychelles warbler and magpie-robin. St. Anne Marine National Park is a marine reserve that features six islands and is a popular spot for snorkelling and diving. Lesotho has Bokong Nature Reserve and Tšehlanyane National Park which anchor a planned Biosphere Reserve found within the Grassland Biome of Southern Africa, which comes after the Savanna Biome (Seleteng-Kose, et al., 2021).

Overall, the state of wildlife tourism in Southern Africa is a complex issue that involves both economic benefits and environmental impacts. The region benefits from the revenue generated by wildlife tourism. This revenue translates to job creation and poverty reduction in local communities. At the same time, the income generated by wildlife tourism supports conservation efforts, mainly through the management of national parks and conservation areas.

6.5 Challenges in Wildlife Tourism in Southern Africa

Wildlife tourism in Southern Africa has experienced both positive and negative changes over time. It creates challenges for conservationists, as they must balance the impact of human interaction with the environment, which includes the preservation of ecosystems and species habitats. The balance between tourism activities and conservation efforts is challenging, particularly in impoverished regions where communities depend on the revenue generated by wildlife tourism. Wildlife tourism destinations run the risk of an overflow of tourists which may lead to ecological deterioration and loss of attractiveness of such places (Roe et al., 1997). Over time, there has been a shift towards more responsible, sustainable, and ethical tourism practices, which prioritize the protection of animal welfare and their habitats. Additionally, conservation efforts, such as stricter regulations, and relevant policies and laws, are gradually emerging. In contrast, some of the negative aspects of wildlife tourism which include issues like poaching, animal cruelty, and illegal trading of animal products, are still rife and ongoing.

Some local people have or will never visit a wildlife reserve in their lifetime because of several reasons, chief among them being poverty. Some wildlife tourism destinations in Southern Africa are financially exclusive and poor people cannot afford them as they are seized with matters of day-to-day survival. This is linked to the question of the use of market-based approaches to conservation in Southern Africa (Diallo, 2015). Tourism becomes a far-fetched leisure activity. Even those close to the parks do not get a chance to enter and enjoy watching the wildlife that sometimes causes havoc rin their neighbourhoods by threatening their lives, and livestock, destroying crops and causing damage. The dynamics of the interactions between the protected areas and the surrounding communities are deep beside the narratives of the historical displacement of people and they need to be understood to manage them better (Child, 2009a; Dondeyne et al., 2012).

The increase in the area under conservation through the expansion of Ruaha National Park in the Usangu Plains in Tanzania resulted in the shift in land use from agriculture to conservation in the displaced and affected communities (Sirima & Backman, 2013). The lack of a deep connection between local communities and national parks in terms of their involvement and deriving some benefits thereof causes tension (Mariki, 2013; Sirima & Backman, 2013). Worldwide it has been proven that areas under indigenous people are well conserved, especially since they derive some livelihoods from the surrounding nature (Ranke et al., 2023). The state and nature of wildlife tourism are an essential topic that mixes environmental and economic aspects to create a balance that will benefit humans and wildlife alike. Although wildlife tourism has brought major benefits to Southern Africa's economy, it requires proper regulation and sustainable approaches to minimise harm to animals and maintain ecosystems' ecological balance.

6.6 COVID-19 and Wildlife Tourism

The impact of the global pandemic on wildlife tourism in Southern Africa was hard-hitting. The global pandemic which started in China in late 2019 spiralled around the world with far-reaching consequences for human life and thus prompted countries to take immediate action including unconventional means to stop its spread (Nhamo et al., 2020; Kamuti, 2022a). For example, lockdowns put a stop to all forms of travel except for essential services and this ground the tourism industry to a halt with devastating effects on employment (Samal & Dash, 2023). This was so significant for the tourism industry as a sector that constitutes the bulk of global travel. The lockdowns that also halted wildlife tourism were also significant because the origins of the global pandemic itself were linked to wildlife. A study in Tanzania showed a slump in visitors to national parks that coincided with the reduced local and international travel due to COVID-19 leading to loss of income, with the parks needing financial support from alternative sources like subsidies (Ranke et al., 2023). For the consumptive wildlife tourism, there was an over-supply of game animals as the hunting seasons had slumped due to low demand caused by the lockdowns that halted global travel (Sommerville et al., 2021). With the easing of COVID-19 restrictions since 2022, wildlife tourism is on a rebound as economies in the SADC regions, just like elsewhere in the world, are roaring back to normal though they will take time to fully recover.

6.7 Natural Disasters and Wildlife Tourism

Natural disasters like tropical cyclones reduce the attractiveness of tourist areas, that includes the tourist destination and its supportive infrastructure and services, and this can affect the sector intermittently or in the long run depending on the extent of the disruption (Nhamo et al., 2021). Wildlife tourism in Southern Africa has not been spared from the vagaries of extreme weather conditions particularly climate change-induced natural disasters like more frequent droughts and tropical cyclones which alter the state of natural ecosystems in the region (Kupika et al., 2021). Supportive infrastructure and services are critical and bring a dilemma for wildlife tourism due to its nature of providing a wilderness experience and thus found in areas that are far away from or that do not reproduce conditions found in towns and cities. Cyclone Idai in March 2019 ravaged and thus disrupted wildlife tourism areas in Madagascar, Malawi, Mozambique, and Zimbabwe (Nhamo, et al., 2021).

The Southern African region is more susceptible to tropical cyclones because of the highly variable climatic conditions and some studies in Angola, Botswana, Lesotho, Malawi, Mauritius, Mozambique, Namibia, South Africa, Eswatini, Tanzania, Zambia and Zimbabwe showed these countries to be more affected by cyclones (Kupika, et al., 2021). Studies on the effect of powerful exogenous factors like tropical cyclones on natural vegetation have paid more attention to those forests found onshore, for example, seagrass meadows in Mozambique, South Africa, and Tanzania as they are the ones that are first exposed since they are close to the sea where the cyclones originate (Bandeira et al., 2021; Chikodzi & Tembani, 2021). However, the increased rate of occurrence and extent of impact has been seen and beginning to ravage the natural vegetation as the cyclones advance into the interior (Chikodzi & Tembani, 2021). For example, "the increasing frequency of these cyclones in the Eastern Highlands of Zimbabwe, which hosts high-value forest plantations and several important forest biodiversities hotspots such as Vumba, Haroni and Chirinda rainforests" (Chikodzi & Tembani, 2021, p. 212). This is in addition to indigenous forests and botanical reserves in Tandai, Ndizwe and Mukurupini, the "Afromontane forests and the low altitude forests of the Rusitu Valley", in Chimanimani District, in the Eastern Highlands of Zimbabwe (Nhamo et al., 2021, p. 249). Flora and fauna in Chimanimani National Park were affected by the impact of Cyclone Idai in March 2019 as much as what had been observed after Cyclone Eline in 2000, and around the Gorongosa area in Mozambique and in Madagascar (Kunedzimwe et al., 2021; Kupika et al., 2021; Nhamo et al., 2021). Chimanimani National Park together with Eland Sanctuary forms the Zimbabwean component of the Chimanimani Transfrontier Conservation Area that is jointly constituted with Chimanimani Nature Reserve from the Mozambican side (Soto, 2009; Dondeyne et al., 2012; Kunedzimwe et al., 2021). Cyclone Freddy in Malawi also negatively impacted wildlife tourism in that country when it made landfall in March 2023.

6.8 Conclusion

This chapter has discussed the state of wildlife tourism in Southern Africa by taking the case of the member states of the SADC. Wildlife tourism is a type of tourism that involves experiencing and observing wild animals in their natural habitat, and if done responsibly, it can create significant benefits for conservation efforts, local economies, and education while minimizing ecologically disruptive impacts on the environment. However, wildlife tourism can also be a cause of ecological disruption and a negative impact on wildlife if not implemented responsibly. Therefore, operators who conduct wildlife tourism must follow strict standards and responsible practices which should prioritize the conservation of wildlife and their natural habitat, while also considering the welfare of the individual animals and minimizing disturbance to the ecosystem. It has also been seen that wildlife tourism creates challenges for conservationists, as they must balance the impact of human interaction with the environment. The balance between tourism activities and conservation efforts is more pressing, particularly in impoverished regions where communities depend on the revenue generated by wildlife tourism. In some instances, there is alienation of local communities from the protected areas, and this casts aspersions on wildlife tourism as extractive and thus exploitative.

References

Bandeira, S., Amone-Mabuto, M., Chitará-Nhandimo, S., Scarlet, M. P., & Rafael, J. (2021). Impact of cyclones and floods on seagrass habitats. In G. Nhamo & D. Chikodzi (Eds.), *Cyclones in southern Africa: Volume three—Implications for the sustainable development goals* (pp. 279–288). Springer Nature.

Bichel, N., & Hart, A. (2023). *Trophy hunting*. Springer Nature.

Buckley, R., & Chauvenet, A. L. M. (2022). Hidden economic counterflows in post-COVID international wildlife tourism. *Tourism Management, 93*, 1–4.

Chidakel, A., Eb, C., & Child, B. (2020). The comparative financial and economic performance of protected areas in the Greater Kruger National Park, South Africa: Functional diversity and resilience in the socio-economics of a landscape-scale reserve network. *Journal of Sustainable Tourism, 1–20*. https://doi.org/10.1080/09669582.2020.1723602

Chikodzi, D., & Tembani, M. C. (2021). The impact of cyclone Idai on natural and plantation forests in the eastern highlands of Zimbabwe. In G. Nhamo & D. Chikodzi (Eds.), *Cyclones in southern Africa: Volume three—Implications for the sustainable development goals* (pp. 211–228). Springer Nature.

Child, B. (2009a). The emergence of parks and conservation narratives in southern Africa. In H. Suich, B. Child, & A. Spenceley (Eds.), *Evolution and innovation in wildlife conservation: Parks and game ranches to transfrontier conservation areas* (pp. 19–33). Earthscan.

Child, G. (2009b). The growth of park conservation in Botswana. In H. Suich, B. Child, & A. Spenceley (Eds.), *Evolution and innovation in wildlife conservation: Parks and game ranches to transfrontier conservation areas* (pp. 51–66). Earthscan.

Cong, L., Wu, B., Morrison, A. M., Shu, H., & Wang, M. (2014). Analysis of wildlife tourism experiences with endangered species: An exploratory study of encounters with giant pandas in Chengdu, China. *Tourism Management, 40*, 300–310.

Conservation Frontlines. (2023). *Community-based conservation*. https://www.conservationfrontlines.org/2023/08/community-based-conservation/?print=pdf. (Accessed 14 Aug 2023).

Diallo, R. N. (2015). Conservation philanthropy and the shadow of state power in Gorongosa National Park, Mozambique. *Conservation & Society, 13*(2), 119–128.

Dondeyne, S., Kaarhus, R., & Allison, G. (2012). Nature conservation, rural development and ecotourism in Central Mozambique: Which space do local communities get? In I. Convery, G. Corsane, & P. Davis (Eds.), *Making sense of place: Multidisciplinary perspectives* (pp. 291–301). The Boydell Press.

Ferreira, S. L. A. (2006). Communities and transfrontier parks in the southern African development community: The case of Limpopo National Park, Mozambique. *South African Geographical Journal, 88*(2), 166–176. https://doi.org/10.1080/03736245.2006.9713859

IUCN ESARO. (2020). The state of protected and conserved areas in Eastern and Southern Africa. State of Protected and Conserved Areas Report Series No. 1. Nairobi, Kenya: IUCN ESARO.

Kamuti, T. (2022a). Counting the cost of COVID-19 on the global tourism industry. *Development Southern Africa, 39*(1), 84–86. https://doi.org/10.1080/0376835X.2021.1912588

Kamuti, T. (2022b). The deepening challenge of governance of wildlife and land issues in the context of rising citizen participation in South Africa. In R. Musavengane & L. Leonard (Eds.), *Conservation, land conflicts, and sustainable tourism in southern Africa: Contemporary issues and approaches* (pp. 33–50). Routledge. https://doi.org/10.4324/9781003188902

Kunedzimwe, F., Gandiwa, E., Kupika, O. L., Muboko, N. N., Taru, P., & Manyuchi, T. (2021). Impact of cyclone Idai on biodiversity and natural resources in Chimanimani District, Zimbabwe. In G. Nhamo & D. Chikodzi (Eds.), *Cyclones in southern Africa: Volume three—Implications for the sustainable development goals* (pp. 229–244). Springer Nature.

Kupika, O. L., Gandiwa, E., Ayuk, J., Bandeira, S., & Kunedzimwe, F. (2021). Evidence of the impact of cyclones and floods on biodiversity and wildlife resources in Southern Africa. In

G. Nhamo & D. Chikodzi (Eds.), *Cyclones in southern Africa: Volume three—Implications for the sustainable development goals* (pp. 265–278). Springer Nature.

Mariki, S. B. (2013). Conservation with a human face? Comparing local participation and benefit sharing from a national park and a state forest plantation in Tanzania. *SAGE Open, 3*(4), 1–16. https://doi.org/10.1177/2158244013512665

Mtahiko, M. G. G. (2004). Wilderness in the Ruaha National Park, Tanzania. *International Journal of Wilderness, 10*(3), 41–44.

Musavengane, R., & Leonard, L. (2022). Land conflicts in southern Africa: The sustainability of tourism and conservation. In R. Musavengane & L. Leonard (Eds.), *Conservation, land conflicts, and sustainable tourism in southern Africa: Contemporary issues and approaches* (pp. 1–13). Routledge.

Nhamo, G., Dube, K., & Chikodzi, D. (2020). *Counting the cost of COVID-19 on the global tourism industry*. Springer Nature.

Nhamo, G., Dube, K., & Saurombe, T. (2021). Impact of tropical cyclone Idai on tourism attractions and related infrastructure in Chimanimani, Zimbabwe. In G. Nhamo & D. Chikodzi (Eds.), *Cyclones in southern Africa: Volume three—Implications for the sustainable development goals* (pp. 245–264). Springer Nature.

Noe, C. (2015). The Selous-Niassa Transfrontier conservation area and tourism: Evolution, benefits and challenges. In R. van der Duim, M. Lamers, & J. van Wijk (Eds.), *Institutional arrangements for conservation, development and tourism in eastern and southern Africa: A dynamic perspective* (pp. 181–201). Springer Nature.

Patterson, M. E., Watson, A. E., Williams, D. R., & Roggenbuck, J. R. (1998). A hermeneutic approach to studying the nature of wilderness experiences. *Journal of Leisure Research, 30*(4), 423–452. https://doi.org/10.1080/00222216.1998.11949842

Ranke, P. S., Kessy, B. M., Mbise, F. P., Nielsen, M. R., Arukwe, A., & Røskaft, E. (2023). The threat of COVID-19 to the conservation of Tanzanian national parks. *Biological Conservation, 282*, 110037. https://doi.org/10.1016/j.biocon.2023.110037

Roe, D., Leader-Williams, W., & Dalal-Clayton, B. (1997). *Take only photographs, leave only footprints: The environmental impacts of wildlife tourism*. International Institute for Environment and Development.

SADC Secretariat. (2013). *SADC Programme for Transfrontier Conservation Areas*. Gaborone, Botswana: SADC Secretariat.

Samal, R., & Dash, M. (2023). Ecotourism, biodiversity conservation and livelihoods: Understanding the convergence and divergence. *International Journal of Geoheritage and Parks, 11*(1), 1–20. https://doi.org/10.1016/j.ijgeop.2022.11.001

Seleteng-Kose, L., Kobisi, K., Pool-Stanvliet, R., & Mohapi, K. (2021). A rapid biodiversity assessment of Lesotho's first proposed Biosphere Reserve: A case study of Bokong Nature Reserve and Tšehlanyane National Park. *African Biodiversity & Conservation, 51*(2).

Sirima, A., & Backman, K. F. (2013). Communities' displacement from national park and tourism development in the Usangu Plains, Tanzania. *Current Issues in Tourism, 16*(7–8), 719–735. https://doi.org/10.1080/13683500.2013.785484

Sommerville, M., Khumalo, L., Kamuti, T., & Brooks, S. (2021). *Game on! Understanding the emerging game meat value chain in South Africa*. Tiny Beam Fund.

Soto, B. (2009). Protected areas in Mozambique. In H. Suich, B. Child, & A. Spencely (Eds.), *Innovation and evolution in wildlife conservation: Parks and game ranches to transfrontier conservation areas* (pp. 85–100). Earthscan.

Tweddle, D. (2000). The role of Nkhotakota wildlife reserve, Malawi, Africa, in the conservation of the Mpasa, *Opsavidium micvolepis* (Günther). In G. Cowx (Ed.), *Management and ecology of river fisheries* (pp. 400–417). Fishing News Books. https://doi.org/10.1002/9780470696026.ch2

Vieira-Alencar, J. P., Bolochio, B. E., Carmignotto, A. P., Sawaya, R. J., Silveira, L. F., Valdujo, P. H., de Campos Nogueira, C., & Nori, J. (2023). How habitat loss and fragmentation are reducing conservation opportunities for vertebrates in the most threatened savanna of the world. *Perspectives in Ecology and Conservation.* https://doi.org/10.1016/j.pecon.2023.02.004

Walsh, M. T. (2007). Island subsistence: Hunting, trapping and the translocation of wildlife in the Western Indian Ocean. *AZANIA: Journal of the British Institute in Eastern Africa, 42*(1), 83–113. https://doi.org/10.1080/00672700709480452

Chapter 7
The Drivers and Barriers of Wildlife-Based Tourism: Cases from Southern Africa

Owen Gohori

7.1 Introduction

The abundance of wildlife resources in Africa makes it one of the world's best destinations for wildlife-based tourism. Africa has various species of flora and fauna, with the southern African region being home to 70% of the remaining African elephant population that is threatened with extinction (World Wide Fund for Nature [WWF], 2018). Over the years, wildlife-based tourism has been growing exponentially both in volume and value for both terrestrial and marine environments (Newsome et al., 2019). It has attracted interest from governments, the tourism industry, and researchers (Higginbottom, 2004). This growth has prompted many developing countries to use it as a strategy to benefit local economies through the sustainable consumptive and non-consumptive use of wildlife resources (Newsome et al., 2019). Wildlife-based tourism directly contributed US$120.1 billion to the global GDP in 2018. Wildlife-based tourism also directly contributed to the creation of 9.1 million jobs worldwide in the same year. Thus, wildlife-based tourism could be key in alleviating poverty in less developed countries (World Travel & Tourism Council [WTTC], 2019).

In South Africa, the Department of Environmental Affairs (2018) states that the wildlife economy drives rural development and prosperity through the sustainable use of wildlife assets, the socio-economic benefits of ecotourism, co-managed conservation areas, and related ancillary services to protected areas. Similarly, the South African National Parks (2019) noted that approximately US$120 million was generated by seven million visitors to national parks. In Zimbabwe, the Communal Areas Management Programme for Indigenous Resources [CAMPFIRE] generates

O. Gohori (✉)
Tourism Research in Economics, Environs and Society (TREES), North-West University, Potchefstroom, South Africa

© The Author(s), under exclusive license to Springer Nature Switzerland AG 2024

L. S. Stone, M. T. Stone (eds.), *Wildlife Tourism Dynamics in Southern Africa*, Geographies of Tourism and Global Change,
https://doi.org/10.1007/978-3-031-57252-4_7

90% of its revenue from consumptive tourism and then allocates 58% of that revenue to communities to fund various development projects and the construction of schools and clinics (Gohori et al., 2022). In Madagascar, lemur viewing and bird watching at Andasibe National Park in Madagascar generates a large amount of revenue for the country (Newsome & Hughes, 2016; Newsome et al., 2019).

However, it is of paramount importance that countries that rely on wildlife-based tourism should gain a better understanding of the drivers and barriers of wildlife-based tourism. This enables them to attain good management of wildlife-based tourism and craft policies that promote drivers of wildlife-based tourism. Therefore, this chapter identifies and discusses the drivers and barriers of wildlife-based tourism by critically evaluating existing literature. The chapter contributes to the knowledge of the drivers and barriers of wildlife tourism. This knowledge may be beneficial to researchers, tourism planners, policymakers, tour operators, and government departments involved with wildlife-based tourism not only in southern Africa but globally.

7.2 Wildlife-Based Tourism

The core of wildlife-based tourism experiences entails the encounter between the visitor and wildlife. According to Newsome et al. (2005, pp. 18–19), "Wildlife tourism is defined as tourism undertaken to view and/or encounter wildlife. It can take place in a range of settings, from captive, semi-captive, to in the wild, and encompass a variety of interactions from passive observation to feeding and/or touching the species viewed". Although the term wildlife-based tourism is usually used to refer to the watching of and interaction with animals, it incorporates both fauna and flora (Newsome & Rodger, 2013). Wildlife-based tourism also includes the consumptive use of wild animals in natural areas (Spenceley, 2008). Consumptive use of wildlife forms a part of the tourism experience since it is sometimes undertaken by tourists (Newsome et al., 2005). Wildlife-based tourism is an area that overlaps nature-based tourism, ecotourism, consumptive use of wildlife, rural tourism, and human relations with animals (Reynolds & Braithwaite, 2001). It shares some of the characteristics of ecotourism when it takes place in natural areas, is educative and/or interpretive, and fosters conservation-supporting practices (Newsome et al., 2005). Wildlife-based tourism is more than travel to enjoy or appreciate wildlife but also entails contributions to conservation and community projects in less developed countries as well as environmental education and awareness for wildlife tourists (Sustainable Tourism Cooperative Research Centre [STCRC], 2009). Thus, wildlife-based tourism experiences are an interaction of elements relating to the natural resource base (i.e. wildlife and associated habitat), the visitor, the operator, host communities, the economy, and any management set in place (STCRC, 2009).

Wildlife-based tourism can be divided into three main experiential categories: (i) general nature-based tours that include a wildlife experience component; (ii) wildlife experience destinations (e.g. breeding colonies and aggregation of wildlife such as coral reefs); and (iii) specialised wildlife tours that target certain groups such as birds, monkeys, or apes (Newsome et al., 2005; Newsome & Rodger, 2013). Nonetheless, other categories such as the semi-natural/captive animal encounters as well as incidental encounters that people may experience while engaging in some other form of activities such as cycling or mountain trekking, are excluded from the described three categories (Newsome & Rodger, 2013). Activities such as scuba diving in a coral reef, going on a wild safari in Africa, whale watching from land or a boat, bird-watching either independently or with a commercial guide, spotting animals while walking in a national park or during a game drive; staying at an eco-resort with neighbouring wildlife; visiting a zoo, aquarium or wildlife sanctuary; and trophy hunting are examples of wildlife-based tourism-related activities (Higginbottom, 2004). Other examples of wildlife-based tourism activities include specialist animal watching; nature-based tours; thrill-seeking tours; artificial wildlife attractions; feeding birds; viewing nocturnal species; snorkelling to view underwater life; and swimming with whales, dolphins, and whale sharks; as well as hunting and fishing tours (Reynolds & Braithwaite, 2001). All wildlife-based tourism activities are characterised by wild animals as the major or significant part of the experience.

Duffus and Dearden (1990) proposed three dimensions of wildlife-human interaction for wildlife-based tourism: (a) hunting and fishing (i.e., consumptive use); (b) zoos and aquaria (i.e., low-consumptive); and (c) wildlife observation and photography (i.e., non-consumptive). The next section discusses the types of wildlife-based tourism.

7.3 Types of Wildlife-Based Tourism

Wildlife-based tourism includes diverse experiences. Sinha (2001) postulates that the categorisation of wildlife tourism is based on the motivation of tourists and the level of interactions of tourists with wildlife. Despite the very diverse experiences, wildlife-based tourism can be categorised into two main broad forms: non-consumptive and consumptive tourism.

7.4 Non-consumptive Wildlife-Based Tourism

This includes recreational activities that neither kill nor catch wild animals (e.g. wildlife watching, photographic safaris, and video recording) (Sinha, 2001). Tourists who participate in wildlife-based activities expect to see or interact with wildlife at the visited destination (purposive motivation of tourists), while tourists who travel without any specific intention to view or interact with wildlife consider their encounter with wildlife to be an added value to their recreational experience (incidental motivation of tourists) (Sinha, 2001). Although non-consumptive wildlife-based tourism activities are considered benign or passive, they can still cause significant negative impacts on wildlife and their habitat. For example, tourists who prefer to get closer to wildlife can cause disturbances to wildlife, while too many tourists at a time on a game drive may damage the environment and disrupt wildlife's reproductive rituals (Sinha, 2001). Likewise, a large number of tourist boats approaching whales during whale watching and close human contact with wild dolphins can affect their feeding and bonding as well as their normal behavioural patterns and survival instincts.

7.5 Consumptive Wildlife-Based Tourism

This involves the capture or killing of target animals (Sinha, 2001). There are three forms of consumptive wildlife-based tourism: (1) recreational hunting of waterfowl and big game (may also be valued for meat); (2) recreational fishing (fish may be released after catching or valued for food); and (3) trophy hunting and fishing (where the trophy itself, as well as the thrill of hunting itself, may be valued) (Freese, 1998). Similarly, there are three different market segments for consumptive wildlife-based tourism: (i) the 'big game' hunters with their various subdivisions (i.e. game ranching, big game, safari, group, indigenous), all of which target the experience, adventure, potential danger, and acquisition of a trophy; (ii) the 'small game' hunters, who are more interested in the hunting experience and skill displayed; and (iii) skill hunting, which specifically uses certain hunting tools (eg. bow, muzzle-loader, and various traps) (Bauer & Herr, 2004, p. 60).

Although some scholars and animal activists dismiss consumptive wildlife-based tourism as unethical and unsustainable (e.g., Newsome et al., 2005), Milner-Gulland and Mace (1998) argue that it is necessary as too much wildlife in a given area may cause damage to the environment due to overcrowding. Nonetheless, uncontrolled hunting and killing of animals through poaching may lead to the extinction of certain threatened wildlife species such as rhinoceros in southern Africa.

However, the distinction between consumptive wildlife-based tourism and non-consumptive wildlife-based tourism is smaller in some cases. The categorisation of wildlife-based tourism activities into consumptive and non-consumptive groupings is limited due to the extent of the overlap between their markets which is not clear. Thus, the appropriate type of classification depends on the purpose of the classification and the person or organisation doing the classification (Higginbottom, 2004).

7.6 Drivers of Wildlife-Based Tourism in Southern Africa

Knowledge of the drivers of wildlife-based tourism is necessary for the quantification of a destination's tourism potential, enabling investments to be channelled to areas that are better suited to wildlife tourism, identifying areas where wildlife-based tourism is likely to succeed, and the development of effective marketing strategies and sustainable management plans that incorporate the needs, interests, and preferences of tourists (Kirkland, 2021).

One of the major drivers of wildlife-based tourism in southern Africa is the abundance of wildlife. It is usually both the quantity and quality of flora and fauna that play a major role in attracting tourists to specific destinations (Newsome et al., 2013). As Africa has the largest proportion of land allocated for protected areas [PAs] (Gohori et al., 2022), most of such protected areas in southern Africa have 'The Big 5' (i.e. lion, elephant, leopard, buffalo, rhino) that are healthy and thus attract visitors to the region. They are found in Hluhluwe-iMfolozi Park (South Africa), Matusadonha National Park (Zimbabwe), Okavango Delta (Botswana), Etosha National Park (Namibia), Majete Wildlife Reserve (Malawi) and the North and South Luangwa National Parks in Zambia. Besides the 'Big 5', the region is home to many other species of both flora and fauna. South Africa is the third most biodiverse country in the world (Driver & Mukhadi, 2019). Similarly, Chobe National Park in Botswana has the highest concentrations and the most diverse range of wildlife in Africa (Moswete et al., 2017; Jagyasi, 2021). Kirkland (2021) asserts that the availability of a wider biodiversity both within the terrestrial and marine realms attracts wildlife enthusiasts. Thus, both land and whale watching are popular in Madagascar, Mauritius, Mozambique, Namibia, and South Africa (O'Connor et al., 2009). Most trophy hunters are attracted to Zimbabwe by the quality of trophies, especially buffalo trophies which are rated among the best (Lindsey et al., 2006).

The availability of abundant wildlife in southern Africa is attributed to good conservation strategies. Tichaawa and Lekgau (2020) consider conservation to be a prominent attribute of wildlife-based tourism. Likewise, Bhatasara et al. (2013) argue that conservation is a core component of wildlife-based tourism since its existence is dependent on the attractiveness and viability of the natural environment. The concept of conservation is not *terra incognita* in the African culture as it has always been a way of life in rural communities for millennia (Gohori, 2020). Most communities in Africa live among wildlife resources that are extinct in many parts of the world (e.g., the African elephant). This is attributed to the indigenous knowledge and culture of the people who have managed to 'maintain a sustainable utilisation and protection of commonly shared natural resources' (Lalonde 1991, p. 4). Research undertaken in Askham, South Africa, revealed that wildlife-based tourism thrives in that area owing to the efforts of the local community to conserve the natural environment (Tichaawa & Lekgau, 2020). In other research conducted in Mahenye village in Zimbabwe, local people boasted of having many elephants in their area due to their conservation abilities and hunting methods that are sustainable (Gohori & van der Merwe, 2022). Consumptive tourism has also been credited with promoting conservation in southern Africa. Lindsey et al. (2006) assert that in Mozambique trophy hunting is conducted on 82,250 km^2 of land and in Zimbabwe, on 65,000 km^2 of land. Most of this land is additional to the protected areas, making trophy hunting a major contributor to conservation. Trophy hunting also contributes to conservation as the offtake is only 2–5% of male populations due to the adherence to the quota setting recommended by the Convention on International Trade in

Endangered Species of Wild Fauna and Flora [CITES] in each country. Thus, trophy hunting is sustainable and low risk if well managed (Bond et al., 2004) while it can play a role in endangered species conservation and the rehabilitation of wildlife species (Lindsey et al., 2007).

Besides the presence of various wildlife species, some scholars (e.g., Loureiro et al., 2012) argue that the landscape is a key element of the wildlife-based tourism experience. Chung et al. (2018) observed that PAs that are highly elevated with diverse land cover and open vegetation are heavily visited. This is due to the opportunity they afford tourists to view wildlife and other natural attractions and a chance to participate in other recreational activities. Protected areas (PAs) at higher elevations may also encompass relatively pristine ecosystems due to low development pressure, making them popular with wildlife-based enthusiasts (Kirkland, 2021). This makes wildlife-based tourism popular in the southern African region as most of the PAs are highly elevated in the mountainous regions (e.g., Drakensberg National Park in South Africa, Nyanga National Park in Zimbabwe, and the Lubombo Transfrontier Conservation Area (TFCA), which straddles the borders of South Africa, Mozambique, and Eswatini). The climate also plays a crucial role as a driver of wildlife-based tourism, as PAs located in areas with cool temperatures and low rainfall have been found to be more heavily visited (Chung et al., 2018). This could be the reason why most wildlife-based tourists visit the region during summer when animals can be easily spotted at waterholes (Jagyasi, 2021).

As wildlife tourists are heterogeneous, they have different preferences for species and other components of wildlife-based tourism experiences (Kirkland, 2021). Thus, there are those who enjoy birdwatching, game viewing, trophy hunting, and whale watching, among others. Duffus and Dearden (1990) suggested a conceptual framework for classifying wildlife tourist typologies based on the experience level and involvement in or commitment to certain wildlife-based tourism activities. The framework argues that initially, a wildlife tourism site attracts exploratory, specialist, enthusiasts (e.g., bird watchers) who are more knowledgeable and environmentally conscious, fewer in number, and require little infrastructure or interpretative facilities. Over time, due to growing awareness, marketing, and development of the site, more tourists begin to patronise the site. When the site has reached this stage, there will be a need to introduce more wildlife-based activities and fauna. The establishment of TFCAs in southern Africa is a good initiative to facilitate the presence of different species of fauna in the countries in the region since the animals can cross the borders freely.

Recently, it was discovered that wildlife-based tourists are also thrilled by the surrounding communities (Tichaawa & Lekgau, 2020). Research in Kgalagadi Transfrontier Park revealed that wildlife enthusiasts who visit the area are not only attracted by the various wildlife species found there but the unique cultural landscape that encompasses one of the oldest traditional communities in southern Africa (Tichaawa & Lekgau, 2020). Wildlife-based tourists are also interested in the culture of the host communities living close to protected areas, especially how they interact with nature. Some recent studies (Black & Cobbinah, 2018) have shown the role that culture plays in wildlife-based tourism. This has prompted the proliferation of community-based tourism projects in the region. Saarinen et al. (2012) claim that the purpose of the establishment of the Kgalagadi Transfrontier Park was to collaborate with local communities and their culture to develop a responsible tourism sector rather than to only conserve wildlife.

7.7 Barriers to Wildlife Tourism in Southern Africa

Despite having several drivers, wildlife-based tourism in southern Africa is facing some notable barriers. Although there is abundant wildlife in the region, poaching is becoming rampant and threatening the extinction of certain species such as rhino and elephant. Poaching decreases wildlife populations and adversely affects ecosystems, which has a deteriorating effect on wildlife-based tourism experiences. Poaching also creates a poor image of a destination country resulting in fewer wildlife-based tourists visiting the country (UNWTO, 2015). Poaching in some parts of southern Africa has been attributed to human-wildlife conflicts (HWC) and a lack of or unequal distribution of revenue from wildlife-based tourism, prompting the locals to connive with poachers. A classic example of this occurred when more than 300 elephants and other animals such as lions, hyenas, vultures, and kudu, among others, were killed by cyanide poisoning when local villagers in Hwange connived with poachers (African Wildlife Foundation, 2013).

In the Okavango Panhandle, around 18,000 elephants are confined into a triangular 8700 km^2 of land, isolated from the rest of KAZA, where there are 13 villages with a population of more than 16,000 people. This has created pressure and HWC (Tan, 2019). In Caprivi, Namibia, 5000 elephants share their habitat with local people, in turn causing much crop damage (Nicole, 2019). Wherever humans and wildlife co-exist, HWC is inevitable (Le Bel et al., 2011). Selier et al. (2014) posit that 80% of the current African elephant range in southern Africa fall outside formally protected areas, leading to increased conflict with local communities. Adams and Infield (2003) conclude that conservation should not compromise poverty reduction. The success of any conservation policy is measured by its impact, not only on species and landscapes but also by the process with which it improves the lives of local people. If wildlife becomes just a cost to local residents through the destruction of crops, eating of livestock, and killing of people, conservation efforts will not be supported by local people and will result in poaching (Adams & Infield, 2003). Hübschle and Shearing (2018, p. 12) assert that "as long as local communities remain on the margins of protected areas and are excluded from the economic benefits of conservation, one should not be surprised when they fail to support the conservation drive, or even take to poaching".

As indigenous and local communities are stakeholders in wildlife-based tourism, the failure to devolve authority and power to grassroots levels present a barrier to wildlife-based tourism in southern Africa. According to Jones and Erdman (2013), formal community-based natural resource management is characterised by the devolution of resource rights by the state to community structures and is mostly based on wildlife utilisation, wildlife-based tourism, and/or forestry. Effective devolution promotes community-based conservation and can facilitate equitable distribution of revenue from wildlife-based tourism as the allocation of benefits is determined by local democratic decision-making processes in which the needs of the local people are known (Brockington et al., 2008). The Zambia Wildlife Authority maintains control over wildlife in communal lands while in Mozambique, ambiguous wildlife legislation has been constraining local influence over wildlife resources (Muchapondwa & Stage, 2015). In Zimbabwe, the Rural District Councils (RDCs) are still reluctant to devolve power and authority further to grassroots levels (Gohori & Van der Merwe, 2021). However, it is argued that the best examples of the devolution of authority and power in the region are found in Botswana (trusts), Eswatini (trusts), Namibia (conservancies), and Makuleke in South Africa (communal property associations) as local residents have received proprietary rights over natural resources and have authority to make rules, approve developments, enter into partnerships with the private sector, receive revenues and distribute benefits (Gohori et al., 2022).

Political instability coupled with bad publicity is another barrier to wildlife-based tourism in southern Africa. Africa popularly known as the 'dark continent', is portrayed by the media as a place of famine, violence, human rights abuse, and genocide (Avraham & Ketter, 2017). Over the years, sub-Saharan Africa has faced various challenges that include political instability, violent conflicts, underdevelopment, natural disasters, epidemics, and extreme poverty (Boniface et al., 2012; Avraham & Ketter, 2017). Such challenges, coupled with what the international media usually writes about the region, have prompted sub-Saharan African countries to gain a negative place image (Muhwezi et al., 2016; Avraham & Ketter, 2017). Political violence during the elections in Zimbabwe, civil war, and rebels in Mozambique as well as the high crime rates in South Africa, all affect the potential of wildlife-based tourism in the region. Several international tourists were killed in South Africa's Kruger National Park in 2022. Perceptions of safety and security play an important role in tourists decision-making, the lack of which would deter them from visiting countries with high levels of crime, conflicts, and unstable political climates (Kirkland, 2021). Nonetheless, problems in other regions of the continent also affect wildlife-based tourism in southern Africa. For example, the outbreak of Ebola in West Africa in 2014 resulted in the cancellation of confirmed bookings in the region, as most overseas tourists perceive Africa as one destination (Zimbabwe Tourism Authority, 2020). Likewise, the current coups in North Africa may also impact wildlife-based tourism in southern Africa negatively.

The regulations of CITES and the actions of some environmental and animal activists who campaign and demonstrate against consumptive tourism affect wildlife-based tourism. The killing of Cecil the lion in Hwange, Zimbabwe, in July 2015 prompted international media attention causing animal conservationists and activists to ban trophy hunting. The CITES species listings of fauna and flora impinge upon communities that live close to them, affecting their harvesting or trading. A former harvester may turn into a criminal (Hübschle & Shearing, 2018), prompting conflicts between communities and wildlife-based tourists. Southern African countries are also contesting the placing of the African elephant (*Loxodonta africana*) as in CITES Appendix II. This has placed limitations on the consumptive use of the elephant as the countries have to be allocated a quota of the animals to be killed while tons of ivory collected from those elephants that are culled or die naturally continue to pile in government warehouses as they cannot be sold until CITES gives permission. Such moves deprive the countries in the region from generating revenues through selling ivory and trophy hunting that can be used to support conservation efforts. Many southern African countries have ventured to such an extent that they have included community representatives in their official delegation list when attending the CITES Conference of the Parties so that they can share their plight with the world (Hübschle & Shearing, 2018).

Unreliable and convenient transport links are other barriers to wildlife-based tourism in southern Africa as most wildlife-based tourism sites are located in remote areas (Jagyasi, 2021; Kirkland, 2021). As most of the countries in the region are less developed, they do not have air connections between major cities and PAs, and the available road networks are in a poor state. In most cases, they do not have scheduled coaches to cater for tourists. These challenges render these PAs not only difficult to reach but also expensive. Besides the unreliable and convenient transport links, most Africans in the region have an unfriendly visa policy. Nonetheless, the region launched the KAZA Uni visa in 2016 to facilitate easy travel within the region. The plan is to make it a single visa for the Southern African Development Community (SADC) region (Myambo & Zengeni, 2019). Although in recent years, the region has followed global trends by adapting the visa on arrival and eVisas, only a few countries in southern Africa score favourably on the visa openness index. These include Mauritius (84.6), Mozambique (70.8), Seychelles (70.6), Madagascar (70.0), and Zimbabwe (67.9) (UNWTO, 2019).

Socioeconomic factors also affect wildlife-based tourism negatively (Willemen et al., 2015). Few or limited amenities and poor infrastructure deter wildlife-based tourists from PAs. The results of several studies (e.g. Hausmann et al., 2019) indicate that PAs with more amenities and well-developed infrastructure are visited more often compared with those with poor amenities and poorly developed infrastructure. This could be the reason why South Africa remains the most visited country in the region, as it has world-class infrastructure (Jagyasi, 2021). Likewise, many PAs and resorts in countries within the region are visited more often than others that are difficult to reach and have poor infrastructure and amenities.

7.8 The Way Forward

This chapter has discussed the factors that promote and deter wildlife-based tourism in southern Africa. The drivers and barriers to wildlife-based tourism have been identified in southern Africa. The drivers include abundant wildlife, good conservation strategies, the landscape, the formation of TFCAs, and local communities. Likewise, poaching and HWC, failure to devolve authority to communities, political instability, CITES regulations, unreliable and convenient transport links, and socioeconomic factors were identified as the barriers to wildlife-based tourism in Southern Africa. The recommendations provided below may assist in mapping the way forward for wildlife-based tourism in southern Africa.

For the region to fully benefit from wildlife-based tourism, there is a need to address the factors that present a deterrent to visitors. In terms of the devolution of power and authority to communities, despite great strides having been made through trusts, conservancies, and community property associations, the crafting of policies that promote the devolution of power and authority to grassroots levels will go a long way. Strategies such as involving communities when setting hunting quotas and having tourism revenues paid directly to communities through trusts are also key in addressing the issue of devolution of power and authority to grassroots levels. Gohori (2020) found that although some communities in Zimbabwe had formed trusts, the government still preferred to pay revenues from trophy hunting through the RDCs who later took time to disburse it to the communities.

In terms of political instability and bad publicity, political leaders should solve their problems amicably through the Southern African Development Community (SADC) which should be more active in promoting regional peace and security. The Regional Tourism Organisation of Southern Africa (RETOSA) should do more to achieve its mandate to promote and market the region to counter negative publicity. This can be carried out by opening offices in the main generating countries of the region as well as campaigning against decisions to ban trophy hunting and counter bad publicity. Of late, the organisation has not been visible.

The governments of the countries in the region must make sure that revenue derived from wildlife-based tourism (e.g., trophy hunting fees, entrance fees) is used for conservation purposes such as fighting poaching or compensating communities affected by HWC rather than being diverted to cater for other issues such as political campaigns. With the KAZA Uni visa now in place, the region must expedite a one-SADC visa to encourage the flow of wildlife-based tourists between countries in the region.

The identified drivers and barriers to wildlife-based tourism as well as the given recommendations may be key to stakeholders in the tourism industry, specifically those involved with wildlife-based tourism. This may assist them with identifying and promoting products and ways that facilitate the growth of the sector. Having a better understanding of both the drivers and barriers offers an opportunity to find ways and methods to maximise the drivers while averting the barriers. This chapter adds to the discussion on the factors that influence and deter the growth of wildlife-based tourism. Such a discussion provides the basis for the crafting of strategies, policies, and the cooperation of countries in the region to promote wildlife-based tourism. The discussion may also benefit countries that promote wildlife-based tourism globally.

References

Adams, W. M., & Infield, M. (2003). Who is on the gorilla's payroll? Claims on tourist revenue from a Ugandan National Park. *World Development, 31*(1), 177–190. https://doi.org/10.1016/S0305-750X(02)00149-3

African Wildlife Foundation. (2013). *Cyanide poisoning wreaks havoc on Zimbabwe elephants, ecosystem.* https://www.awf.org/news/cyanide-poisoning-wreaks-havoc-zimbabwe-elephants-ecosystem

Avraham, E., & Ketter, E. (2017). Destination image repair while combatting crises: Tourism marketing in Africa. *Tourism Geographies, 19*(5), 780–800. https://doi.org/10.1080/14616688.2017.1357140

Bauer, J., & Herr, A. (2004). Hunting and fishing tourism. In K. Higginbottom (Ed.), *Wildlife tourism: Impacts, management and planning* (pp. 57–77). Common Ground Publishing.

Bhatasara, S., Nyamwanza, A. M., & Kujinga, K. (2013). Transfrontier Parks and development in Southern Africa: The case of the Great Limpopo Transfrontier Park. *Development Southern Africa, 30*(4–5), 629–639. https://doi.org/10.1080/0376835X.2013.837377

Black, R., & Cobbinah, P. B. (2018). Local attitudes towards tourism and conservation in rural Botswana and Rwanda. *Journal of Ecotourism, 17*(1), 79–105. https://doi.org/10.1080/14724049.2016.1258074

Bond, I., Child, B., De la Harpe, D., Jones, B., Barnes, J., & Anderson, H. (2004). Private land contribution to conservation in South Africa. In B. Child (Ed.), *Parks in transition* (pp. 29–61). Earthscan.

Boniface, B., Cooper, C., & Cooper, R. (2012). *Worldwide destinations: The geography of travel and tourism*. Routledge. https://doi.org/10.4324/9780429259302

Brockington, D., Duffy, R., & Igoe, J. (2008). *Nature unbound: Conservation, capitalism and the future of protected areas*. Earthscan. https://doi.org/10.4324/9781849772075

Chung, M. G., Dietz, T., & Liu, J. (2018). Global relationships between biodiversity and nature-based tourism in protected areas. *Ecosystem Services, 34*, 11–23. https://doi.org/10.1016/j.ecoser.2018.09.004

Department of Environmental Affairs. (2018). *Unlocking the socio-economic potential of South Africa's biodiversity assets through sustainable use of wildlife resources*. https://static.pmg.org.za/180227DEA.pdf

Driver, A., & Mukhadi, F. (2019). *An initial assessment of biodiversity-related employment in South Africa*. Development Research Unit. University of Cape Town.

Duffus, D. A., & Dearden, P. (1990). Non-consumptive wildlife oriented recreation: A conceptual framework. *Biological Conservation, 53*, 213–231.

Freese, C. (1998). *Wild species as commodities: Managing markets and ecosystems for sustainability*. Island Press.

Gohori, O. (2020). *Towards a tourism and community development framework: An African perspective*. Doctoral thesis, North-West University.

Gohori, O., & Van der Merwe, P. (2021). Barriers to community participation in Zimbabwe's community-based tourism projects. *Tourism Recreation Research*, 1, 91–14, 104. https://doi.org/10.1080/02508281.2021.1989654

Gohori, O., & Van der Merwe, P. (2022). Tourism and community empowerment: The perspectives of local people in Manicaland Province, Zimbabwe. *Tourism Planning & Development, 19*(2), 81–99. https://doi.org/10.1080/21568316.2021.1873838

Gohori, O., Van der Merwe, P., & Saayman, A. (2022). Promotion of pro-poor tourism in southern Africa: Conservation and development critical issues. In L. S. Stone, M. T. Stone, P. K. Mogomotsi, & G. E. J. Mogomotsi (Eds.), *Protected areas and tourism in southern Africa: Conservation goals and community livelihoods* (pp. 245–258). Taylor & Francis Group. https://doi.org/10.4324/9781003193166-22

Hausmann, A., Toivonen, T., Fink, C., Heikinheimo, Tenkanen, H., Butchart, S. H. M., Brooks, T. M., & Di Minin, E. (2019). Assessing global popularity and threats to important bird and biodiversity areas using social media data. *Science of the Total Environment, 683*, 617–623. https://doi.org/10.1016/j.scitotenv.2019.05.268

Higginbottom, K. (2004). Wildlife tourism: An introduction. In K. Higginbottom (Ed.), *Wildlife tourism: Impacts, management and planning* (pp. 1–14). Common Ground Publishing.

Hübschle, A., & Clifford Shearing, C. (2018). *Ending wildlife trafficking: Local communities as change agents*. Global Initiative Against Transnational Organized Crime. https://doi.org/10.13140/RG.2.2.11973.81123

Jagyasi, P. (2021). Complete guide to wildlife tourism. *Travel & Tourism*. https://drprem.com/travel/wild-life-tourism-guide/#Key

Jones, B. T. B., & Erdmann, T. K. (2013). *Community-based natural resource management in Southern Africa: A decision maker's guide*. USAID.

Kirkland, M. (2021). *The global drivers of wildlife tourism and its future potential in a changing world*. Doctoral dissertation, Durham University.

Lalonde, A. (1991, September 26–29). *African indigenous knowledge and its relevance to environment and development activities* [Conference session]. The Common Property Conference, Winnipeg, Manitoba, Canada. https://dlc.dlib.indiana.edu/dlc/bitstream/handle/10535/904/African_Indigenous_Knowledge_and_its_Relevance_to_Environment_and_Development_Activities.pdf?sequence=1

Le Bel, S., Murwira, A., Mukamuri, B., Czudek, R., Taylor, R., & La Grange, M. (2011). Human-wildlife conflicts in Southern Africa: Riding the whirl wind in Mozambique and in Zimbabwe. In J. López-Pujol (Ed.), *The importance of biological interactions in the study of biodiversity* (pp. 283–322). InTech. https://doi.org/10.5772/23682

Lindsey, P. A., Alexander, R., Frank, L. G., Mathieson, A., & Romanach, S. S. (2006). Potential of trophy hunting to create incentives for wildlife conservation in Africa where alternative wildlife-based land uses may not be viable. *Animal Conservation, 9*, 283–291. https://doi.org/10.1111/j.1469-1795.2006.00034.x

Lindsey, P. A., Frank, L. G., Alexander, R., Mathieson, A., & Romañach, S. S. (2007). Trophy hunting and conservation in Africa: Problems and one potential solution. *Conservation Biology, 21*(3), 880–883. https://doi.org/10.1111/j.1523-1739.2006.00594.x

Loureiro, M. L., Macagno, G., Nunes, P. A. L. D., & Tol, R. (2012). Assessing the impact of biodiversity on tourism flows: An econometric model for tourist behaviour with implications for conservation policy. *Journal of Environmental Economics and Policy, 1*(2), 174–194. https://doi.org/10.1080/21606544.2012.692865

Milner-Gulland, E. J., & Mace, R. (1998). *Conservation of biological resources*. Blackwell Science.

Moswete, N., Nkape, K., & Tseme, M. (2017). Wildlife tourism safaris, vehicle decongestion routes and impact mitigation at Chobe National Park, Botswana. In I. Borges de Lima & R. J. Green (Eds.), *Wildlife tourism, environmental learning and ethical encounters ecological and conservation aspects* (pp. 71–88). Springer. https://doi.org/10.1007/978-3-319-55574-4_6

Muchapondwa, E., & Stage, J. (2015). Whereto with institutions and governance challenges in African wildlife conservation? *Environmental Research Letters, 10*, 1–8. https://doi.org/10.1088/1748-9326/10/9/095013

Muhwezi, D. K., Baum, T., & Nyakaana, J. B. (2016). Dealing with negative symbolism of destinations with difficult heritage: Analysis of Uganda's image. *Journal of Hospitality Management and Tourism, 7*(3), 33–42. https://doi.org/10.5897/JHMT2014.0118

Myambo, B., & Zengeni, N. (2019). So Far So KAZA: An assessment of the contribution of the KAZA UNIVISA to tourism performance in Victoria Falls. *Journal of Humanities and Social Sciences, 23*(3), 49–56.

Newsome, D., & Hughes, M. (2016). Understanding the impacts of ecotourism on biodiversity: A multi-scale, cumulative issue influenced by perceptions and politics. In D. Geneletti (Ed.), *Handbook on biodiversity and ecosystem services in impact assessment* (pp. 276–298). Edward Elgar Books. https://doi.org/10.4337/9781783478996.00018

Newsome, D., & Rodger, K. (2013). Wildlife tourism. In A. Holden & D. Fennel (Eds.), *The Routledge handbook of tourism and the environment* (pp. 345–358). Routledge Taylor & Francis Group.

Newsome, D., Dowling, R., & Moore, S. A. (2005). *Wildlife tourism*. Channel View Publications.

Newsome, D., Moore, A. S., & Dowling, R. K. (2013). *Natural area tourism: Ecology, impacts and management* (2nd ed.). Channel View Publications.

Newsome, D., Rodger, K., Pearce, J., & Chan, K. L. J. (2019). Visitor satisfaction with a key wildlife tourism destination within the context of a damaged landscape. *Current Issues in Tourism, 22*(6), 729–746. https://doi.org/10.1080/13683500.2017.1312685

Nicole, B. (2019). An assessment of the human-wildlife conflict across Africa. In M. Feerretti (Ed.), *Wildlife population monitoring* (pp. 1–10). IntechOpen. https://doi.org/10.5772/intechopen.82793

O'Connor, S., Campbell, R., Cortez, H., & Knowles, T. (2009). *Whale watching worldwide: Tourism numbers, expenditures and expanding economic benefits*. Economists at Large.

Reynolds, P. C., & Braithwaite, R. W. (2001). Towards a conceptual framework for wildlife tourism. *Tourism Management, 22*(1), 31–42. https://doi.org/10.1016/S0261-5177(00)00018-2

Saarinen, J., Hambira, W. L., Atlhopheng, J., & Manwa, H. (2012). Tourism industry reaction to climate change in Kgalagadi South District, Botswana. *Development Southern Africa, 29*(2), 273–285. https://doi.org/10.1080/0376835X.2012.675697

Selier, S. J., Page, B. R., Vanak, A. T., & Slotow, R. (2014). Sustainability of elephant hunting across international borders in Southern Africa: A case study of the Greater Mapungubwe Transfrontier Conservation Area. *The Journal of Wildlife Management, 78*(1), 122–132. https://doi.org/10.1002/jwmg.641

Sinha, C. C. (2001, July 27). *Wildlife tourism: A geographical perspective* [Conference session] Geography Curriculum In service Conference, Sydney, Australia. https://www.scirp.org/(S(czeh2tfqyw2orz553k1w0r45))/reference/referencespapers.aspx?referenceid=1504800

South African National Parks. (2019). *Annual report for 2017/2018*. https://www.sanparks.org/assets/docs/general/annual-report-2018.pdf

Spenceley, A. (2008). Impacts of wildlife tourism on rural livelihoods in Southern Africa. In A. Spenceley (Ed.), *Responsible tourism: Critical issues for conservation and development* (pp. 159–186). Earthscan.

Sustainable Tourism Cooperative Research Centre. (2009). *Wildlife tourism: Challenges, opportunities and managing the future*. Sustainable Tourism Cooperative Research Centre.

Tan, J. (2019). Breaking down barriers: Cattle and wildlife compete in Southern Africa. *Mongabay*. https://news.mongabay.com/2019/12/breaking-down-barriers-cattle-and-wildlife-compete-in-southern-africa/

Tichaawa, T., & Lekgau, R. J. (2020). Nature and characteristics of wildlife tourism in the Kgalagadi Transfrontier Park. *AUDŒ, 16*(5), 69–88.

United Nations World Tourism Organisation. (2015). *Towards measuring the economic value of wildlife watching tourism in Africa—Briefing paper*. United Nations World Tourism Organization.

United Nations World Tourism Organisation. (2019). *2019 visa openness report for Africa*. United Nations World Tourism Organisation.

Willemen, L., Cottam, A. J., Drakou, E. G., & Burgess, N. D. (2015). Using social media to measure the contribution of Red List species to the nature-based tourism potential of African protected areas. *PLoS One, 10*(6), e0129785. https://doi.org/10.1371/journal.pone.0129785

World Travel & Tourism Council. (2019). *Global wildlife tourism generates five times more revenue than illegal wildlife trade annually.* https://wttc.org/news-article/global-wildlife-tourism-generates-five-times-more-revenue-than-illegal-wildlife-trade-annually
World Wide Fund for Nature. (2018). *The status of African elephants Loxodonta Africana.* https://www.worldwildlife.org/magazine/issues/winter-2018/articles/the-status-of-african-elephants
Zimbabwe Tourism Authority. (2020). *Tourism trends & statistics report.* http://www.zimbabwe-tourism.net/tourism-trends-statistics/

Chapter 8
Over Reliance on Wildlife Tourism: Impacts and Policy Implications

Lesego S. Stone

8.1 Introduction

Wildlife tourism is important to many countries in Africa. It continues to grow and characterises 80% of the total annual sales of trips to Africa (United Nations World Travel Organisation [UNWTO], 2015). Higginbottom (2004, p. 2) defines wildlife tourism as "tourism based on encounters with non-domestic animals, occurring … in either the animal's natural environment or in captivity. Page and Dowling (2002) posit that wildlife tourism includes nature-based tourism, rural tourism, ecotourism human relations, and consumptive activities such as hunting. According to Rizzolo (2023, p. 1181) non-consumptive wildlife tourism refers to "human recreation that does not remove or permanently alter wildlife" while consumptive tourism refers to "activities that result in wildlife mortality such as hunting, trapping, and fishing". In most Southern and East African countries, non-consumptive and consumptive forms of tourism are frequently practised together in the same destination (Lindsey et al., 2007; Mokgalo & van der Merwe, 2022). Furthermore, according to Mwakiwa et al. (2016), non-consumptive and consumptive wildlife tourists coming to Africa are attracted by the same large mammals, 'the big five' (elephant, rhino, lion, buffalo, and leopard).

Wildlife is the primary source of tourist revenue in Africa through hunting and photographic safaris. It is the main attraction for tourists who visit conserved and protected areas (Tairo, 2020; van der Merwe et al., 2021a). Most wildlife tourists who visit East and Southern Africa are from the Global North from countries such as Australia, Canada, United States of America and from Western Europe (Holechek

L. S. Stone (✉)
Faculty of Business, Department Tourism and Hospitality Management, University of Botswana, Gaborone, Botswana
e-mail: stonel@ub.ac.bw

© The Author(s), under exclusive license to Springer Nature Switzerland AG 2024

L. S. Stone, M. T. Stone (eds.), *Wildlife Tourism Dynamics in Southern Africa*, Geographies of Tourism and Global Change,
https://doi.org/10.1007/978-3-031-57252-4_8

& Valdez, 2018). Wildlife tourists engage in this type of tourism because they want to escape the rush and congestion in their towns and cities (van der Merwe et al., 2021b).

Countries in the southern and eastern parts of Africa are recognised for their variety and abundance of wildlife (Holechek & Valdez, 2018). Southern African countries of Mozambique, Namibia, Zimbabwe, Zambia, Botswana and South Africa are mainly renowned for the "big five" (Holechek & Valdez, 2018). According to Lindsey et al. (2007) and IUCN (2016), since the early 1990s there have been positive changes in wildlife populations in some southern African countries. This has been facilitated by factors such as the reinforcement of international agreements that limit trade in endangered species and animal by-products (particularly rhino horn and ivory); the amplified role of the United Nations Economic, Scientific, and Cultural Organization (UNESCO) in delivering international protection to exceptional wildlife areas (e.g. World Heritage Sites such as the Serengeti); an increase in the number of bioreserves and national parks; extensions of numerous existing national parks; the increased protection of animals and their habitats; the development of game ranching for sport hunting and meat; the formation of private game reserves; private concessions; the growth of community conservancies and ecotourism (Child, 2019; Lindsey et al. 2007; Pack et al., 2013; Holechek & Valdez, 2018; Nattrass, 2021). Although most wildlife tourism sites include attributes of both entertainment and conservation, they vary on the importance they place on these two areas (Fennell, 2012).

Wildlife tourism is an importance part of the economies of many countries in the southern African region. This is because it can contribute to the management of protected areas, biodiversity conservation and local employment (Larm et al., 2018; Mangachena & Pickering, 2023). The significance of wildlife tourism is shown by the amount of land that is protected in many Southern African countries. This includes Namibia and Zambia with 38% of their land protected, 29% for Botswana, 26.6% for Zimbabwe, 17.2% for Mozambique and 8.8% for South Africa (World Bank, 2014). Wildlife tourism mostly occurs in the protected areas found in the region, in game reserves, Transfrontier Parks and national parks (Tichaawa & Lekgau, 2020; UNWTO, 2015).

Due to the significance of wildlife tourism in the Southern African region, some countries heavily rely on it. Although this can be a blessing, it can also have negative impacts on such countries. This is the case in Botswana where wildlife tourism contributes significantly to the country's economy. The country provides a case for this chapter. Botswana's wildlife tourism is characterised by both consumptive (hunting) and non-consumptive (photographic) tourism. The chapter uses available literature to discuss the negative impacts associated with the heavy reliance on wildlife tourism using Botswana as a case study.

8.2 Tourism in Botswana

Botswana's tourism industry heavily relies on wildlife and wilderness tourism and protected and conserved areas (mainly national parks and game reserves). Consequently, 17% of the country has been designated as national parks and game reserves and an additional 20% as wildlife management areas (WMAs) (Government of Botswana, 2000). This is where most of the tourism in the country takes place, mainly in the Northern part of the country, predominantly in the Okavango Delta (the largest inland wetland of its kind in the world) area, the Chobe National Park, and the Moremi Game Reserve (Chatibura, 2023). Protected areas are the main tourist attractions and are standard among international tourists (DWNP, 2014; Stone & Nyaupane, 2016). Botswana receives international tourists from countries such as the United States of America (26.6%), Germany (17.6%) and the United Kingdom (17.3); from the African continent, the country attracts tourists mainly from South Africa (43.5%), Zimbabwe (34.6%) and Zambia (13.9) (Statistics Botswana, 2022).

The country promotes sustainable ecotourism through a high-value low-volume tourism policy to derive revenue, develop the country and for the conservation of natural resources (African Nature-Based Tourism Platform, 2021; Government of Botswana, 1990). Consequently, the country promotes itself as a luxury safari destination, attracting tourists mainly travelling for holiday (19.6%), visiting friends and relatives (11.2%) and business (20.1%) (Statistics Botswana, 2022). Wildlife tourism in the country is not only valuable to the public and private sectors, it is also important to local communities, who also derive benefits from it (Stone & Nyaupane, 2016) through the Community-Based Natural Resource Management (CBNRM) programme. Through the CBNRM programme, the country ensures that sustainable ecotourism development and conservation take place while also developing rural communities and improving their livelihoods (Stone et al., 2021). Wildlife tourism attracts mainly international tourists to the country. This heavy reliance on wildlife tourism and international tourists has led to many benefits for the country including foreign exchange earnings, community benefits, conservation, job opportunities, rural development and business opportunities for many stakeholders (Mbaiwa, 2017, 2018, 2023).

8.3 Consequences of the Over Reliance on Wildlife Tourism in Botswana

The following factors were derived from literature and will be discussed below; the reliance on international markets; a non-existent domestic tourism market and the 2014 hunting ban and the impact of external factors.

8.3.1 The Reliance on International Tourists

Botswana is a popular tourist destination for tourists from the Global North (Westerners), specifically those from North America and Europe (Mbaiwa, 2011a; Statistics Botswana, 2022) who visit the country for holiday purposes at protected areas around the country, especially at popular tourist sites of the Chobe National Park, Okavango Delta and the Moremi Game Reserve. Despite the revenue that the country makes through international tourism, over reliance on international tourists who come for wildlife tourism leads to dependence on foreign skills, investments, and the repatriation of profits (Lepp, 2008). This is very prominent in Botswana's wildlife tourism where there is a huge dependence on international skills, tourists and internationally owned tourism facilities (Stone & Stone, 2022). In 2013, visits to protected areas comprised 71.1% international tourists and only 8.2% domestic tourists (Department of Wildlife and National Parks, 2014). Although wildlife tourism has been lauded for creating employment for local communities in areas such as the Chobe and Okavango Delta regions, most locals hold poor, lower paying jobs with lower salaries, while better paying jobs and managerial positions are held by expatriate workers (Mbaiwa, 2017; Stone et al., 2017). This is because locals lack the high-level technical skills required by the sector (Government of Botswana, 2021). Furthermore, most of the country's international tourist bookings are done in South Africa, resulting on revenue leakages estimated to be between 70% and 90% of transactions (Baatweng, 2014; Mbaiwa, 2017). This is because prime wildlife tourism areas such as the Okavango Delta are largely owned and controlled by multinational tourism companies with headquarters outside Botswana (Mbaiwa, 2023). This results in profits made in Botswana being repatriated back to companies' headquarters leading to large revenue leakages. Furthermore, there are poor linkages between the tourism sector and the local economy. In Botswana, this is often characterised by the tourism sector importing food, and construction materials from outside the country (Mbaiwa, 2017).

According to Mbaiwa (2005a, b, 2012, 2017) and Stone and Stone (2022), foreign owned companies and investors own and control around 82% of accommodation facilities in the Okavango Delta. This has resulted in large revenue leakages and the evasion of tax by these companies in Botswana (Mbaiwa, 2017; Stone & Stone, 2022). According to Mbaiwa (2005a) foreign safari companies own 53.8% of accommodation facilities, Botswana-owned and non-citizen companies jointly own and control around 28%, only 18.5% of facilities are fully owned and controlled by Botswana citizen companies. This indicates that foreign investors and companies have control and influence in almost 82% of accommodation facilities in the popular wildlife tourism destination of the Okavango Delta (Mbaiwa 2005a, b, 2012, 2017). Consequently, according to Mbaiwa (2005a, b, 2017), citizens and local communities perceive the foreign domination of the tourism industry negatively because they believe their natural resources are being 'sold out' or being seized from them by foreign tourism investors (Mbaiwa 2005a, b). This perception can have negative consequences for those resources that tourism relies on.

8.3.2 A Non-existent Domestic Tourism Market

Domestic tourism is the most significant form of tourism, in 2018, approximately 9 billion domestic tourism trips were taken globally, this constitutes six times the number of international tourist arrivals which stood at 1.4 billion that year (UNWTO, 2020).

Data on domestic tourism in Botswana is very scanty (Morupisi & Mokgalo, 2017). Like most developing countries, domestic tourism is neglected in Botswana, with most resources being focused on attracting international tourists. This perception is common in most countries in the Global South where millions of dollars are spent annually enticing international tourists (Scheyvens, 2007). Consequently, the numbers of domestic tourists visiting the wildlife tourism attractions, protected areas, are relatively low. According to Mbaiwa (2011a, b), less focus is given to domestic tourism in Botswana because it is perceived to be unable to bring the same degree of benefits such as employment opportunities, small business development, foreign exchange earnings and development like international tourism (Mbaiwa, 2011b). Despite, the limited focus on domestic tourism, it constitutes over 80% of world tourism flows (Mazimhaka, 2007). Furthermore, domestic tourism is acclaimed for its economic benefits and sustainability (Ghimire 2013; Rogerson & Lisa, 2005; Stone, 2014), especially when epidemics and natural calamities emerge resulting in international and regional tourists' movement restrictions.

Domestic tourism to leisure areas trails in comparison to outbound tourism figures. In 2010, for instance, there were 185,416 domestic leisure trips (Department of Tourism, 2012; Morupisi & Mokgalo, 2017) compared to 634,000 leisure trips taken by Batswana to Namibia, Zimbabwe and South (Business Monitor International, 2014). The above indicates the potential by Botswana residents to travel, this could create a platform for a sustained domestic consumption.

Despite this, domestic tourism has the potential to promote cultural understanding, the use of locally owned facilities and services such as transport, accommodation and food which do not require huge investments, it helps mitigate seasonality during the off season and the tourists interact more with the local economy (Seckelmann, 2002; Shah, 2000; Stone & Nyaupane, 2016). Since domestic tourists spend mostly in locally owned businesses, this could alleviate the huge revenue leakages caused by the over reliance on international tourists and investors. According to Sindinga (1996), domestic tourism can also play a pivotal role in the conservation of natural and cultural and resources (Sindinga, 1996). Domestic tourism is essential, especially in countries such as Botswana since the "absence of locals in their protected areas and the lack of appreciation of their own natural heritage could be a major impediment to the long-term sustainability of these spaces in post-colonial Africa" (Stone & Nyaupane, 2016, 142).

In their study, Morupisi and Mokgalo (2017) found that domestic tourism is not where it should be and that the following obstacles have to be overcome for domestic tourism to succeed in Botswana: the local culture which doesn't recognise the significance of tourism; the undiversified tourism product offering dominated by wildlife and the wilderness which does not appeal to most potential tourists because they seek novelty; the high prices that attract international tourism and deter domestic tourism growth and the belief by most locals that holidays and available time away from work provides prospects for visiting farming areas, cattle posts and home villages. The country has a three-tier settlement system (home/village, lands and cattleposts) where locals move between the three settlements; urban areas are considered as places people go to for economic reasons, to improve their lives and livelihoods and the home, lands and cattle posts are visited when one is away from work; this system can help elucidate on the movements between urban areas, cattleposts, lands and home (Stone & Nyaupane, 2016). Furthermore, the promotion of a high value-low volume tourism policy makes domestic travel to wildlife tourism sites very expensive for most residents and citizens of Botswana (Stone & Nyaupane, 2020).

The impact of the COVID-19 pandemic on the tourism sector showed the importance of domestic tourism to provide a cushion for the tourism sector in Botswana when travellers from source markets in Asia, Europe and North America are affected (Hambira et al., 2022). Domestic and regional tourism are considered necessary to the tourism recovery plan of many African countries (Hambira et al., 2022; Rogerson & Baum, 2020).

8.3.3 The 2014 Hunting Ban and Its Impact on Local Communities

The community-based natural resources management (CBNRM) programme was introduced in the 1990s in Botswana to get community support in natural resources conservation through the promotion of sustainable utilisation of the natural resources using both photographic and hunting tourism by local communities (Blackie, 2019; Government of Botswana, 2007). During its early stages, the focus of the CBNRM programme was on wildlife hunting, with local communities being given hunting quotas in their allocated concession areas (Blackie, 2019). Through their community-based organisations (CBO), local communities have joint venture partnerships (JVP) with private operators. Through the partnerships, communities usually rent their concession areas to private operators and get land rentals from JVP partners rights to conduct tourism operations on their land (Cassidy, 2021; Hambira et al., 2022). Communities also make money from wildlife tourism by selling their wildlife hunting quotas allotted to them by the Department of Wildlife and National Parks (DWNP) (Stone & Stone, 2022).

CBNRM assumes that communities must receive direct benefits from natural resources for communities to manage resources. Furthermore, the benefits must exceed the costs of managing the resources, and if community livelihoods are improved, community members will be more willing to adopt conservation principles (Mbaiwa, 2017; Agrawal & Gupta, 2005). CBNRM was introduced to promote the involvement of local communities in tourism to develop rural areas, involve local communities in tourism, reduce community resentments and alienation, alleviate rural poverty, empowering communities to manage resources sustainably and conserving natural resources (Rozemeijer & Van der Jagt, 2000; Thakadu, 2005). Many communities in the popular wildlife tourism destinations rely on community-based tourism through the CBNRM programme. Communities improve their livelihoods by engaging in both consumptive and non-consumptive tourism. Community-based tourism ventures mainly made money through the selling of their wildlife quota.

In 2014, Botswana banned consumptive tourism (safari hunting), a move that had significant impacts on local communities. The decision was based on suggestions that there was a decline in various wildlife species (Chase, 2011; Blackie, 2019; Department of Wildlife and National Parks, 2012). Despite this, evidence showed that other wildlife species including elephants, buffalos and zebras had been increasing exponentially (Chase, 2011; Department of Wildlife and National Parks, 2012). For instance, between 1992 and 2012, the country's elephant increased by 297% to reach an estimated total of approximately 207,545 elephants (Department of Wildlife and National Parks, 2012). Moreover, Blackie (2019) reports that the great elephant census of 2016 found that 58.9% of the African elephant population and a third (30.1%) of the global elephant population is found in Botswana.

The ban on safari hunting led to many communities losing income from safari hunting and community members losing jobs; this had a negative impact on rural livelihoods (Blackie, 2019; Stone & Stone, 2022). Before the ban, communities involved in safari hunting made huge sums of money through the sale of hunting quotas to professional hunters (Mbaiwa, 2018). For instance, in 2010, the Chobe Enclave Conservation Trust (CECT) made in BWP 63,793,330.80 (see Table 8.1), BWP 4,842,554.52 (76%) of which was from consumptive wildlife tourism (hunting) and only BWP 1,536,776.28 was from photographic tourism (Stone & Stone, 2022). After the hunting ban, communities' lost access to money made from safari hunting tourism. For instance, in 2015 the CECT made losses; the CBO made only

Table 8.1 Revenue made by CECT from 2010 to 2018

Year	Total revenue generated in BWP (Botswana Pula)	Portion of revenue generated from hunting in BWP (Botswana Pula)	Proportion of revenue generated from hunting (in %)
2010	6,379,330.80	4,842,554.52	76%
2011	5,770,992.06	3,435,950.82	60%
2012	5,647,402.14	3,915,385.80	69%
2013	7,097,536.47	4,147,508.52	58%
2014	3,152,725.71	0	0
2015	3,385,782. 88	0	0
2016	4,062,524.57	0	0
2017	4,140,387.75	0	0
2018	4,571,897.92	0	0

Source: Department of Wildlife and National Parks (2020)

BWP 3,385,782.88, about 53% less than what they made in 2010 with safari hunting. Table 8.1 shows the heavy reliance local communities had in wildlife tourism through the hunting quota which contributed 58–76% of revenue generated by community-based ventures such as CECT (Department of Wildlife and National Parks, 2020).

Other communities that experienced losses due to the hunting ban include Okavango Kopano Mokoro Community Trust whose income reduced from BWP4.8 million to BWP2.5 million and 40 job losses; Sankoyo Village whose income declined from BWP3.5 million to BWP1.8 million and 35 job losses and Mababe village with a tourism income decline from BWP3.5 million to BWP500 000 and a loss of 30 jobs (DWNP, 2015). The over reliance on wildlife tourism was especially felt by tourism dependent communities during the COVID-19 pandemic. It was during this time that 44 CBNRM community-based organizations collapsed due to the loss of the hunting quota (Stone et al., 2020). Furthermore, there were increases in human–wildlife conflicts, resulting in increases in the compensation budget which rose from BWP4 million (US$ 292,287.08) in 2014 to BWP21 million (US$ 1,534,507.17) in the 2018/2019 financial year (Stone et al., 2021; Mbaiwa & Hambira, 2023).

In addition to the loss of income and employment, the hunting ban led to increases in poaching and illegal hunting activities, the conversion of marginal areas to photographic tourism sites despite them not being suitable, the collapse of many community projects and trophy hunting investors leaving communities, and some concession areas remained unused because investors were not willing to invest in them (Mbaiwa & Hambira, 2023; Mbaiwa, 2018). Therefore, indications are that the over reliance by communities on wildlife tourism through safari hunting tourism created many problems for communities. Increased human-wildlife conflicts made the situation worse. Based on consultations with community members, the Government of Botswana reintroduced hunting and hunting quotas in 2019 (Stone et al., 2021). This was done to offset the losses suffered and improve the livelihoods of rural communities involved in CBNRM. CBOs rely on international safari hunting operators to buy their hunting quotas (Mbaiwa & Hambira, 2023; Stone et al., 2021).

8.3.4 The Impact of External Factors: The Case of COVID-19

Tourism is often affected by external factors such as disease outbreaks. In 2019, the world was affected by the COVID-19, bringing the sector to a standstill due to measures put in place to reduce the spread of the disease (Khan, 2020). Like other countries, Botswana put in place measures to reduce the spread of the disease. Such included travel restrictions. A s a result of these, the tourism industry in Botswana experienced the cancellation of bookings, the closure of several citizen-owned tourism enterprises and those that were still open operating at low capacity, job and revenue losses and tourism operators not being able to service their loans (Hambira et al., 2022; Presidential COVID-19 Task Force Bulletin, 2020a, b). Tourist flows to Botswana were also affected as a result. Due to movement restrictions, there were declines (from 2019 figures) tourism's contribution to the economy of Botswana. For instance, in 2020 tourism's contribution to GDP stood at 7.8.%, a decline of 48.6%, there was also an 18.4% decline in travel and tourism jobs, 58.7% decline in international visitor spending and a 27.2.% decline in domestic visitor spending (WTTC, 2021).

Tourism dependent communities were especially hit hard by COVID-19 because despite the hunting ban having been lifted in 2019, during the COVID-19 pandemic communities were not able to benefit from the 2020 hunting quota due to travel bans (Hambira et al., 2022). The COVID-19 has shown that the heavy reliance by CBOs in northern Botswana on wildlife tourism and the sale of hunting quotas and land rentals from the private sector is not sustainable. Furthermore, the reliance on the international tourism market also questions the sustainability of tourism in Botswana. This is because without international tourists', private tourism operators find it difficult to pay land concession rentals and CBOs fail to sell their hunting quotas. Consequently, CBOs and their associated communities face challenges during periods of uncertainty (Stone et al., 2021).

8.4 Policy Implications

Wildlife tourism is important to many southern African countries, Botswana included. The chapter has shown the repercussions of the over reliance of Botswana on wildlife tourism. These include the foreign ownership and control of tourism facilities, leading to revenue leakages and dependencies (see Britton, 1982; Weaver, 2006). The heavy reliance on the international market can be attributed to the high value-low volume tourism policy the country promotes (Government of Botswana, 1990, 2021). This has made wildlife tourism to be out of reach for many locals due to the high prices charged and the undiversified products which may be a hindrance for many locals who visit other countries in the region (Morupisi & Mokgalo, 2017; Stone & Nyaupane, 2016; Stone & Stone, 2022).

The chapter also highlighted that the reliance on international tourism has led to the sector ignoring and sidelining the domestic market, which remains very undeveloped in the country. However, research indicates the importance of domestic tourism in promoting sustainability (Kabote et al., 2017) by building resilience and restoring tourism (Woyo, 2021). The chapter highlighted the significance of safari hunting tourism to the livelihoods of local communities participating in the CBNRM programme. Furthermore, it highlighted the need for communities to diversity their income earning opportunities to lessen their risks when the wildlife tourism industry is affected, as was the case during the COVID-19. The COVID-19 pandemic and travel restrictions on the international tourism market showed the significance of a well-developed domestic tourism market. It was during that time that there was a "domestic tourism drive" to wildlife areas through the creation of packages for the local market at reduced prices (Hambira et al., 2022). Other initiatives put in place included the "rediscover Botswana" initiative aimed at showcasing what the country has to offer at a range of budgets from low to high end (Hambira et al., 2022; Presidential COVID-19 Task Force Bulletin, 2020b). Ultimately, results from this chapter indicate the following; firstly, the need for communities to develop various income streams and not to rely on only one source of income. Secondly, it is important for countries to promote both domestic and international tourism markets, so that even during difficult times the tourism industry can still function and lastly, the tourism sector can be affected by external factors and it is important for countries to always be ready and have strategies in place to ensure that tourism can still prosper even during difficult times.

References

African Nature-Based Tourism Platform. (2021). *Botswana, country summary report.*

Agrawal, A., & Gupta, K. (2005). Decentralization and participation: The governance of common pool resources in Nepal's Terai. *World Development, 33*(7), 1101–1114.

Baatweng, V. (2014). Botswana loses up to 90% of tourism revenue. Botswana Sunday Standard, 10 July. www.sundaystandard.info/article.php/email.php?NewsID=20498

Blackie, I. (2019). The impact of wildlife hunting prohibition on the rural livelihoods of local communities in Ngamiland and Chobe District Areas, Botswana. *Cogent Social Sciences, 5*(1). https://doi.org/10.1080/23311886.2018.1558716

Britton, S. (1982). The political economy of tourism in the third world. *Annals of Tourism Research, 9*(3), 331–358.

Business Monitor International. (2014). *Botswana tourism report: A 5-year forecast to 2018.*

Cassidy, L. (2021). Power dynamics and new directions in the recent evolution of CBNRM in Botswana. *Conservation Science and Practice, 3*(1), e205.

Chase, M. (2011). *Dry season fixed wing aerial survey of elephants and wildlife in Northern Botswana.* Elephant Without Boarders.

Chatibura, D. M. (2023). A hedonic analysis of the determinants of hotel room rates in the Greater Gaborone Region (Botswana) using quantile regression. *International Hospitality Review.* https://doi.org/10.1108/IHR-02-2023-0012

Child, B. (2019). *Sustainable governance of wildlife and community-based natural resource management: From economic principles to practical governance.* Routledge.

Department of Tourism. (2012). *Botswana tourism statistics 2006–2010.* Government of Botswana.

Department of Wildlife and National Parks. (2012). *Aerial census of animals in Botswana: 2012 dry season.* Government Printer.

Department of Wildlife and National Parks. (2020). *CECT revenue generation 2010–2018.*

Department of Wildlife and National Parks (DWNP). (2014). *National parks visitor statistics 2010–2013.* Department of Wildlife and National Parks.

Department of Wildlife and National Parks (DWNP). (2015). *Revenue and employment for selected CBNRM villages in Ngamiland 2013–2015.*

Fennell, D. A. (2012). *Tourism and animal ethics.* Routledge.

Ghimire, K. B. (2013). The growth of national and regional tourism in developing countries: An overview. In K. Ghimire (Ed.), *The native tourist: Mass tourism within developing countries* (pp. 1–29). Earthscan.

Government of Botswana. (1990). *Tourism policy: Government paper no. 2 of 1990.* Government Printer.

Government of Botswana. (2000). *Botswana tourism master plan*. Government Printer.
Government of Botswana. (2007). *Community based natural resources management policy*. Government Printer.
Government of Botswana. (2021). *Tourism policy*. Government Printer.
Hambira, W. L., Stone, L. S., & Pagiwa, V. (2022). Botswana nature-based tourism and COVID-19: Transformational implications for the future. *Development Southern Africa, 39*(1), 51–67.
Higginbottom, K. (2004). Wildlife tourism: An introduction. Wildlife tourism: Impacts, management and planning, 1–14.
Holechek, J., & Valdez, R. (2018). Wildlife conservation on the rangelands of eastern and southern Africa: Past, present, and future. *Rangeland Ecology & Management, 71*(2), 245–258.
IUCN (International Union for Conservation of Nature). (2016). *Informing decisions on trophy hunting* (Briefing paper (April 2016)). International Union for Conservation of Nature.
Kabote, F., Mamimine, P. W., & Muranda, Z. (2017). Domestic tourism for sustainable development in developing countries. *African Journal of Hospitality, Tourism and Leisure, 6*(2), 1–12.
Khan, S. (2020). Covid-19: Tourism at crossroads! Where next? *Journal on Tourism & Sustainability, 3*(2), 32–40.
Larm, M., Elmhagen, B., Granquist, S. M., Brundin, E., & Angerbjörn, A. (2018). The role of wildlife tourism in conservation of endangered species: Implications of safari tourism for conservation of the Arctic fox in Sweden. *Human Dimensions of Wildlife, 23*(3), 257–272.
Lepp, A. (2008). Tourism and dependency: An analysis of Bigodi village, Uganda. *Tourism Management, 29*, 1206–1214.
Lindsey, P. A., Roulet, P. A., & Romanach, S. S. (2007). Economic and conservation significance of the trophy hunting industry in sub-Saharan Africa. *Biology Conservation, 134*, 455–469.
Mangachena, J. R., & Pickering, C. M. (2023). Why are some animals popular with wildlife tourists: Insights from South Africa. *Journal of Ecotourism, 22*(2), 312–328.
Mazimhaka, J. (2007). Diversifying Rwanda's tourism industry: A role for domestic tourism. *Development Southern Africa, 24*(3), 491–504.
Mbaiwa, J. E. (2005a). Enclave tourism and its socio-economic impacts in the Okavango Delta, Botswana. *Tourism Management, 26*(2), 157–172.
Mbaiwa, J. E. (2005b). The problems and prospects of sustainable tourism development in the Okavango Delta, Botswana. *Journal of Sustainable Tourism, 13*(3), 203–227.
Mbaiwa, J. E. (2011a). Cultural commodification and tourism: The Goo-Moremi community, Central Botswana. *Tijdschrift voor Economische en Sociale Geografie, 2*(3), 290–301.
Mbaiwa, J. E. (2011b). The effects of tourism development on the sustainable utilization of natural resources in the Okavango Delta, Botswana. *Current Issues in Tourism, 14*(3), 251–273.
Mbaiwa, J. E. (2012). The realities of ecotourism development in Botswana. In *Responsible tourism* (pp. 233–252). Routledge.
Mbaiwa, J. E. (2017). Poverty or riches: Who benefits from the booming tourism industry in Botswana? *Journal of Contemporary African Studies, 35*(1), 93–112.
Mbaiwa, J. E. (2018). Effects of the safari hunting tourism ban on rural livelihoods and wildlife conservation in northern Botswana. *South African Geographical Journal, 100*(1), 41–61.
Mbaiwa, J. E. (2023). Commodification of wildlife resources in the Okavango Delta, Botswana. In M. Bollig, S. Lendelvo, A. Mosimane, & R. Nghitevelekwa (Eds.), *Conservation, markets & the environment in southern and eastern Africa* (pp. 403–430). Boydell & Brewer.
Mbaiwa, J. E., & Hambira, W. L. (2023). Can the subaltern speak? Contradictions in trophy hunting and wildlife conservation trajectory in Botswana. *Journal of Sustainable Tourism, 31*(5), 1107–1125.
Mokgalo, L., & van der Merwe, P. (2022). A revised CBT strategy for Botswana: Reflections from experiences of the ban on trophy hunting. *Cogent Social Sciences, 8*(1), 2081109.
Morupisi, P., & Mokgalo, L. (2017). Domestic tourism challenges in Botswana: A stakeholders' perspective. *Cogent Social Sciences, 3*(1), 1298171.
Mwakiwa, E., Hearne, J., Stigter, J., De Boer, W., Henley, M., Slotow, R., Van Langevelde, F., Peel, M., Grant, C. C., & Prins, H. H. T. (2016). Optimization of net returns from wildlife

consumptive and non-consumptive uses by game reserve management. *Environmental Conservation, 43*(2), 128–139.
Nattrass, N. (2021). Conservation and the commodification of wildlife in the anthropocene: A Southern African history. *South African Historical Journal, 73*(1), 95–116.
Pack, S., Golden, R., & Walker, A. (2013). *Comparison of national wildlife management strategies: What works where, and why?* Wildlife Consulting Associates.
Page, S., & Dowling. (2002). *Ecotourism*. Prentice Hall.
Presidential COVID-19 Task Force Bulletin. (2020a). *Tourism feels the brunt of COVID-19*. Issue 89, 4th August 2020.
Presidential COVID-19 Task Force Bulletin. (2020b). *Tourism sector threatened by COVID-19*. Issue 111, 3rd September 2020.
Rizzolo, J. B. (2023). Wildlife tourism and consumption. *Journal of Sustainable Tourism, 31*(5), 1181–1194.
Rogerson, C. M., & Baum, T. (2020). COVID-19 and African tourism research agendas. *Development Southern Africa, 37*(5), 727–741.
Rogerson, C. M., & Lisa, Z. (2005). 'Sho't left': Changing domestic tourism in South Africa. *Urban forum, 16*(2–3), 88–111. Springer.
Rozemeijer, N., & Van der Jagt, C. (2000). *Community based natural resource management in Botswana: How community based is community based natural resource management in Botswana?* (Occasional paper series). IUCN/SNV CBNRM Support Programme.
Scheyvens, R. (2007). Poor cousins no more: Valuing the development potential of domestic and diaspora tourism. *Progress in Development Studies, 7*(4), 307–325.
Seckelmann, A. (2002). Domestic tourism: A chance for regional development in Turkey? *Tourism Management, 23*, 85–92.
Shah, K. (2000). *Tourism, the poor and other stakeholders: Asian experience* (ODI Fair-Trade in tourism paper). ODI.
Sindinga, I. (1996). Domestic tourism in Kenya. *Annals of Tourism Research, 23*, 19–31.
Statistics Botswana. (2022). *Tourism statistics annual report 2020*. Available at: https://www.statsbots.org.bw/sites/default/files/publications/Tourism%20Statistics%20Annual%20Report%202020.pdf. Accessed 28 Oct 2023.
Stone, L. S. (2014). *Perceptions of nature-based tourism, travel preferences, promotions and disparity between domestic and international tourists: The case of Botswana*. Unpublished PhD dissertation, Arizona State University.
Stone, L. S., & Nyaupane, G. P. (2016). Africans and protected areas: North–South perspectives. *Annals of Tourism Research, 58*, 140–155.
Stone, L. S., & Nyaupane, G. P. (2020). Local residents' pride, tourists' playground: The misrepresentation and exclusion of local residents in tourism. *Current Issues in Tourism, 23*(11), 1426–1442.
Stone, L. S., & Stone, M. T. (2022). Implications of the 'high- value, low- volume' approach in conservation and tourism resources management. In L. S. Stone, M. T. Stone, P. K. Mogomotsi, & G. E. K. Mogomotsi (Eds.), *Protected areas and tourism in Southern Africa: Conservation goals and community livelihoods* (pp. 232–244). Routledge.
Stone, L. S., Stone, M. T., & Mbaiwa, J. E. (2017). Tourism in Botswana in the last 50 years. *Botswana Notes and Records, 49*, 57–72.
Stone, L. S., Mogomotsi, P. K., Stone, M. T., Mogomotsi, G. E. J., Malesu, R., & Somolekae, M. (2020). Sustainable tourism and the SDG's in Botswana: Prospects, opportunities and challenges towards 2030. In S. O. Keitumetse, L. Hens, & D. Norris (Eds.), *Sustainability in developing countries: Case studies from Botswana's journey towards 2030 agenda* (pp. 153–181). Springer.
Stone, L. S., Stone, M. T., Mogomotsi, P. K., & Mogomotsi, G. E. (2021). The impacts of Covid-19 on nature-based tourism in Botswana: Implications for community development. *Tourism Review International, 25*(2–3), 263–278.

Tairo, A. (2020). COVID-19 impact on wildlife conservation in Africa. *eTurbo News*. Available at https://eturbonews.com/571395/covid-19-impact-on-wildlifeconservation-in-africa/. Accessed 20 Oct 2023.

Thakadu, O. T. (2005). Success factors in community based natural resources Management in Northern Botswana: Lessons from practice. *Natural Resources Forum., 29*(3), 199–212.

Tichaawa, T., & Lekgau, R. J. (2020). Nature and characteristics of wildlife tourism in the Kgalagadi Transfrontier Park. *Acta Universitatis Danubius. Œconomica, 16*(5), 69–88.

United Nations World Tourism Organisation. (2020). *UNWTO highlights potential of domestic tourism to help drive economic recovery in destinations worldwide*. Available at https://www.unwto.org/news/unwto-highlights-potential-of-domestic-tourism-to-help-drive-economic-recovery-in-destinations-worldwide. Accessed 1 Aug 2023.

United Nations World Tourism Organisation (UNWTO). (2015). *Towards measuring the economic value of wildlife watching tourism in Africa* (Briefing paper). UNWTO.

van der Merwe, P., Saayman, A., & Jacobs, C. (2021a). Assessing the economic impact of COVID-19 on the private wildlife industry of South Africa. *Global Ecology and Conservation, 28*, e01633.

van der Merwe, P., Slabbert, E., & Koch, M. (2021b). Environmental sustainability of consumptive and non-consumptive wildlife tourists: The case of a game reserve in Namibia. *International Journal of Biodiversity and Conservation, 13*(4), 218–227.

Weaver, D. (2006). *Sustainable tourism: Theory and practice*. Butterworth Heinemann.

World Bank. (2014). *Terrestrial protected areas by country (% of total land area)*. World Bank.

World Travel and Tourism Council (WTTC). (2021). *Travel and tourism: Global economic impact and trends 2020*. World Travel and Tourism Council. https://wttc.org/Research/EconomicImpact. Accessed 20 Oct 2023.

Woyo, E. (2021). The sustainability of using domestic tourism as a post-COVID-19 recovery strategy in a distressed destination. In *Information and Communication Technologies in Tourism 2021: Proceedings of the ENTER 2021 eTourism conference, January 19–22, 2021* (pp. 476–489). Springer.

Chapter 9
Pursuing Sustainable Tourism Development Through Wildlife Tourism: Local Residents' Perceptions

Felix G. Bello and Mathews Lambulira

9.1 Introduction

Tourism has emerged as a popular strategy for promoting economic development and reducing poverty in rural communities of developing nations, particularly in areas that lack opportunities for economic growth. Wildlife reserves and national parks, typically situated in remote regions, have become integral elements of community development systems. Wildlife protected areas create tourism spaces which enable local people to diversify livelihoods by capitalizing on several different economies in sustaining community as well as the tourism industry more generally (Buckley, 2003; Plummer & Fennell, 2009). As a result, the tourism industry has the potential to generate employment and income for local people, reducing poverty and improving their standard of living (Goodwin, 2002; Gordon, 2017). Furthermore, tourism can positively affect social aspects of community development. By participating in tourism, local people can engage with visitors from various regions worldwide, leading to cross-cultural learning and understanding, mutual respect, greater social cohesion, and community spirit (Jiang et al., 2017). Nevertheless, it is crucial to recognize the reciprocal relationship between tourism and community development. While tourism can yield significant advantages, it can also produce harmful

F. G. Bello (✉)
Department of Marketing Management, University of Pretoria, Pretoria, South Africa
e-mail: felix.bello@up.ac.za

M. Lambulira
Department of Tourism, Mzuzu University, Mzuzu, Malawi
e-mail: lambulira.m@mzuni.ac.mw

© The Author(s), under exclusive license to Springer Nature Switzerland AG 2024

L. S. Stone, M. T. Stone (eds.), *Wildlife Tourism Dynamics in Southern Africa*, Geographies of Tourism and Global Change,
https://doi.org/10.1007/978-3-031-57252-4_9

environmental, social, and protected resource effects (Goodwin, 2002; Mbaiwa, 2005; Suntikul et al., 2010). Moreover, poorly managed tourism activities can result in adverse effects on local communities, such as overcrowding, cultural exploitation, and the loss of traditional livelihoods. Thus, it is imperative to carefully design and manage tourism activities to minimize negative impacts on the environment and local communities.

Protected areas play a vital role in the successful implementation of sustainable tourism and sustainable development at large. As the primary global strategy for the preservation of nature, they are intricately linked to the concept of sustainable development (Bushell & Bricker, 2017). However, the role of protected areas in promoting sustainable development goes beyond just safeguarding wildlife and promoting biodiversity. When managed effectively and valued appropriately, protected areas can contribute significantly to achieving several Sustainable Development Goals (SDGs) (Dudley et al., 2017). Protected areas have the potential to promote human welfare and wellbeing by addressing various aspects of sustainable development, including poverty alleviation, food and water security, health, disaster risk reduction, sustainable cities, and climate change strategies. Therefore, protected areas serve as a tool for promoting sustainable development by focusing on conservation and protection, while also intensifying the attainment of SDGs through their outcomes.

The significance of surrounding communities in the management and sustainability of protected areas cannot be overstated. Protected areas are complex integrated systems that comprise the tourism system, park management system, ecological and biophysical systems, and community systems, including resident and indigenous communities. As underscored in the works of Jamal and Stronza (2009) and Farrell and Twining-Ward (2004), local communities are an indispensable element of the management system of protected areas. Consequently, any malfunctioning of the communities can have adverse effects on the efficacy of the sustainable goals of the protected areas (Domínguez & Luoma, 2020). Within the tourism and sustainability literature on protected areas, numerous studies have examined the role of local communities in protected areas and their involvement in sustainable management and tourism development (Bello et al., 2017a; Lee, 2013; Mutanga et al., 2013; Tosun, 2000). The literature has also enormously discussed the role of protected areas in sustainable development of local communities (Bakar et al., 2022; Fairer-Wessels, 2019; Mason et al., 2022; Plummer & Funnel, 2009). The discourse has also extensively explored the role of protected areas in promoting sustainable development in the surrounding communities. The scholarly literature has devoted considerable attention to this topic, scrutinizing the various ways in which protected areas can support and foster sustainable development within local communities.

Protected areas serve as primary attractions in the regions where they are located, but tourism activities in these areas are often attached to the protected areas. However, the categorization of tourism in wildlife protected areas into two distinct forms, i.e., tourism within the protected areas and tourism in the surrounding communities, provides an opportunity to examine the various dimensions of tourism in the surrounding communities. This area is typically overshadowed by the general form of tourism in protected areas, making it imperative to study sustainable tourism in protected areas comprehensively. In this context, it is essential to analyse sustainable tourism development in the surrounding communities as a separate component of such studies. Communities around the wildlife protected areas should not only be viewed as contributors to sustainable tourism in the protected areas but also as a space where sustainable tourism development can take place. This chapter aims to discuss sustainable tourism development in communities around wildlife protected areas, specifically exploring ways to promote sustainable tourism in communities adjacent to protected areas. Based on secondary data, the study draws on a case of two wildlife protected areas, Majete Wildlife Reserve and Liwonde National Park in Malawi managed under a Public Private Partnership (PPP) arrangement between Malawi Government and African Parks.

9.2 Community Development and Tourism

The term "community" is multifaceted and encompasses diverse meanings, including a collection of individuals who share a common interest, as described by Cavaye (2006). However, within the tourism sector, a community is generally defined in terms of its geographic location and pertains to a specific place and its inhabitants. For example, Vogt and Jordan (2015) conceptualize communities in the context of tourism as towns, cities, or villages. In the case of wildlife protected areas, the communities that surround them are typically those located in close proximity to the protected area. However, in most countries, the surrounding communities of wildlife protected areas are predominantly remote villages, as these protected areas are often situated in areas that are economically deprived (Terborgh, 2002). Therefore, in this context, the concept of "surrounding communities" refers specifically to remote villages located in the proximity of wildlife protected areas.

Community development is a process whereby individuals within a community engage in collective action to improve the social, economic, and environmental circumstances of their community (Cavaye, 2006). This approach aims to address shared problems and opportunities within a community. In the tourism sector, community development is focused on using tourism as a means to empower local communities and equip them with the necessary skills to advance their socioeconomic status (Vogt & Jordan, 2015). The process of development that emerges due to tourism activities is referred to as tourism development, although this can also refer to the process of developing tourism within a specific location (Telfer & Sharpley, 2015). In wildlife protected areas, tourism activities occurring within the protected area as well as those taking place in adjacent communities contribute to tourism development (Telfer & Sharpley, 2015; Vogt & Jordan, 2015).

9.3 Sustainable Tourism Development and Local Communities

It is imperative to recognize the significance of sustainable tourism as a concept. It involves the application of sustainable development principles within the tourism sector, ensuring that tourism development meets present needs without compromising future generations' ability to meet their own needs. In this context, the goal is to minimize negative impacts and maximize positive outcomes (Weaver, 2022). The concept of sustainable tourism is underpinned by the three key components of sustainable development, namely economic, ecological, and socio-cultural components (Duran et al., 2015). Scholars have noted the importance of these components in fostering sustainable tourism practices (Duran et al., 2015; Hall, 1998; Sanagustín et al., 2011). Specifically, the economic component of sustainable tourism is concerned with generating economic benefits for local communities, such as job creation, income generation, and entrepreneurship opportunities. Meanwhile, the ecological component of sustainable tourism seeks to protect and conserve natural resources, including biodiversity, water, and energy (Duran et al., 2015). Finally, the socio-cultural component of sustainable tourism recognizes the significance of preserving cultural heritage and promoting cultural exchange (Duran et al., 2015).

The intersection between tourism and sustainable development has been a topic of interest for scholars and practitioners alike. A rich body of literature has emerged, reflecting two primary perspectives. The first perspective considers the potential of tourism to achieve sustainable development (Dłużewska, 2019). This perspective examines the role of tourism in promoting economic growth, poverty reduction, and the attainment of SDGs (Gössling et al., 2012; UNWTO & UNDP, 2017). The second perspective of the literature focuses on the need for the tourism industry to adopt sustainable practices to ensure its long-term viability (Dunets et al., 2019; Coccossis, 2016). This perspective highlights the various forms of tourism, development strategies, and activities that are necessary to sustain the industry. It emphasizes the importance of reducing the negative impacts of tourism on the environment, conserving natural resources, and promoting social and cultural sustainability. Both dimensions of the literature reflect the economic, ecological, and socio-cultural components of sustainable tourism, which are fundamental to the core question addressed in this chapter.

Communities are increasingly encouraged to embrace sustainable tourism as a means to address their developmental needs and aspirations. This sentiment has been echoed by Puczkó and Rátz (2000), who have advocated for sustainable tourism development, particularly for communities utilizing tourism as a tool for local community economic development. Concurrently, community support plays an instrumental role in the successful implementation of sustainable tourism initiatives. As a result, a community participation approach has been widely advocated as an integral component of sustainable tourism development (Carius & Job, 2019; Li & Hunter, 2015; Novelli & Scarth, 2007; Tosun, 2000). Scholars contend that without the active support and participation of local community residents, the development of sustainable tourism is arduous to achieve (Bello et al., 2016; Gursoy & Rutherford, 2004; Lee, 2013). The relationship between communities and sustainable tourism can be conceptualized in two distinct ways. First, communities can leverage sustainable tourism to meet their economic development needs (Gordon, 2017). This perspective views sustainable tourism as a means to generate economic benefits for local communities, such as job creation and income generation, while preserving the natural and cultural resources on which tourism depends. Second, communities are recognized as a strategic component required for the development and management of sustainable tourism (Carius & Job, 2019; Li & Hunter, 2015). This perspective emphasizes the importance of community involvement and participation in decision-making processes related to sustainable tourism development.

In the discourse of sustainable tourism, wildlife protected areas and the communities around them are recognized as interdependent components of a larger ecosystem. Communities not only play an essential role in the sustainable systems of the reserves, but also constitute a crucial segment of the tourism value chain. As Goodwin (2002) posits, communities offer a range of services and products to both tourists and tourism facilities, including cultural tours, handicrafts, and agricultural produce. Despite the significant contributions of communities to the tourism industry, their participation in the tourism value chain is often limited by a host of challenges. Inadequate financial resources, inadequate information, low education levels, unfair distribution of benefits, lack of trained human resource, centralised public administration, lack of coordination and human–wildlife conflicts are just a few examples of the obstacles that impede the ability of communities to effectively participate in protected area-based tourism (Bello et al., 2017a; Goodwin, 2002). Such limitations often result in conflicts between communities and the management of protected areas, further exacerbating their already precarious position in the tourism value chain (Western et al., 2015). Novelli and Scarth (2007) contend that these conflicts can be traced to the historical imposition of nature-protected areas on communities without sufficient regard for the livelihoods of the local residents. As such, the establishment of national parks and other protected areas has frequently resulted in the displacement of communities from their traditional lands and the restriction of access to resources that are vital to their survival. Hough (1988) echoes this sentiment, underscoring the negative outcomes associated with the imposition of national parks on rural communities.

In addition to the aforementioned conflicts between wildlife reserve management and surrounding communities, human wildlife conflicts also constitute a significant issue within the context of protected areas (Dunham et al., 2010; Ghimire, 2019; Mayengo et al., 2017). Appalling acts of poaching by community residents and the destruction of crops and even human fatalities caused by wildlife, are but a handful of the various issues faced by these communities. It is apparent that such limitations and conflicts bear severe implications for the development of sustainable tourism within the communities surrounding wildlife protected areas.

9.4 The Case Study Setting

9.4.1 Majete Wildlife Reserve

The Majete Wildlife Reserve, situated in the southern region of Malawi, was designated a protected area in 1955 (Department of National Parks & Wildlife [DNPW], 2000). The reserve encompasses an expansive area of 715 km^2 within the lower Shire River valley, in proximity to Blantyre, the second most populous city in the country (African Parks, 2022). The reserve is predominantly located within the Chikwawa District, with a small northern section falling within the Mwanza District (Bello, 2021). Throughout the late 1970s, the reserve was afflicted by rampant deforestation and poaching, leading to a substantial decline in the population of wildlife, including the majestic elephant. In 2003, the Malawi Government, through its Department of National Parks and Wildlife, initiated a 25-year public-private partnership (PPP) with African Parks, a non-profit conservation organization, which operates on a business model (Bello, 2021).

This management concession constitutes the first instance of a protected area in Malawi being governed under the PPP model. African Parks successfully undertook extensive measures to restore the depleted wildlife populations of Majete Wildlife Reserve. This involved reintroduction of animals, construction of a perimeter fence, the implementation of effective patrolling programmes, and the improvement of general and tourism infrastructure. As of 2021, approximately 3000 animals from 17 species, including rhinos, elephants, leopards, buffalos, cheetahs, and giraffes, had been reintroduced at Majete, with the majority being translocated from other countries within the region (African Parks, 2022). Majete's case has demonstrated that wildlife reintroduction can be an effective and viable strategy for rejuvenating large carnivores in protected areas in Malawi (Briers-Louw et al., 2019).

9.4.2 Liwonde National Park

The Liwonde National Park (LNP), located in the Upper Shire Valley in the Southern Region of Malawi, spans a vast area of 548 km^2 (DNPW, 1998). The park was designated a national park in 1973, subsequent to a decline in large mammal populations due to sport hunting by colonialists and an exponential rise in the human population around the park (Sievert, 2020). Liwonde was initially declared a controlled shooting area in 1962 and subsequently upgraded to a game reserve in 1969, before eventually attaining national park status in 1973 (Sievert, 2020). Liwonde National Park was officially opened for game-viewing to the public in 1978 (Sievert, 2020). In 2015, following the successful adoption of the PPP model at the Majete Wildlife Reserve, the Government of Malawi entered into a similar arrangement with African Parks for the management of Liwonde National Park. This made Liwonde the second wildlife protected area to be managed by African Parks under the PPP model in Malawi (In Malawi, African Parks also manages Nkhotakota Wildlife Reserve and Mangochi Forest Reserve under the same PPP model). Employing a comparable management approach to that of the Majete Wildlife Reserve, African Parks replicated several successful programmes being implemented at the reserve to LNP.

9.5 Promotion of Sustainable Tourism in the Communities Around Majete Wildlife Reserve and Liwonde National Park

The chapter delves into the critical issue of sustainable tourism development in the two wildlife protected areas, with a focus on promoting sustainable practices in the surrounding communities. A tripartite approach to sustainable tourism development in the adjacent communities is applied, which involves fostering active community engagement in conservation and tourism activities within the reserves, promoting sustainable tourism products and services within the communities, and promoting social and economic development in communities located adjacent to the protected areas. Three central questions guide the analysis of these strategies: How does the reserve management promote community involvement in conservation efforts? How does the management promote sustainable tourism products and services offered by the surrounding communities? and how does the management promote social and economic development in the communities around the protected areas? The data was collected from secondary sources (both published and unpublished), which include policy documents on wildlife and tourism in Malawi, African Parks' annual reports and journal articles. Content analysis was used to analyse the qualitative data from the documents.

9.6 Establishing Community Engagement Mechanisms

African Parks has established effective mechanisms at both Majete and Liwonde to foster engagement with the park management and ensure that the interests of surrounding communities are considered in management decisions. The parks have a dedicated community extension department that serves as the primary liaison point between the park management and local communities. Additionally, Majete and Liwonde each have their management boards, with local representation in place to facilitate decision-making. Community-Based Organizations (CBO) leaders and traditional chiefs serve as representatives for the local communities surrounding the protected areas, which the communities consider to be a fair representation, and they are content with the involvement arrangement (Bello, 2021). However, despite local engagement in decision-making, their participation remains somewhat limited due to a high illiteracy rate among local representatives (Bello, 2021; Kaunda-Khangamwa et al., 2019). Additionally, some communities believe that transparency on certain issues with regard to most management decisions, including issues of tourism development associated with the protected areas is still insufficient (Bello, 2021).

To bolster the relationship between the park management and surrounding communities, the reserves allow the communities to sustainably harvest some resources from the parks under guidance and supervision (Bello et al., 2017b; Gordon, 2017). For example, at Majete, locals are permitted to harvest thatching grass, bamboo, and reeds within the reserve. Additionally, locals are allowed to access traditional medicine derived from plants found in the reserves. Gordon (2017) identified a total of 96 different plant species in Majete which are used by traditional healers in the surrounding areas.

9.7 Providing and Supporting Social Services

African Parks is committed to supporting the provision of social services, particularly in the domains of education and health, both through direct funding of the organization and in collaboration with international development entities. In the realm of education, African Parks constructs schools, refurbishes old school structures, provides teacher housing, and furnishes selected schools (Bello et al., 2017b). For example, in collaboration with Malawi Community Hubs (UK) and Youth for Development and Productivity (YODEP), African Parks constructed two classroom blocks, a teacher's house, and supplied 260 desks at Liwonde (African Parks, 2022). Furthermore, in partnership with the Dioraphte Foundation, African Parks constructed nursery schools and libraries at Majete. Additionally, African Parks provides scholarships for primary, secondary, and tertiary education in the surrounding communities at Majete and Liwonde. In 2021, for example, African Parks spent US$75,765 on scholarships at Liwonde (African Parks, 2022).

The lack of healthcare facilities remains a significant challenge in the surrounding communities of Majete and Liwonde. For example, at Majete, the distance to the nearest healthcare facility ranges from 2.5 to 14.7 km for some households (Kaunda-Khangamwa et al., 2019). In response, African Parks provides support in the realm of healthcare services through a range of initiatives aimed at improving the situation. African Parks offers health support to ensure that marginalized residents in the communities at Majete and Liwonde have access to basic healthcare. This is accomplished through the construction and upgrade of existing structures at government hospitals/clinics in the communities, engagement of the communities in health education, and an active role in research aimed at controlling the spread of common diseases in the communities. African Parks has also upgraded health centres at Liwonde, including the maternity wing at Namanolo Health Centre (African Parks, 2022). They have also constructed a maternity facility at Majete. Additionally,

African Parks has constructed a malaria research and prevention centre at Majete with the goal of reducing the incidence of clinical malaria in the surrounding communities.

9.8 Environmental Education Programmes

African Parks at Majete and Liwonde conducts regular training programmes that aim to educate community residents on the significance of biodiversity and conservation of natural resources (African Parks, 2022). Local environmental education programmes have been established to foster understanding among the local population regarding the necessity of conserving natural resources and utilizing them sustainably. Through these programmes, catchment areas are sensitized on the benefits of natural resource conservation, and how best to effectively practice and observe it. This is particularly crucial in both reserves, but even more so at Liwonde where human-wildlife conflicts, more especially with elephants, hippos, and crocodiles pose one of the most pressing threats to the park. Fatal incidents and unrest between the communities and park management are rampant in this regard (Bello et al., 2016). Some of the programmes target youth and children to instil a value for wildlife and conservation in the community from a grassroots level. For instance, Majete runs an environmental education programme that teaches school children about the significance of the environment. The education programmes do not only involve training but also extend to allowing the residents to experience the beauty of nature in the protected areas. Furthermore, residents are equipped with skills for implementing alternative ways of livelihood, such as making briquettes in place of firewood and charcoal to reduce pressure on resources in the protected areas (African Parks, 2022). At Majete, African Parks supported a community reforestation project that planted 28,343 seedlings in 2021(African Parks, 2022).

9.9 Promoting Sustainable Tourism Development

African Parks regards conservation as a competitive land use and strives to unlock the economic potential of the reserves through a conservation-led economy. At Majete and Liwonde, African Parks provides employment opportunities to community residents and supports the development of local micro-enterprises. In addition to diversifying income streams for the communities, African Parks supports micro-enterprises to alleviate total economic dependence on natural resources in the protected areas (African Parks, 2022). At Liwonde, African Parks is backing beekeeping and Moringa cultivation as part of their community entrepreneurial support initiatives. Beekeeping is also among the community initiatives supported by African Parks at Majete, and other wildlife protected areas in Malawi under African Parks management (African Parks, 2022; Bello et al., 2017b). In 2021, African Parks reported that Majete's beekeeping enterprise grew to 792 beekeepers and produced over 1.8 tonnes of honey (African Parks, 2022). Along with providing necessary equipment to support beekeeping community initiatives, African Parks equips the communities with the necessary beekeeping and basic business management skills through their regular training programmes. In terms of employment, African Parks employs community residents as rangers and guides at both Majete and Liwonde. To increase youth employment opportunities, Majete introduced a technical and skills development programme upscaling youths in cell phone and computer repair, electrical installation, refrigeration, and information technology, with the majority of youths securing jobs using the skills obtained through the training program (African Parks, 2022).

African Parks also has some programmes aimed at promoting the development and provision of tourism services at Majete and Liwonde wildlife reserves, with particularly visible progress at Majete wildlife reserve (African Parks, 2022; Bello et al., 2017b). African Parks introduced a community tour guide programme at Majete that trains young adults in tour guiding, after which they are offered opportunities to work as tour guides in the communities as well as in protected areas. The tour guide training also fosters an understanding of the ecosystem and a sense of pride in the resources found in the reserves, thus strengthening the delivery of tourism services in the communities and ensuring that the youths understand the value of the wildlife resources. African Parks is also supporting tourism-related community-based micro-enterprises, including businesses such as cultural tours, local crafts, and a campsite inside Majete (African Parks, 2017, 2022; Bello et al., 2017b). African Parks developed a tourist campsite within the reserve that is managed by the local communities and provides a direct source of income for the community which is used to finance various community development projects including the purchase of school uniforms and writing materials for needy students in the surrounding villages (Bello et al., 2017b).

Local communities' sharing of benefits of wildlife protected area tourism is an important management requirement as it facilitates local people's positive attitudes towards protected areas (Mutanga et al., 2013). The local communities surrounding both Majete Wildlife Reserve and Liwonde National Park are not yet empowered to the level where they can decide the forms of tourism facilities and wildlife conservation programmes to be developed in their respective areas, mainly due to lack of financial capital (Bello et al., 2016; Bello et al., 2017a). Despite this, the local people still get some benefits, either directly or indirectly from the tourism development associated with the protected areas.

The local communities are benefitting from tourism development through employment in tourism enterprises inside and outside the protected areas, the sale of arts and crafts to tourists, community run tourist facilities, community development projects funded by tourism revenue and income generating activities initiated by management agencies, tourism operators and environmental non-governmental organisations (ENGOs) (African Parks, 2022). African Parks also has a scholarship programme for underprivileged children around both protected areas for both secondary and tertiary education (African Parks, 2022; Bello et al., 2017b).

On employment, only very few local people have had the opportunity to get high paying jobs with either tourism enterprises or African Parks (Bello et al., 2017b). Though local people are employed in tourism operations associated with the protected areas, most of them are not qualified to get high paying tourism related jobs on offer and the limited number of tourism establishments within the protected areas means there are fewer opportunities. Low literacy levels or lack of requisite qualifications and skills for most jobs in the tourism sector is a general problem for most communities around tourism establishments in rural areas in most developing countries (Bello et al., 2017a; Marzuki et al., 2012; Oakley & Marsden, 1984; Tosun, 2000). In most cases the available employment opportunities are also unfairly distributed around the protected area as mainly those people near the park offices and lodges get employed (Bello et al., 2017a).

Furthermore, local communities near entrances and on the tourist' routes have more exposure to tourism activities and a few people have started small businesses (e.g., selling curios, arts, and crafts). The wildlife protected area agency and lodges within and around the protected areas collects or buys arts and crafts from producers in surrounding villages; these are sold at an Arts and Craft Centre located at the protected areas' main entrances and curio shops at the lodges. This is an example of psychological empowerment. A psychologically empowered local community should have faith in its own abilities, be relatively self-reliant, and be proud of its traditions and culture (Scheyvens, 1999, 2000).

The other tourist activities which local people get involved in are village tours. These village tours are organised by either the protected area management agency or tourism operators and only target villages near the park or reserve entrances (Bello et al., 2017b). Although village tours are promoted, the product is not yet well developed, as such very few village tours take place. The benefits to the local communities from such tours are minimal. The local people have little power to determine the rates for the activities which they showcase to tourists; tour rates are set by the tourism operators. As a result, not many local people join such groups engaged in tourists' village tour activities. Earnings for such groups would only be meaningful with an increased number of tours to ensure that they generate enough for members to share. At Liwonde National Park, this cultural experience is more organised through the Njobvu Cultural Village which is run by villagers near the park's Makanga gate. This cultural village offers visitors an opportunity to stay in a traditional Malawian village, sleeping in traditional mud-brick huts and during the day the tourists are invited to take part in the local people's daily lives (Njobvu Cultural Village, 2023). The income from the cultural village is meant to support local community projects (Njobvu Cultural Village, 2023). Tourism development should improve cohesion and integrity of local communities by providing opportunities for community members to work together in social development (Scheyvens, 1999). The Njobvu Cultural Village initiative near Liwonde National Park is a manifestation of social empowerment. As noted by Scheyvens (2000), social empowerment is therefore noticed when tourism directly or indirectly provides resources for a community to embark on social development such as projects for improved water supply and health facilities.

African Parks has been more aggressive in ensuring that the surrounding local communities have alternative means of livelihood by introducing more income generating activities. Interested groups of local people around the protected areas have been supported to start income generating activities such as mushroom production, beekeeping/honey production, pottery, poultry and arts and craft production. The tourism operators in and around the protected areas buy these products and support the local communities in marketing the products. In addition to using tourism revenue, African Parks has also managed to source funds from donors for the introduction of these activities.

9.10 Conclusion

This chapter places its emphasis on the promotion of sustainable tourism development within communities residing near wildlife protected areas. It draws valuable lessons from two highly notable protected areas, Majete Wildlife Reserve and Liwonde National Park, which are managed through a PPP arrangement between the Government of Malawi and African Parks.

It has been ascertained that African Parks is taking an active role in implementing a plethora of programmes aimed at addressing the questions posed in this chapter, in partnership with the Department of Parks and Wildlife and other tourism stakeholders in Malawi. These programmes, which are expounded upon within the chapter, encompass the provision and support of social services, implementation of environmental education initiatives, facilitation of the local economy, and augmentation of tourism services offered by surrounding communities. Upon reflecting on the strategies employed in response to the three central questions and dimensions regarding the promotion of sustainable tourism development within communities, it can be deduced that the endeavours at Majete and Liwonde are heavily geared towards promoting conservation, as well as social and economic growth within the communities. Most of the development initiatives are funded by African Parks' development partners rather than solely by tourism. However, efforts aimed at developing tourism products and services within the surrounding communities appear to be inadequate, and may necessitate additional interventions, such as the development of cultural tourism offerings and the creation of markets for the same.

In both protected areas, it can be concluded that the local communities have been involved in the sharing of tourism benefits to some extent. African Parks has been more aggressive in its efforts to empower local people through tourism, income generating activities related to the reserve and community development projects. The business-oriented management approach used by African Parks attracts investors who see tourism potential of a protected area while its non-profit making management structure attracts individual and corporate donors due to efficient and effective use of their financial gifts.

According to Havel (1996) participation in development related processes should be considered as a continuum along which local people can progressively be empowered. On this continuum, the local communities around the two protected areas have started as mere beneficiaries of tourism development and their participation is mainly in the sharing of the benefits. In the two protected areas, the local communities have participated in the sharing of tourism benefits at different levels. Some communities, more especially those near protected area entrances and on tourism routes, have some understanding of tourism and have started airing their views on how tourism should be developed. This resonates with Goodwin (2002) who found that there is great enthusiasm amongst local people along tourist routes and near park entrances to get involved in tourism, only that they have limited interface with tourists due to the enclave nature of protected area-based tourism. There is a need to create linkages to avoid enclaves to ensure that local small and medium scale entrepreneurs have access to tourists (Ashley & Roe, 1998).

There is some manifestation of psychological and social empowerment (Scheyvens, 1999) among the local people in these communities. The local people have started appreciating that the natural resources in the protected areas and their culture are valuable. They have started working together in groups to get involved in some tourism activities and for this to be sustained the local communities need more help from the other stakeholders to build their capacity in tourism (Ashley & Jones, 2001; Spenceley & Goodwin, 2007; Wells 1996). Private and non-profit partners to local communities are important stakeholders in building capacity for communities to enable local people to run tourism businesses of their own (Jamal & Stronza, 2009). With external help in consolidating the psychological and social empowerment through capacity building and financial assistance in starting up tourism related businesses, the local people could continue to move along the participation continuum. When the local people build their capacity in tourism and access capital, they could eventually reach the far end of the continuum where they would be in a position to own and manage some tourism development activities and be actively involved in decision making. It is at this end of this participation continuum, where the local communities would be expected to be economically and politically empowered.

References

African Parks. (2017). *African Parks annual report 2016*. https://www.africanparks.org/sites/default/files/uploads/resources/2017-05/2016_African_Parks_Annual_Report_Impact_Defined.pdf

African Parks. (2022). *African Parks annual report 2021*. https://www.africanparks.org/sites/default/files/uploads/resources/2022-06/African-Parks-2021-Annual-Report.pdf

Ashley, C., & Jones, B. (2001). Joint ventures between communities and tourism investors: Experience in Southern Africa. *International Journal of Tourism Research, 3*(5), 407–423.

Ashley, C., & Roe, D. (1998). Enhancing community involvement in wildlife tourism: Issues and challenges. *Wildlife and Development Series No. 11*. IIED.

Bakar, A. N., Ali, A., Jangga, R., Yaakob, A. N., Salleh, R. M., & Shukor, M. F. A. (2022, April). Community participation in Gunung Ledang protected area: Supporting the sustainable development goal (SDG). In *IOP conference series: Earth and environmental science* (Vol. 1019, No. 1, p. 012010). IOP Publishing.

Bello, F. G. (2021). Community participation in tourism planning at Majete Wildlife Reserve, Malawi. *Quaestiones Geographicae, 40*(4), 85–100.

Bello, F. G., Carr, N., & Lovelock, B. (2016). Community participation framework for protected area-based tourism planning. *Tourism Planning & Development, 13*(4), 469–485.

Bello, F. G., Lovelock, B., & Carr, N. (2017a). Constraints of community participation in protected area-based tourism planning: The case of Malawi. *Journal of Ecotourism, 16*(2), 131–151.

Bello, F. G., Lovelock, B., & Carr, N. (2017b). Neoliberal conservation and protected area-based tourism development in Malawi. The case of Majete Wildlife Reserve. *Journal of Tourism Challenges and Trends, 10*(1), 45–72.

Briers-Louw, W. D., Verschueren, S., & Leslie, A. J. (2019). Big cats return to Majete Wildlife Reserve, Malawi: Evaluating reintroduction success. *African Journal of Wildlife Research, 49*(1), 34–50.

Buckley, R. (2003). Ecological indicators of tourist impacts in parks. *Journal of Ecotourism, 2*(1), 54–66.

Bushell, R., & Bricker, K. (2017). Tourism in protected areas: Developing meaningful standards. *Tourism and Hospitality Research, 17*(1), 106–120.

Carius, F., & Job, H. (2019). Community involvement and tourism revenue sharing as contributing factors to the UN Sustainable Development Goals in Jozani–Chwaka Bay National Park and Biosphere Reserve, Zanzibar. *Journal of Sustainable Tourism, 27*(6), 826–846.

Cavaye, J. (2006). Understanding community development. *Cavaye Community Development, 1*, 1–19.

Coccossis, H. (2016). Sustainable development and tourism: Opportunities and threats to cultural heritage from tourism. In L. F. Girard & P. Nijkamp (Eds.), *Cultural tourism and sustainable local development* (pp. 65–74). Routledge.

Dłużewska, A. M. (2019). Well-being versus sustainable development in tourism—The host perspective. *Sustainable Development, 27*(3), 512–522.

DNPW. (1998). *Management plan for Liwonde National Park*. Department of National Parks and Wildlife.

DNPW. (2000). *Majete Wildlife Reserve: Park plan*. Department of National Parks and Wildlife.

Domínguez, L., & Luoma, C. (2020). Decolonizing conservation policy: How colonial land and conservation ideologies persist and perpetuate indigenous injustices at the expense of the environment. *Land, 9*(3), 65.

Dudley, N., Ali, N., & MacKinnon, K. (2017). *Natural solutions: Protected areas helping to meet the sustainable development goals*. IUCN.

Dunets, A. N., Vakhrushev, I. B., Sukhova, M. G., Sokolov, M. S., Utkina, K. M., & Shichiyakh, R. A. (2019). Selection of strategic priorities for sustainable development of tourism in a mountain region: Concentration of tourist infrastructure or nature-oriented tourism. *Entrepreneurship and Sustainability Issues, 7*(2), 1217.

Dunham, K. M., Ghiurghi, A., Cumbi, R., & Urbano, F. (2010). Human–wildlife conflict in Mozambique: A national perspective, with emphasis on wildlife attacks on humans. *Oryx, 44*(2), 185–193.

Duran, D. C., Gogan, L. M., Artene, A., & Duran, V. (2015). The components of sustainable development-a possible approach. *Procedia Economics and Finance, 26*, 806–811.

Fairer-Wessels, F. A. (2019). Investigating sustainable development goals, livelihoods and tourism development in a rural protected area: The iSimangaliso Wetland Park, South Africa. In M. T. Stone, M. Lenao, & N. Moswete (Eds.), *Natural resources, tourism and community livelihoods in Southern Africa: Challenges of sustainable development* (pp. 205–217). Routledge.

Farrell, B. H., & Twining-Ward, L. (2004). Reconceptualizing tourism. *Annals of tourism research, 31*(2), 274–295.

Ghimire, P. (2019). Analysis of human wildlife conflict in buffer zone area: A study from Chitwan National Park, Nepal. *International Journal of Natural Resource Ecology and Management, 4*(6), 164–172.

Goodwin, H. (2002). Local community involvement in tourism around national parks: Opportunities and constraints. *Current Issues in Tourism, 5*(3–4), 338–360.

Gordon, C. N. (2017). *People and protected areas: Natural resource harvesting as an approach to support rural communities surrounding Majete Wildlife Reserve, Southern Malawi: A case study*. Doctoral dissertation, Stellenbosch University.

Gössling, S., Hall, C. M., Ekström, F., Engeset, A. B., & Aall, C. (2012). Transition management: A tool for implementing sustainable tourism scenarios? *Journal of Sustainable Tourism, 20*(6), 899–916.

Gursoy, D., & Rutherford, D. G. (2004). Host attitudes toward tourism: An improved structural model. *Annals of Tourism Research, 31*(3), 495–516.

Hall, C. (1998). Historical antecedents of sustainable development and ecotourism: New labels on old bottles? In *Sustainable tourism: A geographical perspective* (pp. 13–24). Longman.

Havel, V. (1996). *The World Bank participation sourcebook*. World Bank.

Hough, J. L. (1988). Obstacles to effective management of conflicts between national parks and surrounding human communities in developing countries. *Environmental Conservation, 15*(2), 129–136.

Jamal, T., & Stronza, A. (2009). Collaboration theory and tourism practice in protected areas: Stakeholders, structuring and sustainability. *Journal of Sustainable Tourism, 17*(2), 169–189.

Jiang, Y., Ramkissoon, H., Mavondo, F. T., & Feng, S. (2017). Authenticity: The link between destination image and place attachment. *Journal of Hospitality Marketing & Management, 26*(2), 105–124.

Kaunda-Khangamwa, B. N., Van Den Berg, H., McCann, R. S., Kabaghe, A., Takken, W., Phiri, K., et al. (2019). The role of health animators in malaria control: A qualitative study of the health animator (HA) approach within the Majete malaria project (MMP) in Chikwawa District, Malawi. *BMC Health Services Research, 19*, 1–16.

Lee, T. H. (2013). Influence analysis of community resident support for sustainable tourism development. *Tourism Management, 34*, 37–46.

Li, Y., & Hunter, C. (2015). Community involvement for sustainable heritage tourism: A conceptual model. *Journal of Cultural Heritage Management and Sustainable Development, 5*(3), 248–262.

Marzuki, A., Hay, I., & James, J. (2012). Public participation shortcomings in tourism planning: The case of the Langkawi Islands, Malaysia. *Journal of Sustainable Tourism, 20*(4), 585–602.

Mason, C. W., Carr, A., Vandermale, E., Snow, B., & Philipp, L. (2022). Rethinking the role of indigenous knowledge in sustainable mountain development and protected area management in Canada and Aotearoa/New Zealand. *Mountain Research and Development, 42*(4), A1–A9.

Mayengo, G., Bwagalilo, F., & Kalumanga, V. E. (2017). Human wildlife conflicts to communities surrounding Mikumi National Parks in Tanzania: A case of selected villages. *International Journal of Environment, Agriculture and Biotechnology, 2*(4), 238852.

Mbaiwa, J. E. (2005). Enclave tourism and its socio-economic impacts in the Okavango Delta, Botswana. *Tourism Management, 26*(2), 157–172.

Mutanga, C. N., Kwanisai, G., & Chitengu, F. (2013). Enhancing community involvement in tourism: An assessment of the interaction between Ngezi Recreational Park and the surrounding community. *IOSR Journal of Business Management, 12*, 74–81.

Njobvu Cultural Village. (2023). *Njobvu Cultural Village—Malawi*. https://www.njobvuvillage.org/#lb22354/njobvu-cultural-village-malawi

Novelli, M., & Scarth, A. (2007). Tourism in protected areas: Integrating conservation and community development in Liwonde National Park (Malawi). *Tourism and Hospitality Planning & Development, 4*(1), 47–73.

Oakley, P., & Marsden, D. (1984). *Approaches to participation in rural development*. International Labour Organization.

Plummer, R., & Fennell, D. A. (2009). Managing protected areas for sustainable tourism: Prospects for adaptive co-management. *Journal of Sustainable Tourism, 17*(2), 149–168.
Puczko, L., & Ratz, T. (2000). Tourist and resident perceptions of the physical impacts of tourism at Lake Balaton, Hungary: Issues for sustainable tourism management. *Journal of Sustainable Tourism, 8*(6), 458–478.
Sanagustín Fons, M., Fierro, J. A. M., & Patiño, M. G. (2011). Rural tourism: A sustainable alternative. *Applied Energy, 88*(2), 551–557.
Scheyvens, R. (1999). Ecotourism and the empowerment of local communities. *Tourism Management, 20*(2), 245–249.
Scheyvens, R. (2000). Promoting women's empowerment through involvement in ecotourism: Experiences from the Third World. *Journal of Sustainable Tourism, 8*(3), 232–249.
Sievert, O. (2020). *Early post-release movements, prey preference and habitat selection of reintroduced cheetah (Acinonyx jubatus) in Liwonde National Park, Malawi*. Doctoral dissertation. Stellenbosch University.
Spenceley, A., & Goodwin, H. (2007). Nature-based tourism and poverty alleviation: Impacts of private sector and parastatal enterprises in and around Kruger National Park, South Africa. *Current Issues in Tourism, 10*(2–3), 255–277.
Suntikul, W., Bauer, T., & Song, H. (2010). Towards tourism: A Laotian perspective. *International Journal of Tourism Research, 12*(5), 449–461.
Telfer, D. J., & Sharpley, R. (2015). *Tourism and development in the developing world*. Routledge.
Terborgh, J. (2002). *Making parks work: Strategies for preserving tropical nature*. Island Press.
Tosun, C. (2000). Limits to community participation in the tourism development process in developing countries. *Tourism Management, 21*(6), 613–633.
UNWTO & UNDP. (2017). *Tourism and the sustainable development goals—Journey to 2030*. UNWTO.
Vogt, C. A., & Jordan, E. J. (2015). Community development, tourism. In J. Jafari & H. Xiao (Eds.), *Encyclopedia of tourism* (pp. 1–4). Springer.
Weaver, D. B. (2022). Sustainable tourism. In D. Buhalis (Ed.), *Encyclopedia of tourism management and marketing* (pp. 317–321). Edward Elgar Publishing.
Wells, M. P. (1996). The social role of protected areas in the new South Africa. *Environmental Conservation, 23*(4), 322–331.
Western, D., Waithaka, J., & Kamanga, J. (2015). Finding space for wildlife beyond national parks and reducing conflict through community-based conservation: The Kenya experience. *Parks, 21*(1), 51–62.

Part IV
Hunting Versus Photographic Tourism: Sustainability Issues

Chapter 10
Consumptive Wildlife Tourism in Southern Africa: Implications for Sustainability

Lyakurwa G. Jerome and Edwin Sabuhoro

10.1 Introduction

During the past 30 years, consumptive tourism has been adopted widely in eastern and southern Africa to promote community support for conservation and finance conservation activities (Lovelock, 2007). The premise underlying consumptive wildlife tourism, mainly in the form of trophy hunting, is the utilitarian perspective which posits that sustainable harvests of a few wildlife species can benefit the conservation of larger remote areas by providing resources to law enforcement activities and enhancing community livelihoods to support conservation (Ghasemi, 2021). Trophy hunting is the selective recreational hunting of specified species of wild animals to obtain part of their bodies as success and memorial. It is distinguished from hunting for survival and traditional or subsistence hunting (Sheikh & Bermejo, 2019). The adoption of trophy hunting in eastern and southern African countries is a response to the reality that most remote areas in those regions need substantial amounts of money for conservation which non –consumptive use through photographic safari cannot generate due to a lack of tourism infrastructures and facilities such as roads and tourist accommodation.

L. G. Jerome (✉)
Department of Recreation, Parks and Tourism Management,
The Pennsylvania State University, University Park, PA, USA
e-mail: gjl5338@psu.edu

E. Sabuhoro
Department of Wildlife Management, The College of African Wildlife Management, Mweka, Moshi, Tanzania
e-mail: eps5479@psu.edu

© The Author(s), under exclusive license to Springer Nature Switzerland AG 2024

L. S. Stone, M. T. Stone (eds.), *Wildlife Tourism Dynamics in Southern Africa*, Geographies of Tourism and Global Change,
https://doi.org/10.1007/978-3-031-57252-4_10

The adoption of hunting as a tool for conservation is a result of multiple factors, including the lack of capacity for government budgets to finance conservation (Lindsey et al., 2020); the protection of remote areas through private company investments in conservation hunting blocks (Lindsey et al., 2006); provision of protein and livelihood support to communities living in wildlife-rich areas (Baker, 1997); and control of wildlife populations that have started creating conflicts with communities (Darimont et al., 2017; Teichman et al., 2016). Irrespective of its diverse social, political, managerial, ecological and ethical critics, trophy hunting is fundamentally premised on supporting the conservation of threatened wildlife population and deteriorating wilderness (Crosmary et al., 2015). In the absence of the trophy hunting industry, the incentive for local people to support wildlife conservation and the capacity of governments to finance conservation activities is likely to face a downward trend (Muposhi et al., 2016) The absence of this stream of revenue generation for conservation is a significant barrier to the sustainable conservation of natural resources in large parts of eastern and southern African countries with vast lands containing threatened megafauna.

In a review of trophy hunting in southern Africa, Di Minin et al. (2016) concluded that the ban of trophy hunting has impacts on community livelihood and consequences for the conservation of game species (Di Minin et al., 2016). Such conservation downfalls have restricted the conservation of remote areas, causing the incursion of livestock to protected areas which in turn has accelerated the decline of wildlife habitat in the region. Hunting bans result in the decline in conservation efforts provided by trophy hunters who provide substantial resources for the protection of remote and wilderness areas.

According to Yeomans et al. (2022), trophy hunting bans are strongly conditioned by media pressure from Western countries who are unaffected by wildlife in the region. Impacts of wildlife such as crop damages does not directly affect the people in the western world and thus they can apply ethical principles to wild animals that cause damage to crops in Africa, leaving people impoverished. Wildlife to westerners is valuable for recreational activities while in Africa they are resources considered for living (Mkono, 2019). The outcome of hunting bans caused by external pressure have negative impacts to communities where trophy hunting is conducted, with less impacts to people living in America and Europe who advocate for trophy hunting bans. Despite the external pressure to ban trophy hunting, evidence shows that trophy hunting bans do not guarantee the recovery of wildlife populations. For instance, the hunting ban in Kenya (in 1977), Tanzania (1973–1973) and Zambia (2000–2003) were associated with more losses of wildlife (Lindsey et al., 2006) while the hunting ban in Botswana in 2014 resulted in more conflicts between people and wildlife and as a result the ban was lifted in 2019 (Mbaiwa, 2018). Zambia is considering uplifting the ban on the hunting of lions and leopards claiming that it has resulted in negative impacts on conservation and people's livelihoods (Mweetwa, 2016). Therefore, the decision to ban hunting in the region is characterised by pressure from the western world; the bans have negative consequences on conservation and community livelihood in the Global South.

The growing pressure from Western countries over the sustainability of trophy hunting in Africa has led to a growing sense among scholars and practitioners that the conservation of wildlife is failing. The resentment over trophy hunting is solely controlled by western pressure, neglecting the role grassroots communities in developing world play on decisions over hunting in their areas (Angula et al., 2018; Dube, 2019). Most communities in which trophy hunting occurs, support hunting for both livelihood and conservation motivations. The decline in hunting has caused communities to lose motivation for conservation and jeopardize the recovery of species and habits (Cooney et al., 2017). Several studies report the positive impacts of hunting to local communities; these include enhancing life and tolerance of losses associated with wildlife (Thomsen et al., 2022).

Our analysis of the trophy hunting industry in east and southern Africa starts with the historical development followed by trophy hunting as a commercial industry, current debates regarding hunting, brief reviews from eastern and southern African countries, the fate of the hunting industry in eastern and southern Africa and lastly, implications for sustainability.

10.2 Trophy Hunting Development in Eastern and Southern Africa

Trophy hunting in East and Southern Africa can be traced back to earlier European settlers and explorers (Akama, 2004). During that time, hunting was uncontrolled, leading to the decline of wildlife species such as elephants and the extinction of others such as blue back (*Hippotragus leocophaus*) and quagga (*Equus quagga* quagga). During the late nineteenth century there was an increasing need to preserve the remaining game stock and by the early twentieth century hunters played a role in the establishment of some of the most famous African protected areas (Munro, 2021). In the early twentieth century formal trophy hunting started in Kenya with wealthy European and American settlers paying settler farmers to guide them on hunting safaris in the areas. From there, the industry spread to other countries in the region. During the 1980s and 1990s, the potential for hunting tourism to create financial incentives for conservation increased and in several nations there was a gradual alignment of trophy hunting with conservation and development with the Community Area Management Program for Indigenous Resources (CAMPFIRE) in Zimbabwe and the Administrative Management Design for Game Management Areas (ADMADE) in Zambia (Lindsey, 2008). Trophy hunting grew into a major industry in Africa, generating significant revenue from and for wildlife over vast areas. From history, consumptive wildlife tourism has been a significant source of income for many communities and has contributed to the conservation of wildlife through the creation of protected areas and the development of sustainable utilization programs in east and Sothern Africa (Lindsey et al., 2006; Saayman et al., 2018).

Over the last decade trophy hunting in Africa has received concern from scientists and the public regarding its legitimacy for conservation and sustainability (Ghasemi, 2021). Much of these concerns are from outside Africa and less on local views. According to recent research (van Houdt et al., 2021) North America and Africa support trophy hunting while Europe and other countries are less interested in trophy hunting. African countries, particularly those in southern Africa are substantially and negatively impacted by the dynamics in the trophy hunting industry because they depend on the business to drive their economies, local development and conservation efforts. Much criticism in this decade came due to public outcries following circulated photos showing the Spain King Carlos posing with his elephant and buffalo trophies (Etherington, 2016) and the hunting of Cecil the lion in 2015 (Carpenter & Konisky, 2019; Macdonald et al., 2016; Mkono, 2018). These events led to the tightening of international legislations and the rise of animal right activist groups (Mkono, 2018). South Africa, however, took a firm stand in developing its trophy hunting industry, owing to its importance for the economy and conservation (Koot, 2019). The downfall of the hunting industry poses two questions. One is whether it is because of the rise in ethical tourism or empirical evidence of the failure of hunting as a conservation tool, or both. The decade witnessed the ban of trophy hunting in Botswana in 2013 (LaRocco, 2016). The increasing resentment about trophy hunting is highly reliant on anti-hunter views, neglecting the opinions of African communities, where hunting takes place, creating fears of increased neo colonialism in trophy hunting.

10.3 Trophy Hunting as a Commercial Industry

The Eastern and Southern Africa region provide a robust consumptive wildlife tourism sector through the trophy hunting industry. The trophy hunting industry comprises commercial operations that involve consumers paying for hunting experiences. It is a lucrative business in which tycoons pay significant amounts of money and support the economies of countries where animals are hunted for leisure. The process not only provides fees but also creates local employment and the trickling down of money to communities, with less environmental impact. Under the theory of wildlife as renewable biological resources, wildlife must pay for their protection, the industry aims to commercialize hunting experiences that pay for people on the ground to work and protect wildlife (Walters, 2009). This theory seems promising in poorer countries with governments that cannot prioritise revenue for the protection of wildlife. Instead of culling to control wildlife populations, wildlife managers can implement trophy hunting to earn more revenue for conservation and maintain carrying capacities of parks. In countries that advocate for trophy hunting, one of the justification is maintaining animal populations and reducing the impacts of problem wildlife on humans through some form of commercialization (Darimont et al., 2017). In other countries like South Africa the model is different; private investors produce wildlife in game ranches and farms supplying trophy to consumers (Saayman et al., 2018). Due to the demand for trophy hunting, farm and ranch owners have protected large land areas and produce wildlife for customers. By doing so, they drive countries' economies and provide employment to people.

Consumptive wildlife tourism is a significant source of income for many communities and has contributed to the conservation of wildlife through the creation of protected areas and the development of sustainable utilization programs in east and Sothern Africa (Lovelock, 2007). The industry was initiated in the seventeenth century but gained traction in East and Sothern Africa in the 1960s to 1970s when more countries in the region gained independence (Baker, 1997). Since then, it has been a growing industry in the region's tourism sector (MacKay & Michael Campbell, 2004). South Africa, Zimbabwe and Namibia are countries that have experienced significant growth in the industry among other countries in the region (Mkono, 2023).

10.4 Current Hunting Debate

Despite the role of the hunting industry in the social, economic and ecological contributions to biodiversity conservation, it is currently facing significant contradictions from the ethical dimension of wildlife (Ghasemi, 2021; Gunn, 2001). The argument for and against hunting is perceived on two major aspects, one on the utilitarian ecological and sustainability sense and the second on animal rights activists/ethical analysist perspectives creating two groups of pro hunters and anti-hunters.

The existing debate on trophy hunting can be traced to the war between pro-hunting and anti-hunting views among western communities, with less influence from trophy hunting hubs in southern Africa (Mkono, 2019). While pro-hunters contend hunting as recreation, conservation tool and population control mechanism, their counterparts view trophy hunting on ethical grounds; some posit that it is causing decline of wildlife populations (Wood, 1997). Western societies including those in Europe and America consider hunting as a recreational pursuit and pay substantial amounts of money for this activity. Hunting in America, has shifted from subsistence for necessity to a leisure pursuit motivated by thrill and trophy seeking, kinship bonding, personal fulfilment and inheritance of tradition (Gunn, 2001). This makes trophy hunting to be both a significant social and economic activity in many countries (Gunn, 2001; Naidoo et al., 2016). Hunting is currently a multibillionaire dollar industry in America and other countries. In 2011, a national survey carried out in the USA reported that 13.7 million people participated in hunting and spent 33.7 billion on trips and equipment (Poudel et al., 2016). The authors evaluated the contribution of hunting in the southern states of USA and found that 31% of the hunting expenditure in America was related to states investment in wildlife management (Poudel et al., 2016). The impacts of America's spending also trickle down to other countries, particular in Africa, where most big game are abundant. The recreational pursuits by developed world have become an opportunity for the developing world to create a business that finds the worthiness of wildlife conservation. Governments in the developing world have created regulations to tap the potential of sustainably harvesting their wildlife populations to generate income (Van der Merwe et al., 2014).

Pro-hunters advocate for trophy hunting as an important conservation tool (Baker, 1997; Lindsey et al., 2006, 2007, 2012; Muposhi et al., 2016). The activities under trophy hunting produce substantial revenues from direct expenditure and taxes to the country where hunting takes place and to the country where the client comes from (Gunn, 2001). It further generates billions of jobs in its chain of activities. The contributions of hunters are fundamental contributions for social and economic support systems for conserving wildlife and habitats for future generations (Naidoo et al., 2016). In the US system, hunting related revenue is channelled to wildlife and habitat conservation through the Federal Aid established by the Wildlife Restoration Act (Heffelfinger et al., 2013). This mechanism provides assurance for wildlife conservation using hunting revenues. Hunting in USA has proven to have significant impact on rural communities in supporting their businesses, job creation and multipliers effects associated with hunting activities (Poudel et al., 2016). Similar models have been used in sub-Saharan Africa to promote wildlife and habitat conservation and community livelihoods. For example, a study in Namibia reported significant contributions of trophy hunting to the economy, wildlife, and habitat in 2016 (Hewitson & Sullivan, 2021). Other countries have also reported evidence of the role trophy hunting plays in the sustainable conservation of wildlife rich habitats (Di Minin et al., 2021). Trophy hunting thus benefits the conservation of wildlife and habitats (Lindsey et al., 2007) if scientific and sustainable management practices as well as laws that govern the investment of revenue for conservation are sufficiently applied (Crosmary et al., 2015).

Focusing on the shrinking of wildlife habitats, pro-hunters believe that conflicts increase the interface between people and wildlife and trophy hunting represents a means for reducing problem animals compared to culling (Macdonald et al., 2016). Further, they apply the principle of wildlife as a renewable resource which can be harvested sustainably under good management. In the USA case, lands are purchased by landowners and wildlife managed for hunting, a mode similar to countries in Africa (such as South Africa) where significant trophy hunting occurs in private land. Under private land management, wildlife is produced for consumption by land owners, including through hunting (Lindsey, 2008; Parker et al., 2020). This mode benefits the conservation habitat for non-game species but also opportunity for harvesting of game species under scientific based optimum yield basis (Chardonnet et al., 2002). Even on public land, if there is good management, wildlife may also be sustainably harvested without jeopardizing the species populations. Large numbers of studies in trophy hunting recommend that under good management, trophy hunting can be a sustainable tool for wildlife conservation (Baker, 1997; Crosmary et al., 2015; Di Minin et al., 2016; Harris et al., 2013; Humavindu & Barnes, 2003). The studies are indicative for the potentialities of sustainability of the trophy hunting industry despite the critics from the anti-hunting views.

Anti-hunters' critique of trophy hunting is grounded on ethical dimensions, positing the claim that it is immoral to kill animals for leisure. They also tend to find no difference between legal hunting and illegal hunting (Ghasemi, 2021). The anti-hunters resentments are heightened by the negative popularisation of hunting cases through media campaigns, the most prominent being the hunting of Cecil the lion by the American hunter, Palmer, in 2015 (Macdonald et al., 2016). Anti-hunters adopt ontological perspective of the industry while pro-hunters adopt utilitarian perspectives (Ghasemi, 2021). The moral dimension of wildlife hunting is complex and problematic in some aspects, it is questioned by its bias towards game species while livestock and chickens are killed by the millions in slaughterhouses (Knezevic, 2009). Thus, the debate needs sufficient scientific driven information to justify position of contradicting parties and to end the debate by leveraging the two contradicting ideologies.

10.5 Tanzania

Tanzania's trophy hunting sector has grown considerably from the early 1990s to 2000s and has become the principal source of income for vast areas of the country (Baldus & Cauldwell, 2004). The industry is an important source of foreign income for the country. Tanzania has gained popularity in hunting owing to its abundance of game species that have attracted famous hunters including President Roosevelt who visited the country in 1905 and Frederic Selous, who fought for African wildlife for most of his life, leading to the establishment of the famous Selous Game reserve in the country. The trophy hunting industry is important because the country provides vast remote areas which provide attractions for wildlife viewing. In these areas, hunting is a viable economic means of using wildlife to provide for conservation and community development. Rich hunting companies have ventured in these areas to establish hunting concessions that have contributed to infrastructure development, support to communities and protection of wildlife and habitats (Musika et al., 2022). With over 130 hunting concessions covering 250,000 km^2 of land available for hunting, Tanzania has set an example, with trophy hunting being used to promote conservation and improve the livelihoods of local communities (Baldus & Cauldwell, 2004). The combination of national parks system which do not allow hunting and hunting concession in the country give assurance of maintaining wildlife populations. National parks have become reservoirs, breeding sites and refuge sites for most wildlife species. Furthermore, the country adheres to strict hunting regulations in which hunting quotas are rigorously set to respect the species biology and ecology (Hurt & Ravn, 2000).

Despite the reputation of Tanzania trophy hunting, there are studies that show a decline of game species as a result of trophy hunting (Brink et al., 2016). Flaws in management remain the driving factor towards unsustainable trophy hunting in Tanzania. For example, using molecular of leopard from 1995 to 2008, Spong et al. (2000) found that 28% of leopard hunted for trophy hunting were females contrary to the Tanzanian hunting regulation which allows only the hunting of old male individuals (Spong et al., 2000). For sustainable trophy hunting, good governance and comprehensive hunting regulations are key requirement in the country. This will allow hunting governed by scientific information and rational distribution of benefits to local communities. For instance, Hurt and Ravn (2000)evaluated trophy hunting in Tanzania and suggested that channelling hunting revenue to communities and good quota setting are essential for the sustainability of the trophy hunting industry (Hurt & Ravn, 2000).

10.6 Namibia

Hunting is an integral part of conservation models in Namibia which benefit communities, wildlife and natural systems (Thomsen et al., 2022). It is coined in the country's constitution that allows the sustainable use and management of natural resources and biodiversity. The constitution allows Namibians to explore both the non-consumptive and consumptive use of wildlife resources. Recognizing the role of the community, Namibia is among countries that have introduced a policy related to wildlife utilisation and tourism in communal areas, linking hunting and community livelihoods.

Though minuscule (0.18% of registered hunting companies in Namibia), the industry has substantial positive impacts on wildlife populations and community development (Barnes & Novelli, 2007). Hunting is also a vital to Namibia's economy, with an estimated 4000 hunters visiting the country each years (Lindsey et al., 2007) and registering a significant growing trend (Humavindu & Stage, 2015). The value of trophy hunting is estimated at USD 18 million per year (Lamprechts, 2009). According to the Game Product Trust fund, trophy hunting accounted for two thirds of the fund needed for conservation in Namibia in 2015. Revenue from hunting has been a key stimulus for wildlife conservation in the country (Lindsey et al., 2007).

It is considered that discouraging hunting by international pressure groups could risk the attainment of the triple benefits of community benefits, wildlife conservation and the maintenance of natural ecosystems. Trophy hunting is supported by both local communities and government in the country; 20% of land is reserved for community conservancy in which hunting occurs (Naidoo et al., 2011; Weaver & Petersen, 2008). A survey from 32 communal conservancies indicated that despite sentiments from western countries, members of these conservancies are happy with continued trophy hunting (Angula et al., 2018). This study tapped the opinions of beneficiaries of trophy hunting who also bear the costs of wildlife in their areas. This is an indication that any attempts to ban hunting because of international pressures could have negative effects on community livelihoods and will lower local support for wildlife conservation. Studies show that the trophy hunting ban in

Namibia reduced the income to many conservancies; this Income could cover their operating costs and support adjacent local communities (Naidoo et al., 2016).

10.7 Botswana

Trophy hunting was re-introduced in 2019 to promote sustainable wildlife harvesting that benefits the national economy, communities, tourism businesses and conservation stakeholders (Mbaiwa & Hambira, 2021).Trophy hunting was banned in 2014 following a national survey conducted in 2013 that showed declines in large mammals in the country (Mbaiwa, 2018). Although the ban shifted the community to photographic tourism, there were significant reductions in benefits from the wildlife sector, leading to negative attitudes towards wildlife and increased poaching (Mbaiwa, 2018). The ban caused job losses and negative impacts on the livelihood of people who were benefiting from hunting through community-based conservation in the country. Before the hunting ban, trophy hunting was significant for the economy of Botswana, ranking second to diamond mining in contribution to GDP.

10.8 Zimbabwe

Trophy hunting in Zimbabwe was affected following the killing of Cecil lion in 2015 (Lindsey et al., 2016). The event created havoc from anti-hunting activists and animal ethics groups through social media. Despite the controversy, trophy hunting is still an avenue for supporting communities adjacent to national parks, and for generating revenue to run conservation activities; to ban hunting there must be alternatives for the benefits to be sacrificed (Makuyana, 2018). The country has a trophy hunting policy that advocates for hunting to support communities and conservation. This led to the establishment of the popular community-based program called CAMPFIRE, that uses trophy hunting to ensure that benefits accrue from conservation to communities. Like other trophy hunting hotspots in Africa, Zimbabwe relies on trophy hunting for rural livelihoods, government revenues and support for conservation activities (Muposhi et al., 2016).Communal areas under CAMPFIRE management are different from national parks, where hunting is not permitted. The model of having trophy hunting sites and strict protected areas is adopted in many African countries to enable the sustainability of wildlife populations.

10.9 Zambia

Local communities gain trophy hunting economic benefits through a revolving fund program that ensures that revenue from hunting reaches local people near protected areas (Lewis & Alpert, 1997). Historically Zambia has provided a good example of a country with a good system of tapping the benefits of hunting for both conservation and sustaining communities close to protected areas. The establishment of game management areas (GMAs) as semi-protected areas to buffer strict protected national parks allowed trophy hunting. Through ADMADE, hunting revenue is distributed to Game Management Areas for local communities' development (Lewis & Alpert, 1997). The revenue from hunting in Zambia has been used to train and hire game scouts for community conservation areas and support community infrastructure such as schools and health canters.

10.10 The Fate of Trophy Hunting in East and Southern Africa

Despite the debate and controversies around trophy hunting, there is a need to revisit the status of the industry following disruptions created by anti-hunters through the media, starting from the Cecil the lion saga to date. Empirical studies in the region are important to establish the magnitude of the effects in relation to conservation and community livelihoods. The studies should be integrative and include social, economic and environmental aspects (Stoddard et al., 2012). For the long-term sustainability of the industry, governments are obliged to initiate discussions from local to international levels, to present context specific to the consumptive utilisation of wildlife for the benefit of wildlife and people. Since both parties of the hunting debate share some common grounds (Clark, 2007) and might have critical reasons for their argument, leverage must be obtained through presentation of pros and cons as well as investing in addressing challenges that tarnish the image of trophy hunting in east and southern African. Most of these are related to the administration, governance, and management of the trophy hunting industry. Efforts are also needed in the management of wildlife populations, to provide numbers that can sustainably be hunted without causing a decline in game species populations.

This is justified because some wildlife management programs such as private farms, ranches and parks have proven their ability to produce wildlife that can be harvested while maintaining the carrying capacity of wildlife (Barnes & Novelli, 2007; Cousins et al., 2008; Taylor et al., 2021). If parks and wildlife reserves are well protected, there is potential for harvesting wildlife while also maintaining the carrying capacities of parks following biological theories. Investing in proper regulatory frameworks and governance will make the industry appealing and feasible to both hunting lovers and the conservation community (Di Minin et al., 2016; Mosimane & Silva, 2015; Nelson et al., 2013). This investment will enable adherence to IUCN guidelines that advocate for the maintenance of wildlife populations and ecosystem functioning (Palazy et al., 2012). Properly crafted hunting guidelines and good administration of the industry should consider the hunting of species showing population concern such as elephants, lions, leopards, and whales. Good regulatory frameworks provide a potential to also ensure equitable benefits sharing from the tourism industry to communities and investments in biodiversity conservation stipulated in the IUCN guidelines.

10.11 Implication for Sustainability

The sustainability of the trophy hunting industry is reliant on the principles of wildlife harvesting and the need for wildlife to pay for its existence, embedded in the concept of wildlife conservation which emphasises wildlife welfare rather than wildlife right. Wildlife conservation is defined by the principles of wildlife protection and wildlife utilisation that are combined holistically (Nasi et al., 2008; Usher, 1986). Under conservation principles, wildlife should be managed and harvested under strict scientifically regulated mechanism. The mechanism includes restrictions on areas to hunt, size, age, physiological status, sex, time, number and the distribution of revenue from trophy hunting. In adherences to hunting regulations, good management of wildlife and equitable benefits distribution, wildlife, as renewable resources, can be harvested sustainably (Ullah & Kim, 2020). Failures in wildlife protection and the mismanagement of wildlife hunting regulations has negative consequences on wildlife populations and has strengthened anti-hunting campaigns against trophy hunting jeopardizing the sustainability of the industry, particularly in developing world.

Wildlife conservation researchers should be regarded as the most important disseminators of scientifically based information, with education in schools and universities being of growing importance. Furthermore, the media should shoulder the social responsibility for the accurate dissemination of conservation information for the industry to appeal to the public.

In conclusion, simplistic media depictions of trophy hunting have the potential to cause negative outcomes for conservation through its impact on public perceptions and political opinions.

References

Akama, J. S. (2004). Neo-colonialism, dependency and external control of Africa's tourism industry. In J. S. Akama (Ed.), *Tourism and postcolonialism: Contested discourses, identities and representations* (pp. 140–152). Routledge.

Angula, H. N., Stuart-Hill, G., Ward, D., Matongo, G., Diggle, R. W., & Naidoo, R. (2018). Local perceptions of trophy hunting on communal lands in Namibia. *Biological Conservation, 218*, 26–31.

Baker, J. E. (1997). Trophy hunting as a sustainable use of wildlife resources in southern and eastern Africa. *Journal of Sustainable Tourism, 5*(4), 306–321.

Baldus, R., & Cauldwell, A. (2004). *Tourist hunting and it's role in development of wildlife management areas in Tanzania.*

Barnes, J. I., & Novelli, M. (2007). Trophy hunting and recreational angling in Namibia: An economic, social and environmental comparison. In *Tourism and the consumption of wildlife* (pp. 177–190). Routledge.

Brink, H., Smith, R. J., Skinner, K., & Leader-Williams, N. (2016). Sustainability and long termtenure: Lion trophy hunting in Tanzania. *PloS One, 11*(9), e0162610.

Carpenter, S., & Konisky, D. M. (2019). The killing of Cecil the Lion as an impetus for policy change. *Oryx, 53*(4), 698–706.

Chardonnet, P., Clers, B. D., Fischer, J., Gerhold, R., Jori, F., & Lamarque, F. (2002). The value of wildlife. *Revue Scientifique et Technique-Office International Des Épizooties, 21*(1), 15–52.

Clark, G. E. (2007). Hunting in context. *Environment: Science and Policy for Sustainable Development, 49*(5), 3–4.

Cooney, R., Freese, C., Dublin, H., Roe, D., Mallon, D., Knight, M., Emslie, R., Pani, M., Booth, V., & Mahoney, S. (2017). The baby and the bathwater: Trophy hunting, conservation and rural livelihoods. *Unasylva, 68*(249), 3.

Cousins, J. A., Sadler, J. P., & Evans, J. (2008). Exploring the role of private wildlife ranching as a conservation tool in South Africa: Stakeholder perspectives. *Ecology and Society, 13*(2).

Crosmary, W., Côté, S., & Fritz, H. (2015). The assessment of the role of trophy hunting in wildlife conservation. *Animal Conservation, 18*(2), 136–137.

Darimont, C. T., Codding, B. F., & Hawkes, K. (2017). Why men trophy hunt. *Biology Letters, 13*(3), 20160909.

Di Minin, E., Leader-Williams, N., & Bradshaw, C. J. (2016). Banning trophy hunting will exacerbate biodiversity loss. *Trends in Ecology & Evolution, 31*(2), 99–102.

Di Minin, E., Clements, H. S., Correia, R. A., Cortés-Capano, G., Fink, C., Haukka, A., Hausmann, A., Kulkarni, R., & Bradshaw, C. J. A. (2021). Consequences of recreational hunting for biodiversity conservation and livelihoods. *One Earth, 4*(2), 238–253. https://doi.org/10.1016/j.oneear.2021.01.014

Dube, N. (2019). Voices from the village on trophy hunting in Hwange district, Zimbabwe. *Ecological Economics, 159*, 335–343. https://doi.org/10.1016/j.ecolecon.2019.02.006

Etherington, N. (2016). *Big game hunter: A biography of Frederick Courteney Selous*. Robert Hale Ltd.

Ghasemi, B. (2021). Trophy hunting and conservation: Do the major ethical theories converge in opposition to trophy hunting? *People and Nature, 3*(1), 77–87.

Gunn, A. S. (2001). Environmental ethics and trophy hunting. *Ethics & the Environment, 6*(1), 68–95. https://doi.org/10.1353/een.2001.0006

Harris, R. B., Cooney, R., & Leader-Williams, N. (2013). Application of the anthropogenic Allee effect model to trophy hunting as a conservation tool. *Conservation Biology, 27*(5), 945–951.

Heffelfinger, J. R., Geist, V., & Wishart, W. (2013). The role of hunting in North American wildlife conservation. *International Journal of Environmental Studies, 70*(3), 399–413.

Hewitson, L. J., & Sullivan, S. (2021). Producing elephant commodities for 'conservation hunting' in Namibian communal-area conservancies. *Journal of Political Ecology, 28*(1), 1–24.

Humavindu, M. N., & Barnes, J. I. (2003). Trophy hunting in the Namibian economy: An assessment: Research article. *South African Journal of Wildlife Research—24-Month Delayed Open Access, 33*(2), 65–70. https://doi.org/10.10520/EJC117175

Humavindu, M. N., & Stage, J. (2015). Community-based wildlife management failing to link conservation and financial viability. *Animal Conservation, 18*(1), 4–13.

Hurt, R., & Ravn, P. (2000). Hunting and its benefits: An overview of hunting in Africa with special reference to Tanzania. In H. H. T. Prins, J. G. Grootenhuis, & T. T. Dolan (Eds.), *Wildlife conservation by sustainable use* (pp. 295–313). Springer. https://doi.org/10.1007/978-94-011-4012-6_15

Knezevic, I. (2009). Hunting and environmentalism: Conflict or misperceptions. *Human Dimensions of Wildlife, 14*(1), 12–20.

Koot, S. (2019). The limits of economic benefits: Adding social affordances to the analysis of trophy hunting of the Khwe and Ju/'hoansi in Namibian community-based natural resource management. *Society & Natural Resources, 32*(4), 417–433.

Lamprechts, M. (2009). *Trophy hunting in Namibia from the 1960s to the present day*. Namibian Professional Hunting Association Report.

LaRocco, A. (2016). The comprehensive hunting ban: Strengthening the state through participatory conservation in contemporary Botswana. *The Politics of Nature and Science in Southern Africa*, 179–207.

Lewis, D. M., & Alpert, P. (1997). Trophy hunting and wildlife conservation in Zambia: Caza Deportiva y Conservación de la Vida Silvestre en Zambia. *Conservation Biology, 11*(1), 59–68.

Lindsey, P. A. (2008). Trophy hunting in sub-Saharan Africa: Economic scale and conservation significance. *Best Practices in Sustainable Hunting, 1*, 41–47.

Lindsey, P. A., Alexander, R., Frank, L. G., Mathieson, A., & Romanach, S. (2006). Potential of trophy hunting to create incentives for wildlife conservation in Africa where alternative wildlife-based land uses may not be viable. *Animal Conservation, 9*(3), 283–291.

Lindsey, P. A., Roulet, P., & Romanach, S. (2007). Economic and conservation significance of the trophy hunting industry in sub-Saharan Africa. *Biological Conservation, 134*(4), 455–469.

Lindsey, P. A., Balme, G. A., Booth, V. R., & Midlane, N. (2012). The significance of African lions for the financial viability of trophy hunting and the maintenance of wild land. *PLoS One, 7*(1), e29332.

Lindsey, P. A., Balme, G. A., Funston, P. J., Henschel, P. H., & Hunter, L. T. (2016). Life after Cecil: Channelling global outrage into funding for conservation in Africa. *Conservation Letters, 9*(4), 296–301.

Lindsey, P., Allan, J., Brehony, P., Dickman, A., Robson, A., Begg, C., Bhammar, H., Blanken, L., Breuer, T., & Fitzgerald, K. (2020). Conserving Africa's wildlife and wildlands through the COVID-19 crisis and beyond. *Nature Ecology & Evolution, 4*(10), 1300–1310.

Lovelock, B. (2007). An introduction to consumptive wildlife tourism. In *Tourism and the consumption of wildlife* (pp. 25–52). Routledge.

Macdonald, D. W., Jacobsen, K. S., Burnham, D., Johnson, P. J., & Loveridge, A. J. (2016). Cecil: A moment or a movement? Analysis of media coverage of the death of a lion, Panthera leo. *Animals, 6*(5), 26.

MacKay, K. J., & Michael Campbell, J. (2004). An examination of residents' support for hunting as a tourism product. *Tourism Management, 25*(4), 443–452. https://doi.org/10.1016/S0261-5177(03)00127-4

Makuyana, P. G. (2018). Trophy hunting: To ban or not to ban? Legal pathways for Zimbabwe in the aftermath of Cecil the lion. *Journal of Sustainable Development Law and Policy (The), 9*(1), 133–167.

Mbaiwa, J. E. (2018). Effects of the safari hunting tourism ban on rural livelihoods and wildlife conservation in Northern Botswana. *South African Geographical Journal = Suid-Afrikaanse Geografiese Tydskrif, 100*(1), 41–61.

Mbaiwa, J. E., & Hambira, W. L. (2021). Can the subaltern speak? Contradictions in trophy hunting and wildlife conservation trajectory in Botswana. *Journal of Sustainable Tourism, 1–19*, 1107–1125. https://doi.org/10.1080/09669582.2021.1973483

Mkono, M. (2018). The age of digital activism in tourism: Evaluating the legacy and limitations of the Cecil anti-trophy hunting movement. *Journal of Sustainable Tourism, 26*(9), 1608–1624.

Mkono, M. (2019). Neo-colonialism and greed: Africans' views on trophy hunting in social media. *Journal of Sustainable Tourism, 27*(5), 689–704.

Mkono, M. (2023). How we can make the trophy hunting debate less fraught. *Nature Human Behaviour, 7*(1), 6–8.

Mosimane, A. W., & Silva, J. A. (2015). Local governance institutions, CBNRM, and benefit-sharing systems in Namibian conservancies. *Journal of Sustainable Development, 8*(2), 99.

Munro, P. (2021). Colonial wildlife conservation and national parks in Sub-Saharan Africa. In *oxford research encyclopedia of African history*.
Muposhi, V. K., Gandiwa, E., Bartels, P., & Makuza, S. M. (2016). Trophy hunting, conservation, and rural development in Zimbabwe: Issues, options, and implications. *International Journal of Biodiversity, 2016*, 1–16. https://doi.org/10.1155/2016/8763980
Musika, N. V., Wakibara, J. V., Ndakidemi, P. A., & Treydte, A. C. (2022). Using trophy hunting to save wildlife foraging resources: A case study from moyowosi-kigosi game reserves, Tanzania. *Sustainability, 14*(3), 1288.
Mweetwa, T. (2016). *Effect of a temporary hunting ban on the demography of African Lions (Panthera Leo) using a protected area.*
Naidoo, R., Stuart-Hill, G., Weaver, L. C., Tagg, J., Davis, A., & Davidson, A. (2011). Effect of diversity of large wildlife species on financial benefits to local communities in Northwest Namibia. *Environmental and Resource Economics, 48*(2), 321–335. https://doi.org/10.1007/s10640-010-9412-3
Naidoo, R., Weaver, L. C., Diggle, R. W., Matongo, G., Stuart-Hill, G., & Thouless, C. (2016). Complementary benefits of tourism and hunting to communal conservancies in Namibia. *Conservation Biology, 30*(3), 628–638. https://doi.org/10.1111/cobi.2016.30.issue-3; https://doi.org/10.1111/cobi.12643
Nasi, R., Brown, D., Wilkie, D., Bennett, E., Tutin, C., Van Tol, G., & Christophersen, T. (2008). *Conservation and use of wildlife-based resources: The bushmeat crisis* (Technical series, 50). Secretariat of the Convention on Biological Diversity/Center for International Forestry Research (CIFOR).
Nelson, F., Lindsey, P., & Balme, G. (2013). Trophy hunting and lion conservation: A question of governance? *Oryx, 47*(4), 501–509. https://doi.org/10.1017/S003060531200035X
Palazy, L., Bonenfant, C., Gaillard, J.-M., & Courchamp, F. (2012). On the use of the IUCN status for the management of trophy hunting. *Wildlife Research, 39*(8), 711–720.
Parker, K., De Vos, A., Clements, H. S., Biggs, D., & Biggs, R. (2020). Impacts of a trophy hunting ban on private land conservation in South African biodiversity hotspots. *Conservation Science and Practice, 2*(7), e214.
Poudel, J., Henderson, J. E., & Munn, I. A. (2016). Economic contribution of hunting expenditure to the southern United States of America. *International Journal of Environmental Studies, 73*(2), 236–254.
Saayman, M., van der Merwe, P., & Saayman, A. (2018). The economic impact of trophy hunting in the South African wildlife industry. *Global Ecology and Conservation, 16*, e00510. https://doi.org/10.1016/j.gecco.2018.e00510
Sheikh, P. A., & Bermejo, L. F. (2019). *International trophy hunting*. Congressional Research Service.
Spong, G., Hellborg, L., & Creel, S. (2000). Sex ratio of leopards taken in trophy hunting: Genetic data from Tanzania. *Conservation Genetics, 1*, 169–171.
Stoddard, J. E., Pollard, C. E., & Evans, M. R. (2012). The triple bottom line: A framework for sustainable tourism development. *International Journal of Hospitality & Tourism Administration, 13*(3), 233–258. https://doi.org/10.1080/15256480.2012.698173
Taylor, W. A., Child, M. F., Lindsey, P. A., Nicholson, S. K., Relton, C., & Davies-Mostert, H. T. (2021). South Africa's private wildlife ranches protect globally significant populations of wild ungulates. *Biodiversity and Conservation, 30*(13), 4111–4135.
Teichman, K. J., Cristescu, B., & Darimont, C. T. (2016). Hunting as a management tool? Cougar-human conflict is positively related to trophy hunting. *BMC Ecology, 16*(1), 1–8.
Thomsen, J. M., Lendelvo, S., Coe, K., & Rispel, M. (2022). Community perspectives of empowerment from trophy hunting tourism in Namibia's Bwabwata National Park. *Journal of Sustainable Tourism, 30*(1), 223–239. https://doi.org/10.1080/09669582.2021.1874394
Ullah, I., & Kim, D.-Y. (2020). A model of collaborative governance for community-based trophy-hunting programs in developing countries. *Perspectives in Ecology and Conservation, 18*(3), 145–160. https://doi.org/10.1016/j.pecon.2020.06.004

Usher, M. B. (1986). *Wildlife conservation evaluation: Attributes, criteria and values*. Springer.
Van der Merwe, P., Saayman, M., & Rossouw, R. (2014). The economic impact of hunting: A regional approach. *South African Journal of Economic and Management Sciences, 17*(4), 379–395.
van Houdt, S., Brown, R. P., Wanger, T. C., Twine, W., Fynn, R., Uiseb, K., Cooney, R., & Traill, L. W. (2021). Divergent views on trophy hunting in Africa, and what this may mean for research and policy. *Conservation Letters, 14*(6), e12840.
Walters, W. H. (2009). Google scholar search performance: Comparative recall and precision. *Portal: Libraries and the Academy, 9*(1), 5–24.
Weaver, L. C., & Petersen, T. (2008). Namibia communal area conservancies. *Best Practices in Sustainable Hunting, 2008*, 48–52.
Wood, F. (1997). *The delights and dilemmas of hunting: The hunting versus anti-hunting debate*. University Press of America.
Yeomans, N., Hare, D., Dröge, E., & Hart, A. G. (2022). Ten years of coverage of trophy hunting in UK newspapers. *Frontiers in Conservation Science, 3*, 133.

Chapter 11
Non-consumptive Wildlife Tourism Resources Utilization Dynamics

Emmanuel Mwakiwa

11.1 Introduction

Wildlife has several uses that generate revenues for the betterment of landowners and society. For example, it is used consumptively, for hunting for trophies, biltong, selling meat and/or hide and recreational fishing. It can also be used non-consumptively for photographic safaris, hiking trails, game drives, wildlife viewing and bird watching. Landowners can choose to concentrate on consumptive or non-consumptive use of wildlife or a mixture of both (Mwakiwa et al., 2016). Game viewing produces several revenue sources for landowners such as entrance fees, accommodation charges, sales from restaurants and shops, trails, rides, tour income, or revenues from permits and fines (Aylward, 2003). Game viewing is the main motivation behind tourists' visits to game reserves, where the number of tourists is positively related to the probability of seeing the "Big Five": elephant, buffalo, rhino, lion, and leopard (Lindsey et al., 2007).

According to the World Travel & Tourism Council [WTTC] (2019), in 2018, the tourism sector accounted for 10.4% of global GDP: wildlife tourism represented 3.9% of this amount ($343.6 billion), which is equivalent to the entire GDP of South Africa or Hong Kong. In addition, 21.8 million jobs or 6.8% of total jobs in 2018 can be attributed to wildlife around the world (WTTC, 2019). Wildlife tourism in Africa plays a tremendous role in the economy than any other continent. This is evident from the regional shares of wildlife tourism within the Travel & Tourism sector which differ quite widely—in Africa it is greatly significant at 36.3%; 8.6% in Latin America; 5.8% in Asia-Pacific; 2% in North America; and just 1.6% in

E. Mwakiwa (✉)
Department of Agricultural Business Development and Economics, Faculty of Agriculture Environment And Food Systems, University of Zimbabwe, Harare, Zimbabwe
e-mail: emakiwa@agric.uz.ac.zw

© The Author(s), under exclusive license to Springer Nature Switzerland AG 2024

L. S. Stone, M. T. Stone (eds.), *Wildlife Tourism Dynamics in Southern Africa*, Geographies of Tourism and Global Change,
https://doi.org/10.1007/978-3-031-57252-4_11

Europe (WTTC, 2019). Furthermore, wildlife tourism provides economic opportunities for communities and an economic justification for the protection of endangered species and associated wildlife habitats.

Non-consumptive wildlife use can have positive and negative economic, socio-cultural, and environmental impacts (Koch, 2021). These impacts affect wildlife dynamics, benefits and costs of those wildlife, and the welfare of both wildlife and society. The type and magnitude of the impacts depend on numerous factors, such as the type of tourists, type of activity pursued by tourists, characteristics of the wildlife, the ecology of the area and the time scale under consideration (Choi et al., 2017). Non-consumptive wildlife tourism can play a great role in fighting against the global wildlife challenges such as the illegal trade in wildlife.

In this chapter, attention will be drawn to what has been happening in the wildlife tourism sub-sector to emphasize the enormous wildlife economic, socio-cultural and environmental opportunities thereby substantiating the need for greater wildlife protection. Focus of the discussions is placed on non-consumptive use by looking at the dynamics, benefits and challenges of wildlife non-consumptive utilization using cases from southern Africa.

11.2 Methodology

To perform the assessment of the wildlife uses, literature review of published, unpublished and grey conservation literature related to both consumptive and non-consumptives uses, with bias towards the latter in southern Africa was conducted. The use of secondary data is also employed to show the dynamics of wildlife tourism. The discussion begins with an overview of the positive and negative impacts of non-consumptive wildlife tourism from the economic, socio-cultural and environment perspectives. Non-consumptive use dynamics is then presented and discussed using case studies from selected southern Africa countries. At the end, conclusions and recommendations are provided.

11.3 Non-consumptive Wildlife Tourism Use

Africa has been experiencing growth within the wildlife tourism industry over the past century, in terms of both consumptive and non-consumptive wildlife tourism (Tremblay, 2001). This growth can be attributed to the attractive landscapes, good weather, distinctive history, abundant and thriving wildlife, multitudes of cultures and the diversity of tourism products offered by the continent (Koch, 2021). The tourism industry can broadly be divided into two types: mass tourism and alternative tourism. Alternative tourism forms part of natural area tourism. In turn, natural area tourism consists of a variety of tourism subsectors, which non-consumptive wildlife tourism forms a part of. Although it can be said that non-consumptive tourism has different types e.g., ecotourism, safari tours and mountain tourism among others, in this chapter we use the term in its broadest sense, that is, viewing and experiencing animals in their natural habitat (WTTC, 2019).

11.4 Benefits and Challenges of Non-consumptive Wildlife Tourism

There are four features that need to be considered to determine whether the economy of an area has been impacted by wildlife tourism: (i) the number of wildlife and tourists in that area; (ii) tourists' spending in that area; (iii) tourists' length of stay in that area; and (iv) the multiplier effect this kind of tourism generates in the area (Saayman, 2000). The economic impacts can further be grouped into direct, indirect, and induced impacts (Lindberg, 2001). Direct impacts are those that result from the wildlife tourists' primary spending in the area, e.g., entry fees, accommodation, and transport. Indirect impacts are secondary and occur when, e.g., wildlife tourism organisations and institutions such as lodges purchase goods and services from other tourism-related businesses, creating an indirect impact; whilst induced impacts occur when wildlife tourism organisations and institutions' employees spend their income on individual goods and services (Lindberg, 2001).

Sociocultural impacts refer to the ways in which tourism alters community and individual values, behaviour, community structure, lifestyle, and general quality of life for both the local community and visitors (Diamantis, 2004). The different economic levels of the local community and visitors should also be considered. Wildlife tourists are mainly from highly developed areas, such as those in Europe and the USA visiting wildlife areas in countries such as Namibia, Zambia, and Zimbabwe where the bulk of the communities are poor. Deliberately or not, the tourists may ignore the fact that the local community do not have the same prospects as they do (Koch, 2021). In addition, dissimilarities in culture, such as the beliefs, values and behaviours between the locals and visitors might have impacts on both parties (Koch, 2021; Diamantis, 2004).

Just like economic impacts, environmental impacts can be categorized as either direct or indirect. Direct impacts are easy to identify, manage and control, since the cause and effect are clear. On the other hand, indirect impacts have more in-depth effect on the environment and are more difficult to measure because the impacts are usually invisible, inconspicuous, and implicit.

11.5 Benefits of Non-consumptive Wildlife Use

11.5.1 Positive Economic Impacts

Wildlife-based tourism can incentivize communities into conserving wildlife and its habitats and as such can gain broad political support for wildlife conservation. Table 11.1 summarizes the main positive economic impacts of non-consumptive wildlife tourism, although some benefits might also apply for wildlife consumptive use.

11.5.2 Positive Socio-Cultural Impacts

Tourism has some positive impacts on culture. Shahzalal (2016) indicates that tourism maintains the undisturbed or closed cultural activities through identifying, developing, rediscovering, and linking them to the international market. Table 11.2 summarises some of the positive impacts that can emerge because of sociocultural contact between wildlife tourists and the host community due to natural area tourism while Table 11.3 summarises some positive environmental impacts of wildlife non-consumptive tourism.

Table 11.1 Positive economic impacts of non-consumptive wildlife tourism

Positive economic impact	Results
Exports	International wildlife tourists' expenditure on local transport, accommodation, food, crafts and curios
Employment opportunities	Creation of both direct and indirect employment opportunities for the small, medium, and micro enterprises
Economic development	Wildlife tourists' expenditure has multiplier effects on the host nation's economy. This leads to improved livelihoods of the local communities and greater diversification of economic activities, thereby reducing economic risks
Decreasing regional economy inequities in marginal areas	Increasing and transferring of wealth and income to vulnerable areas and increased opportunities for wildlife-related businesses and enterprises owned by local people in marginal areas
Improved infrastructure and public services for the local community	Natural area tourism growth supports the local community through the development of new facilities and services, and the enhancement of existing infrastructure
Increased earnings and generation of government revenue	Government income from wildlife tourism reinvested in the local community, such as the enhancement of basic infrastructure, like roads, schools

Source: Adapted from Tisdell (2003); van der Merwe et al. (2021)

Table 11.2 Positive socio-cultural impacts of wildlife non-consumptive tourism

Positive socio-cultural impacts	Results
Development of mutual cultural consciousness	Wildlife tourists and the host community are exposed to each other's cultures, which can help reduce stereotyping and improve appreciation of cultural diversities. Host community self-esteem increases when wildlife tourists are interested in their culture
Conserving history and local heritage	Some of the revenue generated from wildlife tourists contributes to the preservation, conservation and restoration of local heritage which results in the following: Revitalising local customs and traditions; increasing cross cultural interaction; understanding, maintaining, and keeping local culture, arts, crafts and traditions; empowering host communities; and strengthening cultural values
Promotes socio-cultural tolerance	Direct contact between wildlife tourists and local communities help to understand and tolerate each other

Source: Adapted from Shahzalal (2016)

Table 11.3 Positive environmental impacts of wildlife non-consumptive tourism

Positive environmental impacts	Results
Nature conservation	The improvement of degraded and disturbed areas; the improvement and protection of biodiversity; the establishment and expansion of national parks or game reserves; the creation of wildlife reserves
Restore wildlife at the brink of extinction	Restoration of wildlife species which are near extinct; game reserve or farm owners enticed to conserve wildlife found on their land by market demand or just for conservation purposes
Public awareness of environmental problems	Wildlife tourists who come into close contact with nature develop increased public awareness regarding environmental challenges and the appreciation of nature

Adapted from: Koch (2021)

11.6 Challenges of Wildlife Non-consumptive Use

Some argue that consumptive wildlife tourists result in greater environmental impacts on wildlife than non-consumptive wildlife tourists. One of the reasons given is that the impact from consumptive wildlife tourists, although on a smaller scale, is more intense and concentrated (Koch, 2021). However, some might also argue that an influx of tourists to a particular wildlife spot can also cause similar even worse environmental degradation (Prosser (1994). Non-consumptive wildlife tourists who visit a popular wildlife destination, in amalgamation with all their activities, can have severe impacts on the surrounding environment (Ceballos-Lascurain, 1996). We now discuss the challenges posed by non-consumptive wildlife use in detail.

11.6.1 Negative Economic Impacts of Non-consumptive Wildlife Tourism

Non-consumptive wildlife tourism can also bring about negative economic impacts (see Table 11.4) and socio-cultural impacts due to contact between wildlife tourists and the host community due to natural area tourism (see Table 11.5).

11.6.2 Negative Environmental Impacts

Negative environmental impacts occur when the environment use level and demand by tourists is greater than the ability of the environment to cope with this within some tolerable change limits. Subsequently, this causes the environmental resources that wildlife tourism depends on to be eventually depleted. Table 11.6 summarizes the negative environmental impacts of non-consumptive wildlife tourism.

Table 11.4 Negative economic impacts of wildlife non-consumptive tourism

Negative economic impact	Results
Inflationary pressure	Wildlife tourism can cause an increase in cost of living and exorbitant property prices at destination areas
Economic leakages	Money drains out of the economy as goods, services and skills for natural tourism are imported
Economy over-reliance on wildlife tourism	Over-reliance on wildlife tourism can be risky if there is a shock, such as an economic depression, natural disaster, health epidemic, or terrorism
Exclusion of local communities	Exclusion of locals from tourist areas denies them income, employment, and resource utilization opportunities
Seasonal employment	Depressed demand for wildlife tourism in off-season can result in more short-term than stable permanent jobs

Source: Adapted from Tisdell (2003)

Table 11.5 Negative socio-cultural impacts of non-consumptive wildlife tourism

Negative socio-cultural impacts	Results
Import of skilled foreign workers	Local community feel abandoned and presence of economic colonialism
Promotion of tourists' culture	Disruption of the social fabric of the local community which might lead to social unrest
Staging false local culture	Unauthentic local culture package misrepresentation
Diseases transmission	Wildlife tourists may transmit disease to the local community members or vice versa

Source: Adapted from Shahzalal (2016)

Table 11.6 Negative environmental impacts of non-consumptive wildlife tourism

Negative socio-cultural impacts	Results
Overuse of water	Water deficiencies and degradation of freshwater supply due to overutilization of water by the accommodation sector
Local resources put under pressure	Local resources overutilized by wildlife tourism in the form of energy consumption (heating and cooling of rooms, lighting, hot water, food preparation, spas) especially during high tourism season
Land and flora degradation	Disturbance of wildlife and degradation of the minerals in the soil, fertile soils, and water supply in the area due to clearing land of vegetation for the construction of wildlife tourism facilities and infrastructure. Loss of natural biodiversity due to disturbance of seed germination, caused by trampling of vegetation and soil when wildlife tourists use the same trails or roads repeatedly for photographing and wildlife trails
Water pollution	Wastewater and other effluents from accommodation facilities, kitchens and other installations discharged into the water systems in a natural or protected area that hosts wildlife tourists. Excess effluents pollute surface and underground water bodies are detrimental to the wildlife and humans
Air pollution	Carbon monoxide and other air pollutants emitted from transport (air, ground, water, and rail) systems and other sources used by wildlife tourists which contributes to acid rain, climate change, and photochemical pollution
Noise pollution	What about noise pollution—What are its sources in this context. Noise pollution, as a result of aeroplanes, helicopters, recreational vehicles, building operations and wildlife tourist activities such as game drives cause stress and disturbance to wildlife in sensitive and protected areas. This stress and disturbance can further affect the behaviour patterns of wildlife, such as the mating, foraging and resting periods of the animals
Solid waste and littering	Poor waste management and littering, especially in areas of high wildlife tourist concentration due to poor disposal of waste causing damage to the natural environment, rivers and scenic areas; resulting in hazard to both human and wildlife
Aesthetic pollution	Lack of land use planning, building regulations and overdevelopment of wildlife tourists' areas can result in uncontrolled development of tourists' facilities and infrastructure which drastically affects the aesthetic appeal of these destinations resulting in architectural or visual pollution. This can also cause massive deforestation, soil erosion, land degradation, and the loss of wildlife habitats due to destruction of ecosystems and microhabitats especially in sensitive areas
Disturbance of wildlife by tourism systems	Wildlife behaviour patterns (e.g., breeding patterns, foraging and resting periods of the animals) negatively affected by noise pollution from transport (air, ground, water and rail) used by wildlife tourists, building operations, destruction of habitats and wildlife tourist activities (e.g. game drives, picking rare plants)

Source: Adapted from Koch (2021); van der Merwe et al. (2021)

11.6.3 Negative Versus Positive Impacts

One of the requirements in the development of sustainable wildlife tourism is that the economic returns from engaging in it should exceed the costs involved (Mwakiwa et al., 2016). This is possible if the wildlife tourism revenue venture exceeds costs. If the wildlife site is a state protected area, its income may come from the following sources: (a) entry fees, camping fees, and other charges levied on visitors and (b) the allocation of government revenues, (c) sales of services and products at the site, (d) donations by visitors and (e) sales of concessions to others to provide products or services at the site e.g. accommodation, food and tours (Tisdell, 2003). However, resources available to the protected area depend on institutional arrangements. If income raised must be paid into consolidated government revenue, no direct benefits may accrue to the protected area from its income generation activities. If the protected area retains all or a portion of the funds it collects, this will add to conservation resources for the protected area (Tisdell, 2003).

Even when ecotourism development can make available extra resources for wildlife conservation, not all wildlife areas are able to sustain profitable ecotourism ventures. The factors that influence wildlife tourism profitability include accessibility of the area to visitors, the density of wildlife, the availability of complementary services and the cost of visiting the site (Tisdell, 2003).

It is also believed that wildlife tourism can provide direct financial support for local communities where it occurs. This is from the assumption that ecotourism conserves the environment and improve the well-being of local people (Honey, 1999). The local communities involved have limited sources of livelihoods and are often remote from the main centres of economic activity. While benefits from the development of wildlife tourism are possible, wildlife tourism might result in little to no economic benefits to local communities, it may become a drain on finance and can distort the range of species conserved (Tchakatumba et al., 2019; Tisdell, 1995).

11.7 Non-consumptive Wildlife Tourism Dynamics: Case Studies from Southern Africa

In southern Africa, tourism in national parks and game reserve except in privately owned reserves is non-consumptive in nature. Non-consumptive wildlife tourism generates a substantial amount of revenues for countries in southern Africa. According to UNWTO (2015), non-consumptive wildlife tourism for most African countries, represents 80% of the total annual trip sales to Africa, and the proportion is increasing.

11.7.1 Botswana: Non-consumptive Wildlife Tourism

The tourism industry of Botswana remains undiversified and heavily dependent on non-consumptive wildlife use as the major pull factor and as the main driver for international tourist visits (Stone et al., 2021). Botswana is home of the world's largest elephant population (Blackie & Sowa, 2019). The tourism industry contributes about 11.5% to the country's GDP and employs 26,000 people whereas other authors place it at 69,500 jobs (WTTC, 2019; Stone et al., 2017). The tourism industry is Botswana's second largest foreign exchange earner, after diamond mining (Stone et al., 2017). The centre of Botswana's tourism industry is protected and conserved areas, wildlife, and wilderness tourism. It upholds sustainable tourism through a high-value low-volume tourism policy, to promote sustainable generation of revenue and conservation of natural resources (Government of Botswana, 1992). Sustainable wildlife tourism in Botswana supports Community-Based Natural Resource Management (CBNRM) program (Stone et al., 2021). Figure 11.1 shows the number of tourist and associated revenue for Botswana from 1995 to 2020.

In 1995, tourism revenue was US$176 million, about 3.7% of the GNP; there were about 636,000 tourists (see Fig. 11.1). Within a period of 25 years, the country's dependence on tourism increased. Before the outbreak of the COVID-19 pandemic, tourism revenue was US$712.4 million translating to 4.3% of GNP (see Fig. 11.1). Botswana recorded a total of two million tourists in 2018, ranking 73rd in the world in absolute terms (WordData.Info, 2023). By putting the tourist numbers in relation to the population of Botswana: with 0.71 tourists per resident, Botswana ranked 56th in the world (WorldData.Info, 2023). In 2020, Botswana

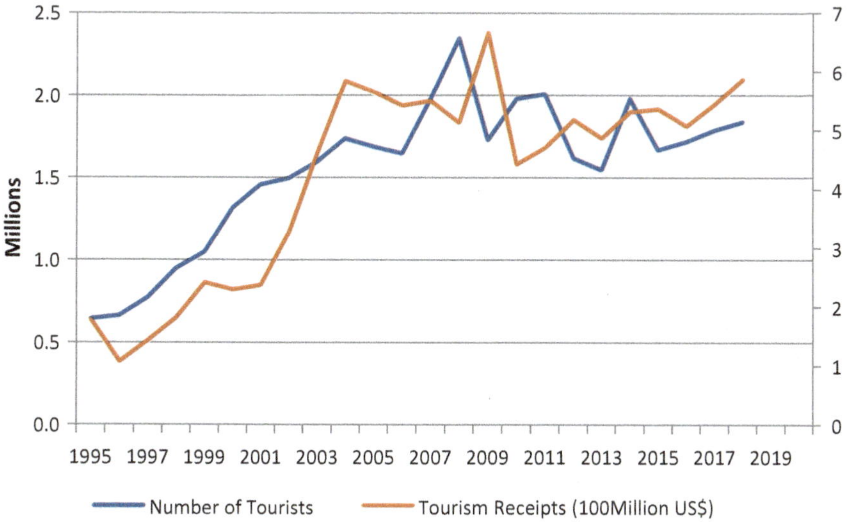

Fig. 11.1 Number of international tourists and revenue from tourism in Botswana. (Source: https://www.worlddata.info/africa/botswana/tourism.php)

generated around US$217 million in the tourism sector alone which corresponds to 1.2% of its GDP (WordData.Info, 2023).

In 2020, tourist receipts tumbled because of the COVID-19 pandemic, with the country making US$217 million, a 70% decrease in revenue (Fig. 11.1). This shows that COVID-19 did not spare the Botswana wildlife tourism sector. Just as elsewhere in the world, the pandemic's reduced global demand due to the travel restrictions and social distancing measures which had an impact on the country's key production and export sectors, including tourism (Blackie & Sowa, 2019).

Botswana's tourism industry is mainly wildlife-based and is carried out in the northern parts of the country, especially in the Okavango and Chobe regions (Stone et al., 2017). About17% of Botswana land area is covered by national parks and game reserves with an additional 22% designated as wildlife management areas (Stone et al., 2017). Botswana's major competitive advantage is its diverse and abundant natural resources especially wildlife and the renowned Okavango Delta and Chobe River Plains in the North to the Kalahari Desert in the South. The resources are sought after by international tourists and compliment the global trends towards greater environmental awareness and the need to experience nature in its unspoiled state (Morupisi & Mokgalo, 2017). Demand for Botswana's wildlife tourism products has increased significantly over time due to greater awareness of Botswana as an attractive destination. This has resulted in the rapid growth of tourism around Kasane and Maun areas as observed through increases in accommodation and other facilities which are dependent on the parks and wildlife areas for their existence and growth. For example, the number of available bed-nights in Kasane grew by 61% between 1998 and 2003 and the number of day visitors in the Chobe National Park rose by 360%, from 4333 to 15,587 over the same period (Leechor & Fabricius, 2006). These developments have placed pressure on the parks' infrastructure (Morupisi & Mokgalo, 2017).

Tourism has driven the development of infrastructure and facilities, such as hotels, lodges and camps, airport, and airstrips; tarred roads and other communication facilities have also been established in Botswana (Stone et al., 2017). Wholesale and retail businesses have also been developed to offer various goods to the tourist industry through backward linkages concept (Stone et al., 2017). Chobe National Park gets about two-thirds of Botswana's parks visitors and most day visitors (Leechor & Fabricius, 2006). Air access is of critical importance to Botswana's tourism development. Most tourists use Air Botswana the main air transport, which might result in uncompetitive fares and limited availability which is a major constraint for tourism growth (Leechor & Fabricius, 2006). Other than the Okavango Delta and the Chobe River Plains, the rest of the attractions in the country are underdeveloped and mostly unknown and thus do not attract many visitors (Morupisi & Mokgalo, 2017).

Although the number of international tourists has been on the rise, the evidence suggests that the situation is rather bad regarding domestic tourism, where only 6.1% of total visits to national parks and reserves in Botswana were made by Batswana (Department of Wildlife and National Parks [DWNP], 2010) in comparison to South Africa's 75% domestic tourists to parks (SANParks, 2008).

11.7.2 Namibia: Non-consumptive Wildlife Tourism

For Namibia, travel and tourism contributed approximately 14.7% of GDP, making it the second most important sector in Namibia, after mining; it also contributed 15.4% to total employment in 2019 (WTTC, 2019). Tourism is the fastest growing economic sector in Namibia and employs over 100,000 Namibians, making this sector an important employer contributing to the reduction in rural poverty; in addition, Namibia's tourism sector is also one of the most competitive sectors in the world (GIZ, 2020; Tervo-Kankare et al., 2018). The increase in tourism to Namibia has led to growth in this sector, including job opportunities, human resource development, and investment in infrastructure (GIZ, 2020). Figure 11.2 shows the number of tourist arrivals registered in Namibia and the associated revenue generated from tourism from 1995 to 2000. Namibia recorded a total of 187,100 international tourists in 2020 (see Fig. 11.3), ranking 159th in the world in absolute terms (WorldData.Info, 2023). A study by Koch (2021) in Namibia revealed that the total number of non-consumptive wildlife tourists visiting the game reserves was much more than the consumptive wildlife tourists. By putting the tourist numbers in relation to the population of Namibia, with 0.074 tourists per resident, Namibia ranked 132nd in the world (WorldData.Info, 2023). In 2020, Namibia generated around US$155 million in the tourism sector alone which corresponds to 1.3% of its GDP (WorldData.Info). In 1995, tourism revenues were US$278 million, about 7.0% of Namibia's GNP. At that time this corresponded to about 421,000 tourists (see Fig. 11.3). Within 25 years, the country's dependence on tourism decreased in GNP terms. Before the outbreak of the COVID-19 pandemic, tourism revenue was

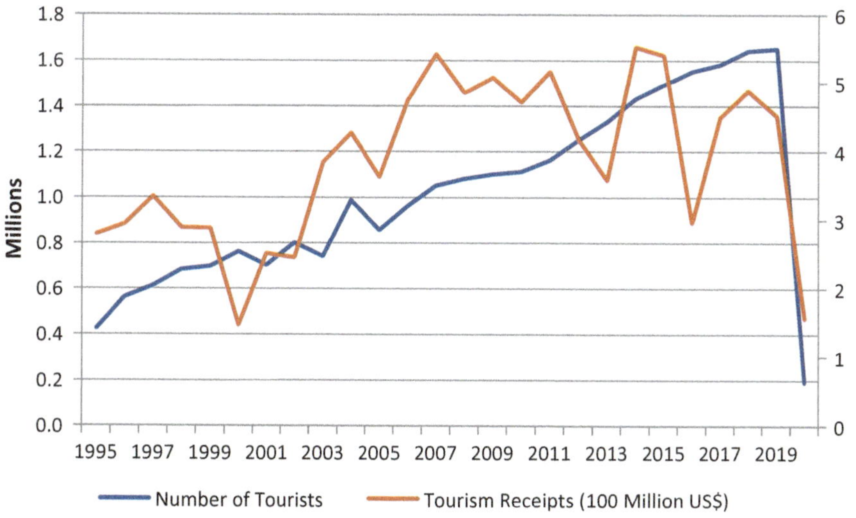

Fig. 11.2 Number of international tourists and revenue from tourism in Namibia. (Source: https://www.worlddata.info/africa/namibia/tourism.php)

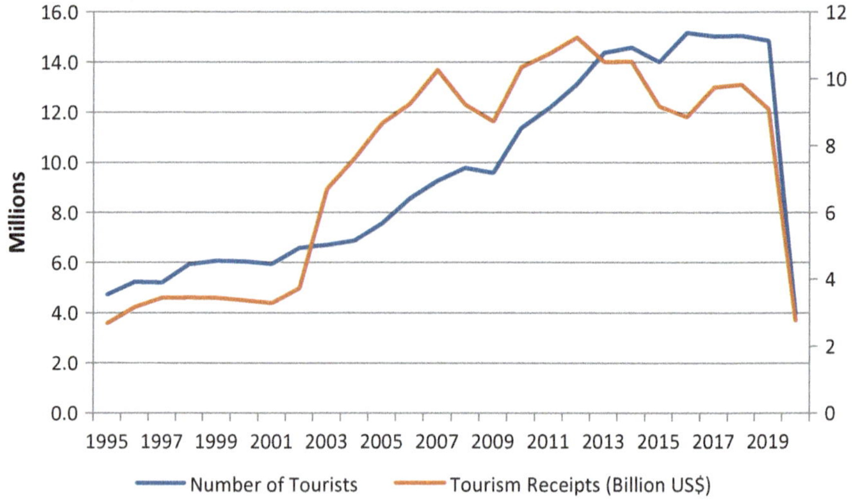

Fig. 11.3 Number of international tourists and revenue from tourism in South Africa. (Source: https://www.worlddata.info/africa/southafrica/tourism.php)

US$451 million, 3.6% of GNP; whilst in 2020, tourist receipts fell due to the COVID-19 pandemic given that the tourism revenue was only US$155.00 million (Fig. 11.3). The top 10 Namibian tourist markets were from Angola, South Africa, Zambia, Germany, Zimbabwe, Botswana, France, United Kingdom, USA and China (GIZ, 2020).

In 2019, the international tourists' arrivals category was made up of 94.9% tourists, 3.3% same-day visitors, 1.4% returning residents and others 0.4% (Ministry of Environment, Forestry and Tourism Namibia, 2019). Apart from the slight increase recorded in 2019, the average growth from 2015 to 2018 has been modest below 4%. The slow growth might be due to various global factors such as world financial crisis and limited marketing efforts—source markets had reached their peak that leave the Namibia Tourism Board to venture into new potential markets such as the Asia, Australia, and Middle Eastern markets (Ministry of Environment, Forestry and Tourism Namibia, 2019). COVID-19 and its associated restrictions in global travel negatively affected the tourism industry, causing mass business closures, lay-offs, and unsustainable debt (Ministry of Environment, Forestry and Tourism Namibia, 2019).

Several features make Namibia an interesting travel destination which include the following: original landscapes; untouched savannah; Namib Desert, the oldest desert in the world; the Fish River Canyon, the second-largest canyon in the world after the USA's Grand Canyon; the world-famous sand dunes at Sossusvlei; the Skeleton Coast with its extraordinary landscapes; Etosha National Park, one of the largest game reserves in Africa; exotic wildlife; and a unique mixture of African and European influences (GIZ, 2020). Etosha National Park and other parks offer world-class safaris with upmarket lodging both inside and near the parks. Protected areas constitute 17% of Namibia's land area and consist of well-known tourist destinations such as the Etosha National Park, the Namib-Naukluft Park and the Skeleton Coast Park; and communal conservancies contribute an additional 20.2%, which leads to a total of at least 37% of Namibia's land area under conservation (GIZ, 2020). In addition, Namibia's superb road infrastructure, potable water, and lower levels of crime than most of its neighbouring countries make it one of the best tourists' destinations in southern Africa.

However, Namibia has one of the driest countries in sub-Saharan Africa (Tervo-Kankare et al., 2018). This, coupled with the impacts of global climate change may have devastating effects on the country's tourism industry (MET, 2015).

11.7.3 South Africa: Non-consumptive Wildlife Tourism

South Africa recorded a total of four million tourists in 2020, ranking 45th in the world in absolute terms. By putting the tourist numbers in relation to the population of South Africa, with 0.065 tourists per resident, South Africa ranked 135th in the world (WorldData.Info, 2023). Figure shows the number of international tourist arrivals and associated revenue in South Africa from 1995 to 2020. In 2020, South Africa generated around US$2.72 billion in the tourism sector alone which represented 0.65% of its the GDP and approximately 88% of all international tourism receipts in Southern Africa (WorldData.Info, 2023). In 2019, the tourism sector's total contribution to the country's GDP was 6.7%, equivalent to US$ $22.1 billion, and contributed 1.5 million jobs (see Fig. 11.3; WTTC, 2021).

In 1995, tourism revenues amounted to US$2.65 billion, corresponding to 1.5% of GNP which corresponded to about 4.68 million tourists at that time (WorldData.Info, 2023). Within 25 years, the country's dependence on tourism has increased. Before the outbreak of the COVID-19 pandemic, tourism revenue was US$9.06 billion (2019), corresponding to a 2.3% of GNP, but in 2020, tourist receipts fell to US$2.72 billion due to the COVID-19 pandemic, a 70% decrease (WorldData.Info, 2023).

South Africa has 1544 protected areas, covering 8% of its terrestrial land and 12% of its marine area (WTTC, 2019). Most protected areas are under the management of private landowners (UNEP-WCMC, 2019). Kruger National Park (KNP) in South Africa is unique because of its wildlife diversity and cutting-edge environmental management techniques and policies (WTTC, 2019). KNP is greatly managed and sustains large and relatively stable wildlife populations (WTTC, 2019).

Unlike other countries in southern Africa, Grünewald et al. (2016) found out that most park visitors to the South African National Parks are locals, with 79% being South African. Visitors spend large fractions of their time viewing predators such as lions, leopards, and cheetahs. Addo Elephant National Park in the Eastern Cape province is the only park where the Big Seven can be viewed, including: the African elephant, Cape buffalo, African lion, African leopard, African rhino as well as whales and Great White sharks (WTTC, 2019). Across the country's protected areas, mega-herbivores and large carnivores are most popular, particularly among first-time and overseas visitors (Lindsey et al., 2007). On the other hand, African visitors and experienced wildlife viewers were interested in bird and plant diversity, scenery, and scarcer, difficult-to-observe and/or less high-profile mammals (Lindsey et al., 2007).

From a study which collected visitor statistics from 19 South African protected areas, Kruger National Park came first in terms of visitation with about 1.5 million visitors annually in 2007; whilst Cape Peninsula National Park recorded an average of 1,462,649 visitors per year from 2002 to 2006 (Balmford et al., 2015). In conducting surveys of South African wildlife tourists, Boshoff et al. (2007) discovered that 23% of respondents visited South Africa's national parks 'frequently', compared with 13% went to its provincial parks and reserves and 4.5% to private reserves. Overall, the sustainability implications of South Africa's wildlife tourism growth prospects appear to be positive given the increase of private wildlife tourism destinations (e.g., private game reserves) which have contributed to the large-scale land-use change of previous agricultural land to wildlife conservation land use (Snyman, 2012).

11.7.4 Zimbabwe: Non-consumptive Wildlife Tourism

Zimbabwe's tourism industry is competitive in the southern Africa region and has contributed a total of 5.8% to the country's GDP in 2019, in addition to providing 8.6% of the total employment (WTTC, 2021). Zimbabwe recorded a total of 639,000 tourists in 2020, ranking 118th in the world in absolute terms. By putting the tourist numbers in relation to the population of Zimbabwe with 0.040 tourists per resident, it ranked 147th in the world (WorldData.Info, 2023). Tourism is Zimbabwe's third largest sector after mining and agriculture and has the potential to play a significant role in Zimbabwe's economic recovery (MFAIT, 2023). Figure 11.4 shows the number of tourist arrivals and associated revenues from 1995 to 2020 for Zimbabwe.

In 2020, Zimbabwe generated around US$66 million in the tourism sector which corresponds to 0.23% of its the GDP (see Fig. 11.4). In 1995, tourism revenues amounted to US$145 million, an equivalence of 2% of the GNP; corresponding to about 1.42 million tourists at that time (4; WorldData.Info, 2023). After 25 years from 1995, the country's dependence on tourism has decreased substantially in terms of contribution to GNP. Before the outbreak of the COVID-19 pandemic, tourism revenue was US$$285 million, 1.3% of GNP which plummeted to US$66 million in 2020, a 77% decrease, due to the COVID-19 pandemic. As a result of the COVID-19 pandemic, the total contributions to GDP and employment dropped in 2020 to 4.1% and 7.8% respectively (WorldData.Info, 2023). Tourist arrivals have fluctuated over the years, in direct correlation to Zimbabwe's economic performance. Most visitors came from within Africa and the Middle East while the country registered declines in arrivals from Europe, the Americas, Oceania, and Asia (ITA, 2023).

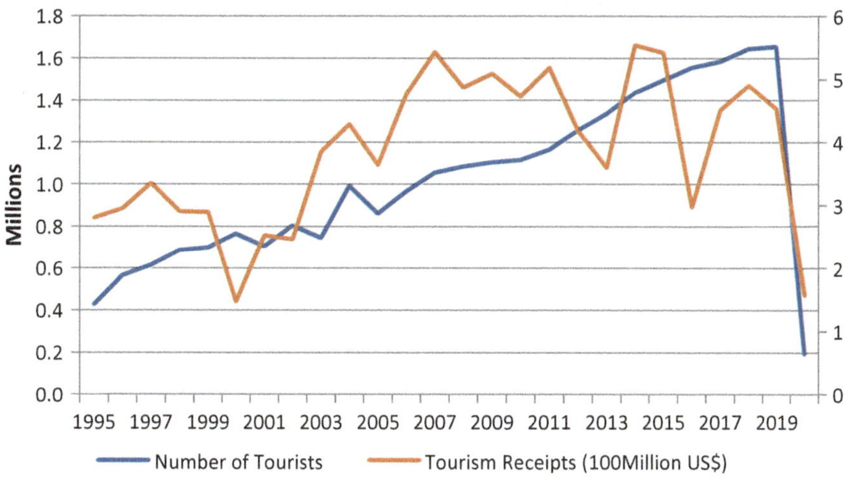

Fig. 11.4 Number of international tourists and revenue from tourism in Zimbabwe. (Source: https://www.worlddata.info/africa/zimbabwe/tourism.php)

Zimbabwe has several national parks and natural attractions such as Hwange, Mana Pools, and Gonarezhou National Parks, Victoria Falls, Lake Kariba, and the Great Zimbabwe National Monument. Wildlife is very essential for tourism in Zimbabwe. From the black and white rhinoceros at the Matobo National Park, the sable and waterbuck at Nyanga National Park, the huge-tasked elephants at the Gonarezhou National Park, Africa's "big five" in Mana Pools, offers opportunity to view animals in their natural habitats (MFAIT, 2023). In recent years, the government of Zimbabwe has made efforts to revamp and develop the tourism industry through various initiatives such as easing visa requirements for tourists, upgrading tourism infrastructure, and promoting the country's unique cultural heritage and wildlife (MFAIT, 2023).

11.8 Conclusions and Recommendations

Non-consumptive wildlife tourism is one of main source of revenue and livelihood for countries across southern Africa. In this study we have discussed the importance of wildlife tourism thereby bringing awareness to its value and the need for wildlife protection. Generally, wildlife tourism has been increasing over time except from 2020 when COVID-19 struck. COVID-19 pandemic presented a threat where it drastically reduced number of tourists and the associated revenues by over 70%.

There are some economic, socio-cultural and environmental benefits and challenges,, which come with non-consumptive wildlife tourism use. There is need to maximize the economic benefits whilst minimizing the costs. However, given that not all wildlife areas are able to sustain profitable ecotourism ventures, the local communities' access to financial support should be prioritized so that they are incentivized to actively contribute to wildlife conservation.

For the socio-cultural impacts of non-consumptive wildlife tourism, it is important to ensure that positive impacts are maximized whilst the negative are minimized. To achieve this, national parks, game reserves and other landowners should be involved in non-consumptive wildlife tourism, continuously monitor wildlife tourism impacts, be involved in joint planning and decision making, provide advice, communicate, and partner with all the stakeholders at different stages of tourism.

For wildlife tourism environmental impact management, it is important to ensure that the natural environment is not affected negatively by non-consumptive wildlife tourism. This can be done using different management tools that address negative environmental impacts, effective and appropriate facility design, site and visitor management strategies and eco-efficient practices that include energy management, waste management and water management.

References

Aylward, B. (2003). The actual and potential contribution of nature tourism in Zululand, considerations for development, equity and conservation. In B. Aylward & E. Lutz (Eds.), *Nature tourism, conservation and development in KwaZulu-Natal, South Africa* (pp. 1–40). The World Bank.

Balmford, A., Green, J. M. H., Anderson, M., Beresford, J., Huang, C., Naidoo, R., Walpole, M., & Manica, A. (2015). Walk on the wild side: Estimating the global magnitude of visits to protected areas. *PLoS Biology, 13*(2), e100.

Blackie, I. R., & Sowa, J. (2019). Dynamics of social ecology of elephant conservation in Botswana and implications on environmental development. *Journal of African Interdisciplinary Studies, 3*(2), 4–25.

Boshoff, A. F., Landman, M., Kerley, G. I., & Bradfield, M. (2007). Profiles, views and observations of visitors to the Addo Elephant National Park, Eastern Cape, South Africa. *African Journal of Wildlife Research, 37*(2), 189–197.

Ceballos-Lascurain, H. (1996). *Tourism, ecotourism and protected areas*. International Union for Conservation of Nature.

Choi, Y., Song, K., Kim, M., & Lee, J. (2017). Transformation planning for resilient wildlife habitats in ecotourism systems. *Sustainability, 9*(4), 1–28.

Diamantis, D. (2004). Ecotourism management: An overview. In D. Diamantis (Ed.), *Ecotourism* (pp. 1–26). Thompson.

DWNP. (2010). *Botswana annual statistical report 2008*. Government of Botswana.

Emerton, L. (2001). The nature of benefits and the benefits of nature: Why wildlife conservation has not economically benefited communities in Africa. In D. Hulme & M. Murphee (Eds.), *African wildlife and livelihood: The promise and performance of community conservation* (pp. 208–226). James Currey.

GIZ. (2020). *Sector brief: Namibia*. Deutsche Gesellschaft für Internationale Zusammenarbeit.

Government of Botswana. (1992). *Wildlife Conservation and National Parks Act No. 2 of 1992: Cap.38*. Government Printer.

Grünewald, C., Schleuning, M., & Böhning-Gaese, K. (2016). Biodiversity, scenery and infrastructure: Factors driving wildlife tourism in an African savannah national park. *Biological Conservation, 201*, 60–68.

Hall, C. M. (2007). *Tourism in Australia*. Pearson Education.

Honey, M. (1999). *Ecotourism and sustainable development: Who owns paradise?* Island Press.
ITA. (2023). *Zimbabwe – Country commercial guide*. The International Trade Administration, U.S. Department of Commerce.
Koch, M. H. (2021). *A comparison of the perceived environmental impacts of consumptive and non-consumptive wildlife tourism*. MSc thesis, North West University.
Leechor, C., & Fabricius, M. (2006). *Developing tourism in Botswana: Progress and challenges*. World Bank & BIDPA.
Lindberg, K. (2001). Economic impacts. In D. Weaver (Ed.), *The encyclopaedia of ecotourism*. CABI.
Lindsey, P. A., Alexander, R., Mills, M. G. L., Romanach, S., & Woodroffe, R. (2007). Wildlife viewing preferences of visitors to protected areas in South Africa: Implications for the role of ecotourism in conservation. *Journal of Ecotourism, 6*, 19–33.
MET. (2015). *Namibia third national communication to the United Nations framework convention on climate change*. Government of Namibia.
MFAIT. (2023). *Wildlife: At the heart of Zimbabwe's tourism*. Ministry of Foreign Affairs and International Trade, Zimbabwe.
Ministry of Environment, Forestry and Tourism Namibia. (2019). *Tourist statistical report*. Ministry of Environment, Forestry and Tourism Namibia.
Morupisi, P., & Mokgalo, L. (2017). Domestic tourism challenges in Botswana: A stakeholders' perspective. *Cogent Social Sciences, 1*. https://doi.org/10.1080/23311886.2017.1298171
Mwakiwa, E., Hearne, J., Stigter, J. D., de Boer, W. F., Henley, M., Slotow, R., van Langevelde, F., Peel, M., Grant, C. C., & Prins, H. H. T. (2016). Optimization of net returns from wildlife consumptive and non-consumptive uses by game reserve management. *Environmental Conservation, 43*(2), 128–139.
Panos, B. (1995). *Ecotourism* (Panos Media Briefing No. 4). Panos Institute.
Prosser, R. (1994). Societal change and the growth in alternative tourism. In E. Cater & G. Lowman (Eds.), *Ecotourism: A sustainable option*. Wiley.
Saayman, M. (2000). *En route with tourism*. Leisure.
SANParks. (2008). *Economic impact assessment report*. Retrieved April 9, 2023, from South African National Parks Authority. http://www.sanparks.org/about/economic_impact_study_sept08.pdf
Shahzalal, M. (2016). Positive and negative impacts of tourism on culture: A critical review of examples from the contemporary literature. *Journal of Tourism, Hospitality and Sports, 20*.
Snyman, S. L. (2012). The role of tourism employment in poverty reduction and community perceptions of conservation and tourism in southern Africa. *Journal of Sustainable Tourism, 20*, 395–416.
Stone, S. L., Stone, T. M., & Mbaiwa, J. E. (2017). Tourism in Botswana in the last 50 years: A review. *Botswana Notes and Records, 49*, 57–72.
Stone, L. S., Stone, M. T., Mogomotsi, P. K., & Mogomotsi, G. E. (2021). The impacts of Covid-19 on nature-based tourism in Botswana: Implications for community development. *Tourism Review International, 25*(2–3), 263–278.
Tchakatumba, P. K., Gandiwa, E., Mwakiwa, E., Clegg, B., & Nyasha, S. (2019). Does the CAMPFIRE programme ensure economic benefits from wildlife to households in Zimbabwe? *Ecosystems and People, 15*(1), 119–135.
Tervo-Kankare, K., Saarinen, J., Moswete, N., & Kimaro, E. (2018). Nature-based tourism operators' responses to changing environment and climate in Uis, Namibia. *African Geographical Review, 37*, 273–282.
Tisdell, C. (2003). Economic aspects of ecotourism: Wildlife-based tourism and its contribution to nature. *Sri Lankan Journal of Agricultural Economics, 5*, 1–14.
Tremblay, P. (2001). Wildlife tourism consumption: Consumptive or non-consumptive? *International Journal of Tourism Research, 3*(1), 81–86.
UNEP-WCMC. (2019). *Protected area profile for South Africa from the world database of protected areas*. www.protectedplanet.net

UNWTO. (2015). *Towards measuring the economic value of wildlife watching tourism in Africa—Briefing paper*. World Tourism Organization (UNWTO).

Van der Merwe, P., Slabbert, E., & Koch, M. (2021). Environmental sustainability of consumptive and nonconsumptive wildlife tourists: The case of a game reserve in Namibia. *International Journal of Biodiversity and Conservation, 13*(4), 218–227.

WorldData.Info. (2023). *Tourism in [Various countries]*. Eglitis-media.

WTTC. (2019). *The economic impact of global wildlife tourism: Travel & tourism as an economic tool for the protection of wildlife*. World Travel & Tourism Council.

Chapter 12
Contemporary Issues Affecting Wildlife Tourism: Crises and Disasters

Chiedza N. Mutanga and Edson Gandiwa

12.1 Introduction

The demand for tourism is extremely vulnerable to disasters, crises, and hazards, including wars, disease outbreaks, currency instability, energy crises, natural disasters, terrorism, plane and railway crashes, ship wreckages, and crime (Cornelissen, 2006; Phoofolo, 2019; Sun et al., 2022; Wakeford, 2006). As a result, there are many different forces and factors that shape tourism, including exogenous ones that are unrelated to the tourism industry. Examples include natural disasters and unforeseen events that have a significant impact on people's lives and society, potentially having a significant impact on tourism flows (Rosselló et al., 2020). Additionally, the extraordinary globalisation of the tourism industry means that small-scale crises in one region of the world can have a big effect on others. Due to this, contemporary crises do not respect national borders or confine themselves to a certain sector of policy, like health or energy (Maditinos & Vassiliadis, 2008). Tourism destinations that are unable to adequately prepare for certain crises and disasters as well as recover from their effects are therefore vulnerable and risk losing a sizable portion of the tourist market.

A crisis is a disturbance that physically affects an entire system and imperils its underlying presumptions, sense of self, and fundamental existence (Pauchant & Mitroff, 1992). On the other hand, a disaster is a circumstance that significantly disrupts the usual social order, overwhelms local capacity to withstand, cope with,

C. N. Mutanga (✉)
Okavango Research Institute, University of Botswana, Maun, Botswana
e-mail: mutangac@ub.ac.bw

E. Gandiwa
Scientific Services, Zimbabwe Parks and Wildlife Management Authority, Harare, Zimbabwe
e-mail: egandiwa@zimparks.org.zw

© The Author(s), under exclusive license to Springer Nature Switzerland AG 2024

L. S. Stone, M. T. Stone (eds.), *Wildlife Tourism Dynamics in Southern Africa*, Geographies of Tourism and Global Change,
https://doi.org/10.1007/978-3-031-57252-4_12

and recover from, requires external assistance, and involves a variety of stakeholders (Coppola, 2015). Both crises and disasters attract public and media attention and pose a threat to public trust. As such, they are anomalous circumstances that pose significant risks to enterprises and may lead to quick changes in public policy (Al-Dahash et al., 2016). Disasters, crises, and threats, therefore, influence how tourists view destinations, making some places either appropriate or inappropriate for visitation. Crises and disasters are sudden changes that shock the system in which tourism is ingrained because they have the potential to economically threaten, weaken and erode the competitiveness of tourist destinations due to declines in visitor numbers, hotel reservations, and bookings for cultural, sporting, and service activities (Miller, 2008). The nature and scope of effects determine the type of shock and the affected system's resilience (Organisation for Economic Co-operation and Development [OECD], 2014).

Crises and disasters negatively impact people, households, businesses, and institutions all over the world by causing injuries and fatalities as well as the loss of infrastructure for households and tourist destinations. For instance, the coronavirus (COVID-19) pandemic confirmed more than 500,000 deaths worldwide in 2020, and international travel restrictions and lockdowns severely reduced economic activity, reducing the global output of emerging economies (Rondeau et al., 2020). According to the World Bank (2020a), the COVID-19 pandemic's impacts were threatening to push 70–100 million more people into poverty. This demonstrates how crises and natural catastrophes may have an impact on both tourism arrivals and food security. Barrett et al. (2011) posit that food insecurity and poverty are the main causes of biodiversity loss, which is the biggest danger to wildlife tourism in Southern Africa.

Wildlife protection activities and funding levels are related to tourism (Smith et al., 2021). Sustainable tourism programmes offer greater chances for wildlife protection. They frequently lead to positive conservation outcomes, including enhanced conservation perceptions and methods, decreased poaching, involvement in conservation initiatives, higher species numbers and/or ranges, habitat improvement, and community development (Lindsey et al., 2020). When funding for patrols declines or anti-poaching programs are scaled back because of a lack of tourism caused by crises and disasters, conservationists contend that poaching activities pose the greatest threat to critically endangered animal species (Newsome, 2021). The efficiency of wildlife protection is primarily dependent on the resources available to handle the various threats and conservation problems protected areas (PAs) are facing.

While some crises and disasters, like the September 11 attacks on New York's twin towers in 2001, the global financial crisis of 2007–2008, and the Ebola virus outbreak beginning in 2013, generally tend to cause short- and medium-term disruption to tourism, Cheer et al. (2021) argue that the unprecedented climate crisis is the most detrimental situation that will be felt by future generations. However, COVID-19 has proven to be more severe, to last longer, and to be more challenging for global communities to deal with. As Cheer et al. (2021) deduced from the COVID-19 pandemic, the most difficult element of crises is that they can evolve and adopt the shape of yet unanticipated versions of themselves, or dramatically alternative pathways, largely nullifying futures planning and scenario creation. This puts increasing pressure on managers and planners of the tourism sector to consider the effects of disasters, crises, and dangers on the industry and to devise strategies to deal with those consequences in order to safeguard the sector and society at large (Wakeford, 2006). Despite their complexity, understanding crises and disasters can aid in the development of policies and strategies to deal with such incidents and stop or lessen the severity of their effects on business and society (Maditinos & Vassiliadis, 2008). This chapter discusses some of the crises and disasters that have affected wildlife tourism in Southern Africa; these include COVID-19 and other disease outbreaks, and natural disasters. Finally, the chapter considers the lessons learnt and implications for policy and practice.

12.2 Methods

Literature from peer-reviewed journal articles, books, edited book chapters, academic theses, policy documents, and reports by conservation and welfare non-governmental organisations (NGOs) and non-profit organisations, was used to search articles about crises and disasters in Southern African countries from the year 2000. Google Scholar, Web of Science, Scopus, and Google were used to find the literature. The key search terms and phrases used for the search were 'COVID-19', 'diseases', 'epidemics', 'pandemics', and 'natural disasters'. The search was conducted from December 2022 to February 2023, and this yielded 184 articles that provided initial data for analysis. While the initial literature search showed many crises and disasters that may have affected the Southern African region on different scales, only two that have affected the region at a larger scale, i.e., epidemics and pandemics, and natural disasters, became the major focus of this chapter. Since the initial data set was too big to handle, further refinements were made (Dube, 2022), and the search was mainly inclined toward articles focusing on COVID-19 and tropical cyclones. Botswana and Zimbabwe which constituted much of the literature became the main reference countries for the analysis. Articles in which the impacts of the crisis or disaster were not adequately explained were excluded at the screening stage 62 articles were finally used in the analysis. Following guidelines from Roberts et al. (2019), content analysis was used to analyse the articles. Each article was first read through to familiarise researchers with its content and an initial list of ideas about what is in each article was generated. We then identified texts carrying ideas about the chapter objectives and these then informed the ensuing discussion.

12.3 Results and Discussion

12.3.1 Impacts of Epidemics and Pandemics on Wildlife Tourism

The COVID-19 pandemic was the deadliest of all pandemics and tourism was among the top of its economic casualties (Ozbay et al., 2022). Due to the pandemic's effects on wildlife populations as well as the distorted views that result from the zoonotic origins of the pandemic, forms of tourism that entail interactions with wildlife were severely impacted (Rosselló et al., 2020). COVID-19 had a significant negative impact on many countries' tourism industries, which made the nature-based tourism sector more unpredictable and compromised its ability to quickly resume normal operations. This section discusses the impacts of epidemics and pandemics on wildlife tourism drawing mainly from the COVID-19 pandemic.

Epidemics and pandemics cause decreased destination competitiveness as they are frequently accompanied by the cancellation of flights to afflicted nations and the publication of travel advisories to destinations from the tourism-producing nations. As an illustration, the development of Ebola has lowered Africa's attractiveness as a travel destination (Maphanga & Henama, 2019). A common pattern has been noticed in which a tragic event in one African nation or region is frequently connected to the entire continent. As Maphanga and Henama (2019) note, during the period of Ebola, 50% of tour operators in Africa experienced cancellations due to fears of the virus, and Tanzania and South Africa experienced declines in tourism business due to fears of Ebola. The tourism sector made numerous layoffs in response to the Ebola outbreak in keeping with the declining business.

In the case of COVID-19, in response to the effects of the pandemic, governments restricted human movement, which reduced economic activity (Lindsey et al., 2020). Travel bans and other restrictions intended to stop the spread of COVID-19 had a significant negative economic impact on Botswana's tourism industry where, for example, some tourism businesses were closed while some, like the accommodation sector, were operating at less than 10% of their capacity (Hambira et al., 2022; Stone et al., 2021). Booking cancellations cost the worldwide tourist industry around US$80 billion in the first quarter of 2020. The effects were widely felt in developing countries like Zimbabwe where the tourism sector contributed 6.3% of the GDP in 2018 and aimed to contribute US$7 billion annually by 2030 (World Bank, 2020b).

Due to the decline in tourism, the COVID-19 pandemic had several detrimental effects on workplaces, including lower income for wildlife management. Wildlife resource protection was significantly impacted by the nearly 100% drop in tourism revenue and the lack of funding to replenish habitats (Ndlovu et al., 2021). This is because most developing nations' national parks and reserves are primarily funded by wildlife tourism, which accounts for more than 10% of the economies of Tanzania, South Africa, and Namibia (Coke-Hamilton, 2020). As a result of this situation, threats to biodiversity like poaching, wildlife trafficking, and deforestation have unavoidably escalated (Lindsey et al., 2020; Spenceley et al., 2021).

The World Bank (2020b) estimates that COVID-19-related tourism losses caused an 80% drop in revenue for Zimbabwean parks (ZimParks) between April and June 2020. Additionally, COVID-19 had a significant negative impact on contributions from the private sector and NGO donors, leaving ZimParks with a major budget vacuum to sustain essential operations including anti-poaching surveillance and wildlife population monitoring. In essence, the decline in tourism coupled with employment losses, stagnant or declining income levels, and insufficient support from revenue-generating businesses put conservation at the forefront of worries and expectations for intensified poaching trends. The potential for a sustained decline in revenue from wildlife tourism poses a threat to reverse the advancements made in development work for the creation of sustainable community-based conservation initiatives.

COVID-19 also became a new addition to the list of diseases that affect rural households (Bruce-Lockhart, 2020). The disease impaired working individuals and increased mortality, which lowered household income. Another possible contributing reason is a general rise in labour supply which in turn increased human threats to the environment. Developing nations with biodiversity hotspots are particularly vulnerable when one considers the link between poverty and illegal hunting activities. The World Bank (2020b) notes that due to an increase in poachers' attempts to cross the Zambezi River border into Zimbabwe, border patrols by boat and drone are now more necessary. Therefore, considering the pandemic that has decreased the income of all parties involved, the Zimbabwe Parks and Wildlife Management Authority expressed worries regarding the growing need for conservation activities (Mudzengi et al., 2022).

Tourism losses affect the rural population in marginalised regions, many of whom are close to PAs where safari revenues often make up a larger percentage of their livelihood. For instance, during the COVID-19 pandemic, communities that relied on tourism in Botswana were negatively impacted by the postponement of the hunting season (Stone et al., 2021). The losses put the provision of services like healthcare, food, and water at risk. When local communities do not benefit financially from wildlife, retaliatory killings are anticipated to increase in cases of human-wildlife conflict. Lindsey et al. (2020) report that during the period of full lockdown the trend of local poachers and illegal miners being arrested increased. For instance, Bumi Hills Anti-Poaching Unit (BHAPU) in Zimbabwe reported a rise in the number of snares they tracked since the start of COVID-19 (African Bush Camps, n.d.). Similarly, income losses may push people into cooperating with poachers, further increasing poaching and illegal wildlife trade (Singh et al., 2021). In Botswana, wildlife crimes including poaching escalated during lockdown, for example, 17 deaths of poachers were reported in 2020, while 275 elephant deaths were reported in the same year (Stone et al., 2021). Also, in 2020, the Forestry Service in Zimbabwe reported an increase in deforestation as people increasingly turned to charcoal production and biomass fuel as they could no longer afford electricity (World Bank, 2020b).

The reduction of wildlife tourism revenue due to COVID-19 directly affects the conservation of wildlife. Wardens and conservation officers are affected, both in terms of catching the disease and in carrying out their obligations, thereby reducing conservation work and enforcement against illegal resource extraction (Bockler, 2022; International Fund for Animal Welfare, 2020). Many wardens and conservation officers also lost their jobs, while some even lost their lives. Stone et al. (2021) report that due to COVID-19, many jobs were lost in Botswana which may have contributed to widespread poverty in many households. According to the International Fund for Animal Welfare (2020), the sudden loss of revenue led Zimbabwe and other nations to lay off wardens and rangers. On another note, governments have been seen to redirect funds intended for wildlife protection to more urgent social needs when a disease outbreak occurs, especially one of significant magnitude (Bang & Khadakkar, 2020). The COVID-19 pandemic has also restricted the activities of conservation organisations, and hampered collaborations with conservation partners to address pressing conservation issues like human-wildlife conflict through conservation education, awareness, social engagements, and work that necessitates close human contact. Additionally, support for coping with the crisis's aftermath and its broader impact on wildlife conservation and the wildlife economy has been drastically diminished (Ndlovu et al., 2021).

Further, the unexpected drop in tourism has been blamed for the increase in poaching and a rise in illicit logging (Rondeau et al., 2020). Mutanga and Gandiwa (2023) argue that the COVID-19 outbreak has made it clear how important tourism is for deterring illegal hunting and establishing wildlife sanctuaries. The rise in poaching incidences during the COVID-19 period in the Okavango region in Botswana has also been attributed to the absence of safari tourists during that period (Hambira et al., 2022). As such, the absence of tourists effectively modifies the poachers' production function by expanding their range and raising the stock of the targeted resource that is available.

It is also important to note some positive environmental outcomes that can be realised from crises like the COVID-19 pandemic. These include reductions in emissions and air pollution (due to reduced industrial activity and mechanised transportation), limited trade in wildlife products due to transportation restrictions, and provision of a breather for PAs that suffer negatively from the effects of tourist congestion (Lindsey et al., 2020; Stone et al., 2021). However, Lindsey et al. (2020) contend that when travel restrictions are relaxed and economies resume normal operations, the positive environmental effects of crises are typically transient and susceptible to reverse.

12.3.2 Impacts of Natural Disasters on Wildlife Tourism

Earthquakes, tsunamis, floods, bushfires, hurricanes, droughts, and heatwaves have always occurred and have contributed to a larger "riskscape" that people have learnt to manage and adapt to (Rossello et al., 2020). However, due to climate change, intense and severe weather events are becoming common in Southern Africa, having serious negative effects on tourism (United Nations Educational, Scientific and Cultural Organization [UNESCO], 2022). Tropical cyclone Idai of March 2019 was the strongest and most destructive tropical cyclone ever to hit Africa and the Southern Hemisphere. This section discusses the impacts of cyclone Idai and other natural disasters on wildlife tourism.

In Southern Africa, droughts and the effects of floods have had a disastrous effect on tourist arrivals and the industry at large (Dube & Nhamo, 2020). Tourists may avoid a specific destination if there is a disaster there or if one is thought to be imminent, which could result in revenue losses for the tourism industry (Fitchett & Hoogendoorn, 2019; Tsai et al., 2016). Natural disasters can cause tourist sites to lose their beauty and economies, with the extent of the damage largely depending on the severity of the natural disaster. While natural disasters can lead to a reduction in visitation to the impacted area, adjacent areas frequently see a reduction in tourism even when they are not affected by the event (Bhati et al., 2016). This is because disaster damage inhibits the affected areas from taking part in tourism-related activities.

Disasters frequently lead to reduced economic activity and capacity by imposing considerable physical restrictions on the provision of tourism services, significantly restricting the industry's supply side (Shaw et al., 2012). Critical infrastructure like airports and ports, infrastructure for land transportation, and networks for telecommunications and energy may be compromised or dysfunctional, depending on the nature and scope of the disaster (Ghobarah et al., 2006; Parajuli & Haynes, 2006). For instance, the Manicaland province of Zimbabwe experienced damage to roads and bridge infrastructure as a result of cyclone Eline in 2000 (Nhamo et al., 2021). Nhamo et al. (2021) further state that besides natural scenes, in 2019, cyclone Idai destroyed the tourism infrastructure in Chimanimani, including tourism accommodation as well as water sources (weirs) and piping system. In addition, some highways were demolished and became unusable due to the harm done to bridges. Tourist goods and other services were cut off from the main routes that feed into Chimanimani, and as a result, tourists were unable to access the goods and services they used to enjoy in the area (Kunedzimwe et al., 2021). When tropical cyclone Idai cut off the power supply, all ICT services were lost, and wild animals and birds were also killed, all of which had a negative influence on tourism (Nhamo et al., 2021).

Natural disasters can also have a negative impact on tourists' perceptions of the wider region where the calamity occurred. As an alternative, some people can decide to alter their trip plans and go to places where they face fewer personal risks (Phoofolo, 2019). The way the accountable authorities handle a crisis may have an impact on these safety perceptions. Successful disaster management reduces the negative effects on people's perceptions of the damaged area, while disaster management gone wrong has the opposite effect. Media coverage of disasters frequently portrays the ensuing human suffering, loss of life, destruction of public and private property, and disruption of the economy and social order. The immediate aftermath of a tragedy, which lasts until complete recovery is accomplished, is frequently characterised by the accompanying unfavourable publicity (Sönmez et al., 1999). Gandiwa et al. (2014) detail how the spillover of frames occurs in the media in times of natural disasters and how this can influence how a disaster is addressed and handled.

Human capital is lost as a result of the death of skilled workers. Kabonga et al. (2021) state that cyclone Idai killed about 172 people. The authors also state that about 800,000 hectares of agricultural land had crops devastated by cyclone Idai. Since natural disasters affect agricultural production in the affected region, the problems encountered by the agricultural sector often spill over to the tourism business which negatively impacts the tourism sector's image, sustainability, and competitiveness. For instance, in South Africa, drought is considered a calamity that has the potential to affect how tourists view the country as a tourism destination due to the resultant unattractive and degraded landscapes, severe water and food shortages, and accompanying losses of wildlife (Phoofolo, 2019). Moreover, due to the impact on agriculture, local communities may also lose another source of income, necessitating their turn to illegal and unsustainable activities like poaching to survive.

Ecosystem resources such as wildlife are threatened by both natural and anthropogenic factors (Mohan et al., 2020). Climate change is expected to bring about risks like rising sea levels, more frequent and stronger storms, and significant rainfall events in the Southern African region (Kunedzimwe et al., 2021). The diversity and distribution of species will change as a result of these natural disasters, particularly in poorly maintained PAs (Gizaw & Gan, 2017). The natural resources of a nation may also be impacted by hurricanes that decimate forests and droughts that reduce soil fertility (Deraniyagala, 2016). Tropical cyclones and floods brought on by climate change in Southern Africa have had a devastating impact on a variety of ecosystem products and services that both terrestrial and aquatic ecosystems supply (Kunedzimwe et al., 2021). In Mozambique, for example, significant floods resulted in some species' extinctions in the floodplain of Gorongosa National Park and other low-lying rift valley regions close to the national park (Phiri et al., 2021). Cyclone Idai killed more than half of the species that had been replenished in the Gorongosa National Park after animal numbers drastically decreased during the country's civil war (Cambaza et al., 2019; Phiri et al., 2021). UNESCO (2022) reports that in Chimanimani, Zimbabwe, cyclone Idai destroyed wildlife habitats, contaminated water sources, and brought tourism to a halt.

12.4 Implications for Policy and Practice

The volume of international visitors has a direct impact on the performance of the national tourism industry and, ultimately, wildlife tourism (Massidda & Mattana, 2013). Therefore, it is crucial for policymakers to increase their comprehension of how crises and natural catastrophes impact visitor demand. Furthermore, the

Chapter has illuminated that perceptions of safety are a crucial component of destination image and that various risks, crises, and disasters have diverse effects on tourists' perceptions. A lesson learnt from COVID-19 is that crises can undermine visitor confidence. Cheer et al. (2021) point out that a delay in resolving this issue could result in the spread of fear and false information, both of which could be very challenging to reverse. Therefore, it is essential to restore public perceptions of safety and attractiveness after a crisis or disaster to attract and reassure potential tourists to travel to the destination. This will also assist the destination in regaining functionality and economic recovery (World Travel and Tourism Council [WTTC], 2018). For effective disaster response and recovery, decision-makers must also address the risk perceptions and behaviours of tourism stakeholders at the national or destination level because they have a significant role to play in this (Kozak et al., 2007; Park & Reisinger, 2010). Nguyen et al. (2021) emphasise the significance of stakeholder collaboration in developing crisis destination marketing strategies to combat negative destination perceptions that disasters are responsible for creating. Since the stakeholders also have a great deal to lose if a pandemic breaks out, it is crucial to involve them at the national, provincial, and local governmental levels.

Destination managers should concentrate their efforts on restoring the essential commercial infrastructure. This effort could be accelerated by proactive planning, such as that related to business continuity, business support networks, and recovery aid programs (Hystad & Keller, 2008). Effective strategies must be put into place, including emergency preparedness and training for key stakeholders, to minimise the effects of crises and disasters on the destination. Investments in preparedness are likely to generate positive returns in the long term (Rosselló et al., 2020).

Another lesson learnt from the COVID-19 situation is that there is a need to equip local communities to lessen their reliance on nature and wildlife-based tourism. This is because communities have the propensity to revert to unlawful and unsustainable behaviours when they lack viable sources of income for their livelihoods. A study by the World Bank (2020b) suggests that, in addition to regularly educating local communities about the value of preserving wildlife and enhancing community benefits from conservation and wildlife tourism, the introduction of conservation tax credits as an incentive will help maintain natural lands even outside PAs.

Lindsey et al. (2020) suggest that donors could band together to establish an emergency fund for struggling wildlife authorities, communities, private landowners, and conservation NGOs because the majority of governments in Southern African countries are cash-strapped and lack specific mechanisms for fostering conservation. They also stress the significance of supporting important sectors underpinning conservation efforts, such as tourism, through tax incentives and other forms of direct financial aid, provided those sectors can show continued commitment to protecting the wildlife on which they depend. Furthermore, to prevent layoffs, NGOs in Africa should prioritise salaries for staff members where possible since there are typically no salary protection programs in the region.

12.5 Conclusion

Understanding crises and disasters, and their potential impacts is important in the development of policies and strategies to stop or reduce the severity of the impacts on business and society. The broader objective of the chapter was to shed more light on the main crises and disasters that have affected wildlife tourism in Southern Africa and their impacts. Our findings point to two main contemporary issues which are COVID-19 and other disease outbreaks, and natural disasters whose frequencies have been exacerbated by climate change. Mainly drawing from the impacts of COVID-19 in Botswana and Cyclone Idai in Zimbabwe, we posit that the net impacts of the COVID-19 crisis and natural disasters in Southern Africa have been strongly negative. These include reduced economic activity, negative tourist perceptions, destruction of infrastructure, less revenue for wildlife management, reduced conservation capacity, biodiversity loss and disturbance, reduced household income, and reduced agricultural production.

The chapter highlights some of the lessons learnt from the discussed crises and disasters, and implications for policy and practice. The chapter underscores the need for re-establishing public perceptions of destination safety and attractiveness following a crisis or a disaster; addressing risk perceptions and behaviours of tourism stakeholders; stakeholder cooperation in formulating crisis destination marketing; capacitating local communities to reduce over-reliance on nature and wildlife-based tourism; continuously educating the local communities about the importance of preserving wildlife; introducing conservation tax credits; creating emergency wildlife funds; providing support to key industries underpinning conservation efforts, such as tourism; and prioritising salaries for staff where possible, to avoid laying off employees.

The Chapter concludes that risk and crisis management is crucial even if some crises, like COVID-19, are complicated because they can change into whole different forms. To minimise the negative effects on visitors and the overall tourism business, it is important for tourist destinations to prepare for crises and disasters well in advance. Having a well-prepared destination and tourism industry allows for speedy planning and implementation of post-disaster recovery measures, which in turn aids in the conservation of wildlife.

References

African Bush Camps. (n.d.). *Safari talks: COVID-19 tourism, conservation & community*. Retrieved February 9, 2023, from https://africanbushcamps.com/safari-talks-covid-19-tourism-conservation-community/

Al-Dahash, H., Thayaparan, M., & Kulatunga, U. (2016, August). Understanding the terminologies: Disaster, crisis and emergency. In *Proceedings of the 32nd annual ARCOM conference, ARCOM 2016* (pp. 1191–1200). https://openresearch.lsbu.ac.uk/item/872x5

Bang, A., & Khadakkar, S. (2020). Biodiversity conservation during a global crisis: Consequences and the way forward. *Proceedings of the National Academy of Sciences, 117*(48), 29995–29999. https://doi.org/10.1073/pnas.2021460117

Barrett, C., Travis, A., & Dasgupta, P. (2011). On biodiversity conservation and poverty traps. *Proceedings of the National Academy of Sciences, 108*(34), 13907–13912. https://doi.org/10.1073/pnas.1011521108

Bhati, A., Upadhayaya, A., & Sharma, A. (2016). National disaster management in the ASEAN-5: An analysis of tourism resilience. *Tourism Review, 71*, 148–164. https://doi.org/10.1108/TR-12-2015-0062

Bockler, Z. (2022). *The global impact of COVID-19 and tourism on conservation rangers' guardianship capabilities* (Publication NO. 1233). Honors undergraduate thesis, University of Central Florida. https://stars.library.ucf.edu/honorstheses/1233

Bruce-Lockhart, A. (2020, May 25). *Africa 'least affected' by COVID-19; hydroxychloroquine trial suspended—WHO briefing*. Retrieved February 9, 2023, from https://www.weforum.org/agenda/2020/05/covid19-africa-hydroxycholoroquine-world-health-organization/

Cambaza, E., Mongo, E., Anapakala, E., Nhambire, R., Singo, J., & Machava, E. (2019). Outbreak of cholera due to cyclone Kenneth in northern Mozambique, 2019. *International Journal of Environmental Research and Public Health, 16*(16), 2925. https://doi.org/10.3390/ijerph16162925

Cheer, J., Lapointe, D., Mostafanezhad, M., & Jamal, T. (2021). Global tourism in crisis: conceptual frameworks for research and practice. *Journal of Tourism Futures, 7*(3), 278–294. https://doi.org/10.1108/JTF-09-2021-227

Coke-Hamilton, P. (2020) *Impact of covid-19 on tourism in small island developing states*. Retrieved December 28, 2022, from http://www.ipsnews.net/2020/05/impact-covid-19-tourism-small-island-developing-states/

Coppola, D. (2015). *Introduction to international disaster management* (3rd ed.). Elsevier. https://www.sciencedirect.com/book/9780128014776/introduction-to-international-disaster-management#book-description

Cornelissen, S. (2006). Tourism development and policy in cape town: Patterns, issues and lessons for other African cities. Seminar paper presented at University of Stellenbosch, 25 August.

Deraniyagala, S. (2016, June 24). *Economic recovery after natural disasters. The United Nations Chronicle*. The Magazine of the United Nations. https://doi.org/10.18356/e93f67b0-en

Dube, K. (2022). Nature-based tourism resources and climate change in southern Africa: Implications for conservation and development sustainability. In L. S. Stone, M. T. Stone, P. K. Mogomotsi & G. E. J. Mogomotsi (Eds.), *Protected areas and tourism in Southern Africa* (pp. 161–174). Routledge.

Dube, K., & Nhamo, G. (2020). Vulnerability of nature-based tourism to climate variability and change: Case of Kariba resort town, Zimbabwe. *Journal of Outdoor Recreation and Tourism, 29*, 100281. https://doi.org/10.1016/j.jort.2020.100281

Fitchett, J., & Hoogendoorn, G. (2019). Exploring the climate sensitivity of tourists to South Africa through TripAdvisor reviews. *South African Geographical Journal = Suid-Afrikaanse Geografiese Tydskrif, 101*(1), 91–109. https://doi.org/10.1080/03736245.2018.1541022

Gandiwa, E., Sprangers, S., van Bommel, S., Heitkönig, I., Leeuwis, C., & Prins, H. (2014). Spillover effect in media framing: Representations of wildlife conservation in Zimbabwean and international media, 1989–2010. *Journal for Nature Conservation, 22*(5), 413–423. https://doi.org/10.1016/j.jnc.2014.03.009

Ghobarah, A., Saatcioglu, M., & Nistor, I. (2006). The impact of the 26 December 2004 earthquake and tsunami on structures and infrastructure. *Engineering Structures, 28*, 312–326. https://doi.org/10.1016/j.engstruct.2005.09.028

Gizaw, M., & Gan, T. (2017). Impact of climate change and El Niño episodes on droughts in sub-Saharan Africa. *Climate Dynamics, 49*(1–2), 665–682. https://doi.org/10.1007/s00382-016-3366-2

Hambira, W., Stone, L., & Pagiwa, V. (2022). Botswana nature-based tourism and COVID-19: transformational implications for the future. *Development Southern Africa, 39*(1), 51–67. https://doi.org/10.1080/0376835X.2021.1955661

Hystad, P., & Keller, P. (2008). Towards a destination tourism disaster management framework: Long-term lessons from a forest fire disaster. *Tourism Management, 29*(1), 151–162. https://doi.org/10.1016/j.tourman.2007.02.017

International Fund for Animal Welfare. (2020, April 30). *Ranger FAQ: Protecting wildlife in Africa during covid-19*. Retrieved February 9, 2023, from https://www.ifaw.org/news/faq-rangers-africa-protect-wildlife-covid-19

Kabonga, I., Mhembwe, S., & Dziva, C. (2021). The reconstruction of livelihoods by survivors of Cyclone Idai in the Chimanimani district of Zimbabwe. In G. Nhamo & D. Chikodzi (Eds.), *Cyclones in Southern Africa: Volume 3: Implications for the sustainable development goals* (pp. 143–156). Springer. https://doi.org/10.1007/978-3-030-74303-1_10

Kozak, M., Crotts, J., & Law, R. (2007). The impact of the perception of risk on international travellers. *International Journal of Tourism Research, 9*, 233–242. https://doi.org/10.1002/jtr.607

Kunedzimwe, F., Gandiwa, E., Kupika, O., Muboko, N., Taru, P., & Manyuchi, T. (2021). Impact of Cyclone Idai on biodiversity and natural resources in Chimanimani district, Zimbabwe. In G. Nhamo & D. Chikodzi (Eds.), *Cyclones in Southern Africa: Volume 3: Implications for the sustainable development goals* (pp. 229–244). Springer. https://doi.org/10.1007/978-3-030-74303-1_15

Lindsey, P., Allan, J., Brehony, P., Dickman, A., Robson, A., Begg, C., et al. (2020). Conserving Africa's wildlife and wildlands through the COVID-19 crisis and beyond. *Nature Ecology & Evolution, 4*(10), 1300–1310. https://doi.org/10.1038/s41559-020-1275-6

Maditinos, Z., & Vassiliadis, C. (2008, July). Crises and disasters in tourism industry: Happen locally, affect globally. In *Management of international business and economics systems, MIBES conference* (pp. 67–76). chrome-extension://efaidnbmnnnibpcajpcglclefindmkaj/. http://mibes.teilar.gr/ebook/ebooks/maditinos_vasiliadis%2067-76.pdf

Maphanga, P., & Henama, U. (2019). The tourism impact of Ebola in Africa: Lessons on crisis management. *African Journal of Hospitality, Tourism and Leisure, 8*(3), 1–13.

Massidda, C., & Mattana, P. (2013). A SVECM analysis of the relationship between international tourism arrivals, GDP and trade in Italy. *Journal of Travel Research, 52*(1), 93–105. https://doi.org/10.1177/0047287512457262

Miller, D. (2008). Disaster tourism and disaster landscape attractions after Hurricane Katrina: An auto-ethnographic journey. *International Journal of Culture, Tourism and Hospitality Research, 2*(2), 115–131. https://doi.org/10.1108/17506180810880692

Mohan, V., Hardee, K., & Savitzky, C. (2020). Building community resilience to climate change: the role of a population-health-environment programme in supporting the community response to cyclone Haruna in Madagascar. *Jàmbá: Journal of Disaster Risk Studies, 12*(1), 1–4. https://doi.org/10.4102/jamba.v12i1.730

Mudzengi, B., Gandiwa, E., Muboko, N., & Mutanga, C. (2022). Innovative community ecotourism coping and recovery strategies to COVID-19 pandemic shocks: The case of Mahenye. *Development Southern Africa, 39*(1), 68–83. https://doi.org/10.1080/0376835X.2021.1980375

Mutanga, C., & Gandiwa, E. (2023). Natural heritage: wildlife and nature preserves in the African tourism landscape. In D. Timothy (Eds.), *Cultural heritage and tourism in Africa* (pp. 269–283). Routledge. https://www.taylorfrancis.com/chapters/edit/10.4324/9781003153955-16/

Ndlovu, M., Matipano, G., & Miliyasi, R. (2021). An analysis of the effect of COVID-19 pandemic on wildlife protection in protected areas of Zimbabwe in 2020. *Scientific African, 14*, e01031. https://doi.org/10.1016/j.sciaf.2021.e01031

Newsome, D. (2021). The collapse of tourism and its impact on wildlife tourism destinations. *Journal of Tourism Futures, 7*(3), 295–302. https://doi.org/10.1108/JTF-04-2020-0053

Nguyen, D., Kumakura, M., Kudo, S., Esteban, M., & Onuki, M. (2021). Overcoming negative disaster images: How Fukushima's sake breweries challenged negative stereotypes and rebuilt its regional brand. *Journal of Tourism Futures, 7*(3), 390–405. https://doi.org/10.1108/JTF-11-2020-0197

Nhamo, G., Dube, K., & Saurombe, T. (2021). Impact of Tropical Cyclone Idai on tourism attractions and related infrastructure in Chimanimani, Zimbabwe. In G. Nhamo & D. Chikodzi (Eds.), *Cyclones in Southern Africa: Volume 3: Implications for the sustainable development goals* (pp. 245–264). Springer. https://doi.org/10.1007/978-3-030-74303-1_16

Organisation for Economic Co-operation and Development. (2014). *Guidelines for resilience systems analysis*. OECD Publishing. Retrieved December 26, 2022, from chrome-extension://efaidnbmnnnibpcajpcglclefindmkaj/. https://www.oecd.org/dac/Resilience%20Systems%20Analysis%20FINAL.pdf

Ozbay, G., Sariisik, M., Ceylan, V., & Çakmak, M. (2022). A comparative evaluation between the impact of previous outbreaks and COVID-19 on the tourism industry. *International Hospitality Review, 36*(1), 65–82. https://doi.org/10.1108/IHR-05-2020-0015

Parajuli, J., & Haynes, K. (2006). The earthquake impact on telecommunications infrastructure in Nepal: A preliminary spatial assessment. *Regional Science: Policy and Practice, 8*, 95–109. https://doi.org/10.1111/rsp3.12075

Park, K., & Reisinger, Y. (2010). Differences in the perceived influence of natural disasters and travel risk on international travel. *Tourism Geographies, 12*, 1–24. https://doi.org/10.1080/14616680903493621

Pauchant, T., & Mitroff, I. (1992). *Transforming the crisis-prone organization: Preventing individual, organizational, and environmental tragedies*. Jossey-Bass.

Phiri, D., Simwanda, M., & Nyirenda, V. (2021). Mapping the impacts of cyclone Idai in Mozambique using Sentinel-2 and OBIA approach. *South African Geographical Journal, 103*(2), 237–258. https://doi.org/10.1080/03736245.2020.1740104

Phoofolo, T. (2019). *Tourism disasters, crises and the risk-fear nexus: Tourists perceptions of Durban as a tourism destination*. Doctoral thesis, University of KwaZulu-Natal. https://researchspace.ukzn.ac.za/handle/10413/20117

Roberts, K., Dowell, A., & Nie, J. B. (2019). Attempting rigour and replicability in thematic analysis of qualitative research data; a case study of codebook development. *BMC medical research methodology, 1*(19), 66.

Rondeau, D., Perry, B., & Grimard, F. (2020). The consequences of COVID-19 and other disasters for wildlife and biodiversity. *Environmental and Resource Economics, 76*, 945–961. https://doi.org/10.1007/s10640-020-00480-7

Rosselló, J., Becken, S., & Santana-Gallego, M. (2020). The effects of natural disasters on international tourism: A global analysis. *Tourism Management, 79*, 104080. https://doi.org/10.1016/j.tourman.2020.104080

Shaw, G., Saayman, M., & Saayman, A. (2012). Identifying risks facing the South African tourism industry. *South African Journal of Economic and Management Sciences, 15*(2), 190–206. https://hdl.handle.net/10520/EJC121523

Singh, R., Galliers, C., Moreto, W., Slade, J., Long, B., Aisha, H., et al. (2021). Impact of the Covid-19 pandemic on rangers and the role of rangers as a planetary health service. *Parks, 27*, 119–134. https://doi.org/10.2305/IUCN.CH.2021.PARKS-27-SIRS.en

Smith, M., Smit, I., Swemmer, L., Mokhatla, M., Freitag, S., Roux, D., & Dziba, L. (2021). Sustainability of protected areas: Vulnerabilities and opportunities as revealed by COVID-19 in a national park management agency. *Biological Conservation, 255*, 108985. https://doi.org/10.1016/j.biocon.2021.108985

Sönmez, S., Apostolopoulos, Y., & Tarlow, P. (1999). Tourism in crisis: Managing the effects of terrorism. *Journal of Travel Research, 38*(1), 13–18.

Spenceley, A., McCool, S., Newsome, D., Báez, A., Barborak, J., Blye, C., et al. (2021). Tourism in protected and conserved areas amid the COVID-19 pandemic. *Parks, 27*, 103–118. https://doi.org/10.2305/IUCN.CH.2021.PARKS-27-SIAS.en

Stone, L., Stone, M., Mogomotsi, P., & Mogomotsi, G. (2021). The impacts of Covid-19 on nature-based tourism in Botswana: Implications for community development. Tourism Review International, 25(2–3), 263–278. https://doi.org/10.3727/154427221X16098837279958, 263.

Sun, S., Zhong, L., Law, R., Zhang, X., Yang, L., & Li, M. (2022). A proposed DISE approach for tourist destination crisis management. *Sustainability, 14*(17), 11009. https://doi.org/10.3390/su141711009

Tsai, C., Wu, T., Wall, G., & Linliu, S. (2016). Perceptions of tourism impacts and community resilience to natural disasters. *Tourism Geographies, 18*(2), 152–173. https://doi.org/10.1080/14616688.2016.1149875

United Nations Educational, Scientific and Cultural Organization. (2022, August 31). *From recovery to resilience in the Zimbabwean highlands*. Retrieved January 31, 2023, from https://www.unesco.org/en/articles/recovery-resilience-zimbabwean-highlands

Wakeford, J. (2006). *Risks to global trade and implications for South Africa's economy and policy* (DPRU Working Paper No.6/111 July 2006). Development Policy Research Unit. University of Cape Town. https://doi.org/10.2139/ssrn.943995

World Bank. (2020a, May 19). *World bank group: 100 countries get support in response to covid-19 (coronavirus)*. Retrieved February 9, 2022, from https://www.worldbank.org/en/news/press-release/2020/05/19/world-bank-group-100-countries-get-support-in-response-to-covid-19-coronavirus

World Bank. (2020b, July 30). *Risking lives to protect wildlife and wildlands: Stories from rangers in the field*. Retrieved February 8, 2023, from https://www.worldbank.org/en/news/feature/2020/07/30/risking-lives-to-protect-wildlife-and-wildlands-stories-from-rangers-in-the-field

World Travel and Tourism Council. (2018). *Caribbean resilience and recovery: Minimising the impact of the 2017 hurricane season on the Caribbean's tourism sector*. https://www.wttc.org/-/media/files/reports/2018/impact-of-the-2017-hurricane-season-on-the-caribbean-spread.pdf

Chapter 13
Indigenous Communities and Wildlife Tourism: Issues of Sustainability in Botswana's Tourism Industry

Bongani G. Gumbo

13.1 Introduction

Globally, the tourism industry has been heralded as a key sector and an important driving force for the regeneration of rural areas in developing countries (Okello & Novello, 2014; Rasool et al., 2021). Tourism is viewed as critical to rural development and poverty alleviation for local communities in developing countries, becoming the latter's major growth area. Many scholars contend that tourism is one of the cornerstones of development for rural communities that are mostly found in remote areas where there is little or no Foreign Direct Investment (FDI) to generate income for livelihoods of indigenous communities in the Global South (Hyytia & Kola, 2013; Moswete & Mavondo, 2003; Mowforth & Munt, 1998). Thus, international tourism is viewed as a source of foreign exchange so dearly needed especially by developing economies. Goodall (1987, p. 69) captured it aptly: 'one person's holiday is another person's livelihood.' In southern Africa, tourism in most countries is largely nature-based, boasting of a rich biodiversity that includes different species of wildlife and scenic beauty, all of which draw international tourists to the region to satisfy their 'wanderlust'.

This chapter focuses on the place and nature of the roles of indigenous communities in wildlife tourism in Botswana. It seeks to move beyond the usual examination of the impact of tourism on local communities to also assess the actual participation of local people in the wildlife tourism through the various processes of planning, development, and management of wildlife tourism in Botswana. As Bello et al. (2018, p. 310) put it, "meaningful involvement of local communities in the tourism planning process entails empowerment of the local

B. G. Gumbo (✉)
Department of History, University of Botswana, Gaborone, Botswana
e-mail: gumbo@ub.ac.bw

© The Author(s), under exclusive license to Springer Nature Switzerland AG 2024

L. S. Stone, M. T. Stone (eds.), *Wildlife Tourism Dynamics in Southern Africa*, Geographies of Tourism and Global Change,
https://doi.org/10.1007/978-3-031-57252-4_13

residents to enable them to set their own goals, and to identify their hopes and concerns for tourism, in order that tourism benefits to the community are maximized". Indigenous communities have lived with wildlife from time immemorial and there is a historical relationship between local communities in Africa, Botswana included, and wildlife, "a relationship that is resilient and one that has existed from time immemorial and through the generations to the present and hopefully to posterity" (Sifuna, 2012, p. 31). This relationship, as Sifuna further explains, encompasses utility value in economic, social and cultural spheres of lives of indigenous communities. This manifested in, for example, where some animals have been revered in some communities for their totemic, ritualistic, and symbolic values (Sifuna, 2012). Given this history, indigenous communities are thus sine qua non to wildlife tourism development.

Botswana is recognised as one of the prime tourist destinations in the region (Saarinen & Manwa, 2008). Tourism in Botswana is the second largest foreign exchange earner, after diamonds (Mbaiwa, 2005, Mogomotsi et al., 2018). The country's wild landscapes are host to one of the largest concentrations of prolific wildlife species such as the famous 'big five' that include elephants, buffaloes, lions, leopards, and rhinos, and many other species, as well as a mosaic of floodplains with channels that snake their way around islands waterways (Gumbo, 2010). These are found in the country's tourist bastions, in the northern regions, namely the Chobe District, and Okavango Delta in the Ngamiland District, making Botswana a wildlife-based tourism destination (Mbaiwa, 2018). Against the broader goal of sustainable development through tourism, this chapter considers the scope and significance of indigenous community involvement in wildlife tourism and explores pathways to sustainable human-wildlife coexistence in the tourism industry in Botswana. It examines the extent of indigenous communities' involvement in the management of wildlife, the nature of benefits and challenges in the management of the wildlife resource over time and the prospects for sustainable development. It is important to understand from the onset, that sustainable development in tourism cannot be achieved without collaboration between government officials and local communities in the management of wildlife tourism.

13.2 Methodology

The chapter adopts the qualitative research methodology, drawing on both primary and secondary sources. Primary sources include material gleaned from the Botswana National Archives and Records Management Services (BNARS) and oral interviews. Among records at the National Archives are original authoritative documents such as correspondences between government officials including letters, memoranda, and policy documents. Secondary sources comprise a review of published books, journal articles, academic theses, newspaper articles and internet sources.

13.3 Definition of Key Concepts

Three concepts are key in the understanding of the subject of this chapter, namely 'indigenous communities', 'wildlife tourism', and 'sustainability' in the context of tourism development. There is no universal definition of any of the three terms but rather a modicum of convergence on the elements that constitute each of these. Albeit briefly, these concepts are explained below.

The World Bank defines indigenous communities from a socio-cultural perspective, with a major leaning towards relations to economic resources. In the Bank's words, 'Indigenous Peoples are distinct social and cultural groups that share collective ancestral ties to the lands and natural resources where they live, occupy or from which they have been displaced' (The World Bank, 2022). The socio-economic nexus is amplified further: 'The land and natural resources on which they depend are inextricably linked to their identities, cultures, livelihoods, as well as their physical and spiritual well-being' (The World Bank, 2022). Like The World Bank, Ashley and Roe (1999, p. 7) recognise the "common interest in resources of an area by a heterogeneous set of local people". Historically, the term 'indigenous peoples' in the southern African region, including Botswana, was used to refer to the San-speaking indigenous groups who occupied territories in the region over a period dating back to over 70,000 years (Sifuna, 2012). However, with influx of the Bantu-speaking peoples, the term is generally used as the nomenclature for the communities settled in the areas within the Botswana state. For example, in the Chobe and Ngamiland Districts where most protected areas teeming varied wildlife endowment are found, the San were the first inhabitants to settle but were later joined by a host of migrating Bantu-speakers who are now Batswana and together regarded as the indigenous communities.

'Wildlife Tourism' is another concept that needs unpacking. According to Higginbottom (2004, p. 1), wildlife tourism is tourism that is based on '...encounters with non-domesticated (non-human) animals. Such encounters can occur in either the animals' natural environment or captivity.' Echoing a similar sentiment, Jordan (2021) is emphatic on the aspect of viewing the wildlife in its natural habitat. Tourist activities in wildlife tourism are very diverse. They include, among others, game viewing, photography, hunting, viewing the wilderness, and sport fishing also known as angling. Wildlife tourism is of great economic value to indigenous communities as it provides economic opportunities for local people and further motivates them to conserve game animals knowing that benefits can accrue to them in the present and for future generations (Ashley & Roe, 1999). Arguably, wildlife tourism is extremely important to the African continent which is home to a variety of species of wildlife and 'it is no wonder that many tourists seek Africa as a place to get closer to nature' (Jordan, 2021). Bragagnolo et al. (2016) argue that despite the growing narratives on tourism potential, wildlife tourism has both positive and negative impacts on indigenous communities as shall be seen with the chapter unfolding.

'Sustainability' is one of the most contested concepts with no agreement over what constitutes sustainability, its nature or even the content. The meanings of the term are many and varied. The concept is used in many realms—sometimes environmental, other times economic as is commonly applied to 'sustainable tourism' or 'sustainable development' (Mowforth & Munt, 1998). The debate on sustainability is protracted, partly because, as Stone et al. (2022, p. 3) suggest, it is "facilitated by the lack of solutions in striking a balance between conservation and livelihood development sustenance." In this chapter, among other things, we critique the 'wildlife tourism-indigenous communities' development' nexus, in the context of the concept of sustainable development among communities living adjacent to protected areas in northern Botswana. We examine the extent to which the wildlife conservation trajectory synchronises with the development needs of the indigenous communities on whose land protected areas [PAs] have been established and, in the same vein, strive for protection and sustainable utilisation of biodiversity. Gumbo (2002) argues that local communities play a pivotal role in managing wildlife and its habitat advocating for similar reciprocal recognition of the local communities by state authorities, to incentivise local people for the conservation of wildlife resources. The chapter takes a historical perspective to understanding the extent to which the promotion of wildlife tourism has been able to create a balance between sustainability and community development. This trajectory, from the roles of both government officials and local communities (every Motswana), in conservation strategies through to other structures and stages of tourism development in the area.

13.4 Indigenous Communities, Wildlife Tourism and Sustainability: The Issues

13.4.1 Botswana Tourism—'Lauded and Reproved in Equal Measure'[1]

Before we delve into the complex functional relationships between wildlife tourism and local communities and the question of sustainability, it is important to provide a synopsis of tourism development in Botswana in general in the study area. In the development literature, it is universally accepted that tourism is increasingly being recognised as a new engine for growth, especially for poor communities in rural areas of the Global South (Hambira, 2020, Mowforth & Munt, 1998, Ndege, 1992). It is hailed as an important '…strategy for economic upliftment, community development and poverty relief' (Binns & Nel, 2002, p. 235). Its benefits to local communities have been viewed as wielding potential for incentivising rural communities into wildlife conservation. As with many southern African countries, Botswana

[1] Expression borrowed from David Magang, *Delusions of Grandeur: Paradoxes and Ambivalences in Botswana's Macroeconomic Firmament,* Vol. 2, (Gaborone: Print Media Consult, 2015), p. 182.

adopted tourism as a tool for local economic development among rural communities and, most importantly, as an alternative to existing livelihood strategies (Saarinen & Manwa, 2008).

Nestling in the heart of the tourist sought-after unique ecosystem of the Okavango Delta and the aesthetic vegetation comprising dense woodland forests, shrub savannah and a unique floodplains and small water channels meandering around small islands on the Chobe River, tourism in Botswana is anchored on both consumptive wildlife utilisation such as safari hunting, and non-consumptive that include the viewing of wildlife and scenic beauty of the wilderness, as well as photographic tourism (Gumbo, 2002).

Scholarly analyses on global tourism development oscillate between perceived successes and challenges of the industry in relation to the extent to which, in the quest for sustainability, the industry has been able to balance wildlife conservation with community development (Honey & Gilpin, 2009; Moscardo, 2008; Richards & Hall, 2000). Others have derided it for, among others, its enclave nature especially its tendency to be dominated by a few global finance magnates (Mogalakwe & Nyamnjoh, 2017, Reiser, 2003). In Botswana, these sentiments notwithstanding, it is universally agreed that tourism plays a vital role in increasing government revenue, through *inter alia*, taxation and park fees; the industry also wields huge potential for poverty alleviation and economic diversification from the country's overdependence on diamonds, which are non-renewable and very vulnerable to global economic shocks (Gumbo, 2022). Thus, the sector is a niche specialty (Magang, 2015). As the second largest foreign exchange earner, tourism's direct impact on economic development includes being a significant source of foreign exchange. According to the World Travel and Tourism Council, tourism's contribution to Botswana's Gross Domestic Product (GDP) in 2021 for example, amounted to BWP 15,664.4 m/(US$412.8 m), representing 8% of the GDP (Gumbo, 2022).

Regarded as a pro-poor strategy, tourism has brought some measure of economic development to rural areas, creating job opportunities especially in sectors such as hospitality—hotels, lodges, and camps; and directly from wildlife related jobs and activities created in the conservation and wildlife management in the Department of Wildlife and National Parks sub-sector, transport services by airlines and tour operators, and tour guides. Apart from generating income for workers, tourism related job opportunities in rural areas also provide checks on rural-urban migration thus relieving economic pressure on cities such as Gaborone and Francistown (Magang, 2015). Tourism has also been hailed for luring Foreign Direct Investment (FDI) to far flung remote areas that hitherto were either unknown to or shunned by foreign capital. Such investments indirectly increased employment opportunities and subsequent development and honing of skills. Added to the array of positive contributions is also the development of infrastructure such as roads and increased government services that include security through police stations in an otherwise remote and hitherto undeveloped area (Gumbo, 2002).

Equally, critics of tourism development have frowned at a variety of what they perceive as negative developments in the industry, citing among others the enclave phenomenon as evidenced by foreign domination, with only very few citizens participating as entrepreneurs while most locals serve as unskilled employees in sub-sectors of the tourism industry (Bolaane & Kanduza, 2008). Criticism has also been levelled against tourism because of its association with loss of land to the industry by indigenous communities—land that was appropriated to tourism infrastructure and protected areas for the conservation of fauna and flora (Mogalakwe & Nyamnjoh, 2017). Loss of land to tourism imperatives such as protected areas. This is palpably illustrated by, among other cases, the Chobe District, for example. While the entire district measures 20,800 km^2, a massive 11,700 km^2 is occupied by the Chobe National Park, and, additionally, an extra 4198 km^2 is host to six forest reserves, leaving only 4902 km^2 for settlement and economic activities by the local communities (Republic of Botswana, 2003). More of the apprehensions associated with wildlife tourism will be discussed in detail as the chapter unfolds.

13.5 The Genesis of Protected Areas in Northern Botswana: A Brief History

A discussion on wildlife tourism is incomplete without the historicisation of protected areas that served and continue to serve as habitats for game animals which draw tourists to wildlife tourism destinations. In the early 1960s, to protect the wildlife from marauding hordes of trophy hunters from Europe and South Africa, and to ensure sustainable utilisation of the tourism resources, authorities established Protected Areas (PAs) over most of the wildlife habitat areas in northern Botswana. European settlement at the Cape in the seventeenth century and their subsequent dispersal into the hinterland in the eighteenth and nineteenth centuries resulted in unchecked hunting and killing of substantial populations of large mammals for trade in skins, hides and ivory by these Europeans (McCormick, 1989). During this period, Botswana was a British protectorate, known as the Bechuanaland Protectorate, a status that spanned between 1885 and 1965, with the country gaining independence in 1966. The colonial administration established protected areas such as national parks, game reserves, wildlife management areas and forest reserves for the conservation of wildlife resources.

In 1962, the first protected area to be created was the Chobe National Park in the Chobe District, which the post-colonial government maintained to date (Gumbo, 2010). It must be noted that almost invariably everywhere in Africa, the creation of protected areas for the conservation of wildlife throughout the nineteenth century was an exercise in colonial landscaping in which local opinion was excluded—eliciting resentment from indigenous communities (Murombedzi, 2009). This was the case with the establishment of the Chobe National Park in 1962. This unilateralist action by the colonial administration was fiercely contested by the indigenous communities in the Chobe District, arguing that they had not been consulted (BNA S.163/4, 1959–1961). A plethora of game laws was promulgated, still without local input, effectively limiting access to some wildlife species which they had historically utilised for subsistence. The entire Chobe District had been designated 'Crown Lands', which allowed the administration to put the land to any use they deemed necessary without recourse to the indigenous people (Hailey, 1953). Gumbo (2015) has argued that successive republican administrations in Botswana inherited and adopted the existing Western structures of wildlife management with only very few modifications to assuage the indigenous communities with a view to securing their support for the wildlife conservation agenda.

However, in the case of the Moremi Game Reserve in Ngamiland, the creation of this game sanctuary was the brainchild of the local BaTawana elites. Bolaane (2005), reveals a new dynamic, illustrating how local initiatives in collaboration with other external actors, who were not necessarily the colonial administration, resulted in the creation of the game reserve, demonstrating local management strategies, as early as 1963. The colonial government benefitted rather fortuitously by virtue of being the superior political authority. In Bolaane's words:

The key actors in the foundation of Botswana's Okavango/Moremi National Park in the 1960s were not state officials but local BaTawana chiefs and a network of hunters and adventurers turned conservationists. The initiative was conceived as a means of protecting wildlife from the depredation of illegal South African hunting parties and ensuring future local use and was initially opposed by the colonial state (Bolaane, 2005, p. 241).

Therefore, the conception and establishment of the Moremi Game Reserve by indigenous BaTawana chieftaincy belies the narrative in colonial history that insinuated that game reserves were always initiated by colonial administrators. This was far from the Eurocentric sentiments that reverberated in the colonial world that local people saw game simply as food and not a national asset to be conserved for posterity (BNA 568/13/5, 1960). In essence, indigenous were alive to the importance of wildlife as a natural heritage not only for the present but also for future generations with beneficiary uses of wildlife resources and the need for its conservation as exemplified by the Batawana of Ngamiland in the then Bechuanaland Protectorate.

13.6 The Political Economy of Wildlife Tourism, Indigenous Communities, and the Ambiguities of the Notion of Sustainability

Like many tourist destinations in Africa, Botswana is richly endowed with diverse wildlife species, fascinating wilderness and ecosystems that lure international tourists from Western Europe, United States of America, Asia, Southern Africa and other parts of the world, making the country a wildlife tourism destination (Republic of Botswana, 2022). Wildlife tourism thrived on wildlife resources as the main attraction for tourists to Botswana, prompting government to set aside chunks of land for the establishment of protected areas for the conservation of wildlife resources. In South Africa, European sportsmen almost decimated wildlife populations in the then Transvaal and blamed it on Africans (Carruthers, 1995). In Botswana too, the protection of game animals was necessitated by the near extermination of wild animals on the Chobe and Caprivi floodplains in the nineteenth and early twentieth century by European trophy hunters, traders and the unrelenting game poaching for financial gain in the post-colonial period (Fisch, 1999). As elsewhere in Africa, these protected areas were established on indigenous peoples' territories, bringing both benefits and costs to the host communities. Additionally, to bolster the conservation regime, the state promulgated laws and policies to protect wildlife. These included the *National Parks Act* of 1968, the *Fish Protection Act* of 1975, the Fauna Conservation Act of 1979 (amending that of 1961); *The Tourism Policy* of 1990, the *National Conservation Strategy* of 1990, and the *Wildlife Conservation and National Parks Act* of 1992.

Pundits on tourism development aver that a successful tourism industry is one that embraces the principles of sustainable development (Bragagnolo et al., 2016; Moscardo, 2008). However, the concept of sustainable development is a contested space in academic literature with some questioning if it can be achieved in tourism, notably the balancing of conservation and development (Stone et al., 2022). Sustainable development is a subject with many dynamics. Hinz (2003) notes that some environmentalists advanced neoliberal orientations that emphasized the conservation of the environment and its resources, namely wildlife, vegetation, and aquatic life, unfortunately at the exclusion of developing indigenous communities in developing countries. Fervent supporters of environmental conservation such as Buhalis and Fletcher et al. (1995) argue that wildlife tourism conflicts with the environment as it results in the over-utilisation of the environment. Such a trajectory though, is at variance with reality as wildlife habitats were established adjacent to land and resources that local communities have historically utilised and subsisted on (Rodary, 2008).

Achieving sustainable development in wildlife tourism needs a holistic approach that considers not only the conservation of natural resources but also a keen interest in uplifting the communities in the vicinity. Moscardo (2008, p. 6), frowns at the conceptualisation of sustainable tourism being '...much more about the continuity of tourism than it is about the contribution to sustainable outcomes', implying unfettered contribution to both conservation and livelihoods. A similar balancing act is proffered by Richards and Hall (2000), who posit that an ideal sustainable wildlife tourism trajectory ought to enable local communities to participate in all the processes of tourism development including decision-making in activities which affect their lives including the management of wildlife conservation. The balancing act would serve as a mechanism for inclusive poverty reduction and global conservation (The World Bank, 2018). Furthermore, according to my oral informants on field research in Pandamatenga village in the Chobe District, local communities state that they have a long history of co-existence with wild animals and possess indigenous knowledge systems that could enhance sustainability in modern conservation regimes. They are unequivocal in having been conscious of the need to conserve wildlife as part of their heritage from time immemorial.

Despite the imperfections inherent in the traditional conservation regimes of local communities, there was, at the minimum, a modicum of commitment to conservation as seen in the adoption of totemism as a tenet of conservation. This is corroborated by some researchers on wildlife conservation such as Spinage (1991), Hinz (2003) and other informants lament exclusion from tourism planning as well as in the management of wildlife conservation. The exclusion from tourism structures irks the communities because wildlife tourism affects their lives both positively and negatively. For example, wild animals roam their village day and night causing despicable destruction to property and human life, clearly suggesting that government and other partners in tourism planning ought to enlist the support of local communities. The geographical position of Pandamatenga village does not help the situation either. The village is sandwiched between two huge national parks, namely the Chobe National Park to the west in Botswana, and the Hwangwe National Park to the east in Zimbabwe, allowing wild animals to move from one park to the other, passing through the village creating human-wildlife conflict. Informants' testimonies are a microcosm of what many communities feel about the western conservation regimes. There are other misgivings from indigenous communities about wildlife tourism, which compromise sustainable development as far as wildlife tourism is concerned.

Elsewhere, Ashley and Roe (1999), aver that protected areas for the conservation of wildlife have resulted in limiting access for local communities to natural resources through restrictive legislation as well as demarcated parks. Cases of forced evictions were reported in many developing countries due to wildlife conservation imperatives. As Ashley notes, "in many cases communities have been forcibly displaced from land or denied access to resources, to make way for tourism developments. For example, the Maasai in Kenya lost access to significant portions of their land with the gazettement of parks and reserves.' In Botswana the Bhugakhwe Khoisan communities of Mababe and Sankuyu villages of northern Ngamiland were evicted from their ancestral land to give way to the creation of the Moremi Game Reserve (Taylor, 2000). Many communities were disenfranchised of the right to harvest firewood for fuel energy, wild fruits, building material such as poles and thatch grass, herbs and certain animal portions for medicinal purposes, leading to the collapse of traditional medicine practices, according to informants in the Chobe District (Gumbo, 2015).

However, like other forms of tourism, wildlife tourism has significantly contributed to both the livelihoods of indigenous communities and the conservation of wildlife and biodiversity, contributing to a modicum of sustainably run wildlife tourism in rural areas in developing countries (The World Bank, 2018). Arguably, almost invariably everywhere in wildlife tourism destinations, communities living adjacent protected areas are usually very poor and tourism has been found to contribute not only to their upliftment but also enhancing wildlife conservation because with the flow of tourism benefits, communities also see the value of national parks and wildlife therein (The World Bank, 2018). Besides, as Ashley (1999, p. 13) puts it, "wildlife tourism can be a way to complement existing livelihood strategies of rural communities, by diversifying off-farm income and spreading risk." One of the most important contributions of wildlife tourism is the potential for local employment opportunities, and poverty alleviation as the sector is relatively labour intensive (Ashley & Roe, 1999). Job opportunities are ranked high because wildlife tourism destinations are usually in remote rural areas in which there are not many chances for income generation serve for agriculture which is only a subsistence regime that is susceptible to vagaries of weather and destruction by animals. Employment opportunities range from the service provision as rangers, tour guides, the hospitality sub-sector, security, cultural performances, and many others (United Nations World Tourism Organisation [UNWTO], 2015). Jobs thus provide the local communities with much needed purchasing power in an ever-changing global economy. Other benefits include the establishment of infrastructure such as access roads, service institutions and tourism concessions which also provide potential for tourism business by local communities (UNWTO, 2015).

In Botswana, wildlife tourism comprises consumptive and non-consumptive activities, namely trophy hunting and photographic tourism respectively. Albeit at different levels, both trajectories create opportunities for improved livelihoods for local communities and conservation of wildlife, arguably a step towards sustainable tourism development. Notwithstanding the value of employment creation, according to a local newspaper the *Sunday Standard* (2016), most of the jobs for locals are menial, low paying and with poor working conditions for many. 'Be that as it may, controversial as it, consumptive tourism through trophy hunting is one regime that has been adopted with a view to benefitting both government and communities and to some extent contributing to the conservation of wildlife' (Mbaiwa & Mogende, 2022, p. 134). The duo argue that rich trophy hunters can inject huge income into the communities in whose concession the trophy animal is killed, and this can '…boost local economies and further encourages local communities to participate in conservation,' further stating that local communities tend to support the conservation agenda where they are highly rewarded. Besides, communities can also access game meat through licensed hunting which also persuades local communities to see the value of conserving wildlife. Thus, the ability to involve indigenous communities to meaningfully participate in the various structures and processes of wildlife tourism planning, development, and management and benefits, is an important measure of a successful wildlife tourism.

Lekgau and Tichaawa (2021, p. 139), succinctly state that community participation is "central to obtaining the benefits conceptualised in wildlife tourism." So crucial is communities' participation that, in a review of an article entitled 'Enhancing Community Involvement in Wildlife Tourism: Issues and Challenges' Buhalis (2002) identified three essential goals of community involvement in tourism. These are rural development, conservation and industry development arguing that these can be achieved through community members' active involvement in the industry, stating that such participation has the potential to maximise benefits as well as potentially minimising the negative impacts on indigenous. Wildlife tourism is one area that cannot avoid involvement of indigenous communities—they have a long-standing historical association with wildlife and view them as a natural heritage to be proud of (Sifuna, 2012). Besides, through indigenous knowledge systems acquired over time, local communities generally understand the behaviour of wildlife such that they can add immense value in the processes of planning, development, and management of wildlife tourism, as well as the conservation of wild animals (Gumbo, 2002). Nevertheless, as indicated earlier, in most African countries, Botswana included, indigenous communities are usually marginalised and treated as adjuncts in wildlife tourism development. In most cases, communities have historically been excluded from the stages and structures of tourism development which have been dominated by government and to some extent, investors in the industry.

Benchmarking on Zimbabwean government's recognition of the importance of 'persuading' communities into wildlife conservation through the Communal Areas Management Programme for Indigenous Resources (CAMPFIRE), The Botswana government established the Community-based Natural Resource Management (CBNRM) programme which "...recognizes the rights of local communities in managing and deriving benefits from natural resources and wildlife conservation" (Mogomotsi & Mogomotsi, 2022, p. 119). Funded by the USAID and introduced in Botswana in 1992, the aim of CBNRM programme is to incentivise local communities into conservation of biodiversity while improving rural livelihoods, thus bringing in inclusivity of local communities in tourism development. The programme also seeks to reduce the enduring human wildlife conflicts as well as alleviate poverty (Centre for Applied Research, 2016). In western Chobe the local CBNRM is known as the Chobe Enclave Community Trust (CECT), but there have since been many of such programmes throughout northern Botswana where wildlife tourism abounds. In Mababe village, about 125 km northeast of Maun in Ngamiland, the local community established the Mababe Zokotsama Community Development Trust (MZCDT) and they state government gave them usufruct rights over a tourism concession NG 41, where they undertake tourism business that has provided the community with jobs and community development (*The Patriot on Sunday*, 2023). The newspaper report explains that the trust has employed over eighty people from the village, almost a quarter of the village population. Among the jobs created and provided are those of escort guides '...to undertake wildlife monitoring in the concession' primarily to check poaching crimes in the area. Furthermore, according to the report, MZCDT conducts community education, sensitising the community on the importance of wildlife and the benefits accruing therefrom. The community also receives land rental concessions from those who utilise their campsites. That said, suggest some requisites that are essential to achieve full and active community participation namely, among others, 'public awareness and education, capacity building, creation of linkages, use of appropriate participation methods, use of appropriate local community organisations.'

13.7 Conclusion

Wildlife tourism, indigenous communities, and issues of sustainability in Botswana speaks to the conservation-development nexus, a highly contested space. According to some analysts, the debate is protracted because for some, sustainable development means a conservation trajectory that is exclusive to the environment and nature resources excluding livelihoods and rural development. It cannot be that communities' livelihoods can be ignored yet wildlife habitat surrounds human settlement. Indigenous communities' lives are affected by wildlife both positively and negatively and community involvement is paramount in addressing sustainable wildlife tourism development and achieving a measure of sustainability. We have indicated how tourism developers can learn from history, notably the communities' indigenous knowledge systems that could be infused into the western conservation regimes to strengthen wildlife conservation. Notwithstanding, wildlife tourism has both costs and benefits to indigenous communities. The fact that wildlife tourism destinations are found in far flung remote areas where income generating sources are limited, tourism is very crucial for the inhabitants of such areas. It creates employment opportunities, improves infrastructure, and facilitates the delivery of services that would otherwise not reach such rustic settlements.

Achieving a modicum of sustainable wildlife tourism development in Botswana as elsewhere, essentially requires a reciprocal relationship between the drivers of the tourism industry and indigenous communities, a balance that takes into cognizance the importance of biodiversity conservation and improving livelihoods of indigenous communities that have sacrificed so much in the form of land loss, human-wildlife conflict, limited access (sometimes no access altogether) to resources that they historically subsisted on. Active involvement of indigenous communities in wildlife tourism cannot be overemphasised.

References

Ashley, C., & Roe, D. (1999). *Enhancing community involvement in wildlife tourism: Issues and challenges*. Russel Press.

Belo, F., Brent, L., & Carr, N. (2018). Enhancing community participation in tourism planning associated with protected areas in developing countries: Lessons from Malawi. *Tourism and Hospitality Research, 18*(3), 309–320.

Binns, T., & Nel, E. (2002). Tourism as a local development strategy in South Africa. *The Geographical Journal, 168*(3), 235–247.

Bolaane, M. (2005). Chiefs, hunters and adventurers: The foundation of the Okavango/Moremi National Park, Botswana. *Journal of Historical Geography, 31*(2), 241–159.

Bolaane, M. M. M., & Kanduza, A. M. (2008). Critical factors in cultural tourism in Botswana. *Botswana Notes and Records, 39*, 54–61.

Botswana National Archives, BNA S.163/4, 1959–1961 'Game Reserves and Protection of Wildlife, File No. M.H/2/60, 1959–1961.'

Botswana National Archives, BNA 568/13/5, 1960.

Bragagnolo, C., Malhado, A. C. M., Jepson, P., & Ladle, R. J. (2016). Modelling local attitudes to protected areas in developing countries. *Conservation and Society, 14*(3), 163–182.

Buhalis, D. (2002). Book review. *International Journal of Tourism, 4*, 253–269.

Buhalis, D., & Fletcher, J. (1995). Environmental impacts of tourist destinations: An economic analysis. In H. Coccossis & P. Nijkamp (Eds.), *Sustainable tourism development* (pp. 3–24). Avery.

Carruthers, J. (1995). *The Kruger National Park: A social and political history*. University of Natal.

Centre for Applied Research. (2016). *2016 review of community based natural resources management in Botswana: Report prepared for SAREP*. https://www.car.org.bw/wp-content/uploads/2016/05/Botswana-CBNRM-2016-Review.pdf. Accessed 13 Feb 2023.

Fisch, M. (1999). *The Caprivi Strip during the German colonial period, 1890 to 1914*. Out of Africa.

Goodall, B. (1987). Tourism and regional development. *Built Environment, 13*(2), 68–72.

Gumbo, G. B. (2002). *The political economy of development in the Chobe: Peasants, fisherment and tourists, 1960–1995*. Unpublished Masters thesis, Department of History, University of Botswana.

Gumbo, G. B. (2010). *Economic and social change in the communities of the wetlands of Chobe and Ngamiland, with special reference to the period since 1960*. Unpublished PhD thesis, Department of Historical Studies, University of Cape Town.

Gumbo, B. (2015). *In whose interest? Conservation and livelihoods in Botswana's tourism industry: The case of the Chobe District* (African Economic Review) (pp. 131–146). Shanghai Joint Publishing House.

Gumbo, G. B. (2022). The negative impacts of tourism in Africa: The case of Botswana. *African Journal of Hospitality, Tourism and Leisure, 11*(2), 1750–1764.

Hailey, M. H. (1953). *Native administration in the British African Territories, part V. The High Commission Territories: Basutoland, the Bechuanaland Protectorate and Swaziland*. Her Majesty's Stationery Office.

Hambira, W. L. (2020). A review of community social upliftment practices by tourism multinational companies in Botswana. In M. T. Stone, M. Lenao, & N. Moswete (Eds.), *Natural resources, tourism and community livelihoods in Southern Africa: Challenges of sustainable development* (pp. 52–63). Routledge.

Higginbottom, K. (Ed.). (2004). *Wildlife tourism: Impacts, management and planning*. Common Ground Publishing.

Hinz, M. O. (2003). *Without chiefs there would be no game: Customary law and nature conservation*. Out of Africa.

Honey, M., & Gilpin, R. (2009). Tourism in the Developing World: Promoting Peace and Reducing Poverty. *United States Institute of Peace: Special Report*. https://www.usip.org/sites/default/files/tourism_developing_world_sr233_0.pdf

Hyytia, N., & Kola, J. (2013). Tourism policy as a tool for rural development. *Applied Economic Perspectives and Policy, 35*(4), 708–730.

Jordan, J. (2021). *How wildlife in Africa reduces poverty*. https://borgenproject.org/wildlife-tourism-in-africa/. Accessed 28 Feb 2023.

Keakabetse, B. (2023, March 26). Mababe -where people and wildlife live side by side. *The Patriot on Sunday*.

Lekgau, R. J., & Tichaawa, T. M. (2021). Tourism in Kgalagadi transfrontier park. *Tourism Review International, 25*(2–3), 139–155.

Magang, D. (2015). *Delusions of grandeur: Paradoxies and ambivalences in Botswana's macroeconomic firmament* (Vol. 2). Print Media Consult.

Mbaiwa, J. E. (2005). Enclave tourism and its socio-economic impacts in the Okavango Delta, Botswana. *Tourism Management, 26*(2), 157–172.

Mbaiwa, J. E. (2018). Effects of the safari hunting tourism ban on rural livelihoods and wildlife conservation in Northern Botswana. *South African Geographical Journal, 100*(1), 41–61.

Mbaiwa, J. E., & Mogende, E. (2022). Lifting of the hunting ban and the elephant debate in Botswana: Implications for conservation and development in Southern Africa. In L. S. Stone,

M. T. Stone, P. F. Mogomotsi, & G. E. J. Mogomotsi (Eds.), *Protected areas and tourism in Southern Africa: Conservation goals and community livelihoods* (pp. 131–145). Routledge.

McCormick, J. (1989). *The global environmental movement: Reclaiming paradise*. Belhaven.

Mogalakwe, M., & Nyamnjoh, F. (2017). Botswana at 50: Democratic deficit, elite corruption and poverty in the midst of plenty. *Journal of Contemporary African Studies, 35*(1), 1–14.

Mogomotsi, P. K., & Mogomotsi, G. E. J. (2022). Protected areas and community-based tourism: The effectiveness of current mitigation techniques in human wildlife conflicts. In L. S. Stone, M. T. Stone, P. F. Mogomotsi, & G. E. J. Mogomotsi (Eds.), *Protected areas and tourism in Southern Africa: Conservation goals and community livelihoods* (pp. 114–128). Routledge.

Mogomotsi, P. K., Kolobe, M., Raboloko, M., & Mmopelwa, G. (2018). The economic contribution of tourism to local communities: The case of Khumaga and Moreomaoto villages, Botswana. *PULA: Botswana Journal of African Studies, 32*(1), 84–99.

Moscardo, G. (2008). Sustainable tourism innovation: Challenging basic assumptions. *Tourism and Hospitality Research, 8*(1), 4–13.

Moswete, N., & Mavondo, F. T. (2003). Problems facing the tourism industry in Botswana. *Botswana Notes and Records, 35*, 69–77.

Mowforth, M., & Munt, I. (1998). *Tourism and sustainability: Development and new tourism in the third world*. Routledge.

Murombedzi, J. (2009). Pre-colonial and colonial conservation practices in southern Africa and their legacy today. In W. Hwande, T. Kepe, & M. Murphree (Eds.), *Local communities, equity and conservation in southern Africa: A synthesis of lessons learnt and recommendations from a southern African technical workshop* (pp. 23–24). Programme for Land and Agrarian Studies (PLAAS).

Ndege, G. O. (1992). Tourism in Kenya: Genesis, growth and impact. In W. R. Ochieng & R. M. Maxon (Eds.), *An economic history of Kenya*. East African Educational Publishers.

Okello, M. M., & Novello, M. (2014). Tourism in the East African Community (EAC): Challenges, opportunities, and ways forward. *Tourism & Hospitality Research, 14*(1–2), 53–66.

Online editor. (2016, August 30). Tourism workers refute claim they are well paid. *Sunday Standard*. https://www.sundaystandard.info/tourism-workers-refute-claim-they-are-well-paid/

Rasool, H., Maqbool, S., & Tarique, M. (2021). The relationship between tourism and economic growth among BRICS countries: A panel cointegration analysis. *Future Business Journal, 7*(1), 9–11.

Reiser, D. (2003). Globalisation: An old phenomenon that needs to be rediscovered for tourism? *Tourism and Hospitality Research, 4*(4), 306–320.

Republic of Botswana. (2003). *Chobe District development plan 6, 2003–2009*. Chobe District Council.

Republic of Botswana. (2022). *International visitor arrival, stats brief quarter 1*. Statistics Botswana.

Richards, G., & Hall, D. (2000). The community: A sustainable concept in tourism development? In G. Richards & D. Hall (Eds.), *Tourism and sustainable community development* (pp. 1–13). Routledge.

Rodary, E. (2008). From common to banal tourism in Southern Africa. *Botswana Notes and Records, 39*, 89–97.

Saarinen, J., & Manwa, H. (2008). Tourism as a socio-cultural encounter: Host-guest relations in tourism development in Botswana. *Botswana Notes and Records, 39*, 43–53.

Sifuna, N. (2012). The future of traditional customary uses of wildlife in modern Africa: A case of Kenya and Botswana. *Advances in Anthropology, 2*(1), 31–38.

Spinage, C. (1991). *History and evolution of the Fauna Conservation Laws of Botswana*. The Botswana Society.

Stone, M. T., Stone, L. S., Mogomotsi, G. E. J., & Mogomotsi, P. K. (2022). Protected areas and tourism dynamics in Southern Africa: An overview. In L. S. Stone, M. T. Stone, P. K. Mogomotsi, & G. E. J. Mogomotsi (Eds.), *Protected areas and tourism in Southern Africa: Conservation goals and community livelihoods* (pp. 1–11). Routledge.

Taylor, M. (2000). *Life, land and power: Contesting development in Northern Botswana*. Unpublished PhD thesis, University of Edinburgh.

The World Bank. (2018). *Growing wildlife-based tourism sustainability: A new report and Q & A*. http://www.worldbank.org/en/news/feature/2018/03/01/growing-wildlife-based-tourism-sustainibility-a-new-report. Accessed 27 Jan 2023.

The World Bank. (2022). *Indigenous peoples*. https://www.worldbank.org/en/topic/indigenous-peoples Accessed 24 Feb 2023.

World Tourism Organisation. (2015). *Towards measuring the economic value of wildlife watching tourism in Africa* (Briefing paper). WTO.

Part V
Protected Areas and Wildlife-Based Tourism

Chapter 14
International Laws and Wildlife Tourism in Southern Africa

Selma Lendelvo, Mfundo Mlilo, Margaret N. Angula, Jona Heita, and Raili Hasheela

14.1 Introduction

Over the past decade, research into tourism in Southern Africa has significantly expanded. This research has predominantly centred on the tourism potential in Africa (Christie et al., 2013), the nature of tourism in Africa, and the examination of the extent to which local communities benefit from wildlife tourism (Kalvelage et al., 2022). Furthermore, a substantial body of research delves into nature-based tourism, specifically focusing on its influence on economic development and the involvement of minority communities (Bush et al., 2008). In light of the abundant wildlife in the Southern Africa region and the various threats posed by human activities, there has been a noticeable shift toward adapting international governance structures to protect wildlife within domestic contexts. Nevertheless, the extent of the effectiveness and impact of these international instruments on conservation and tourism development remains an area of limited understanding (Holechek & Valdez, 2018).

S. Lendelvo (✉) · M. N. Angula
School of Science, Department of Environmental Science, University of Namibia, Windhoek, Namibia
e-mail: slendelvo@unam.na; mangula@unam.na

M. Mlilo
Institute of Geography, University of Cologne, Koln, Germany
e-mail: mmlilo@uni-koeln.de

J. Heita
University of Namibia (Main Campus), Windhoek, Namibia
e-mail: jheita@unam.na

R. Hasheela
Ministry of Environment, Forestry and Tourism, Windhoek, Namibia

© The Author(s), under exclusive license to Springer Nature Switzerland AG 2024

L. S. Stone, M. T. Stone (eds.), *Wildlife Tourism Dynamics in Southern Africa*, Geographies of Tourism and Global Change,
https://doi.org/10.1007/978-3-031-57252-4_14

This chapter aims to delve into and discuss the influence of international legislation on the wildlife tourism industry in Southern Africa. The chapter focuses on pivotal legislation at both the regional and international levels, with particular attention paid to the treaties and protocols of the Southern African Development Community (SADC). We also explore international conventions such as the Rio Conventions, the Convention on International Trade in Endangered Species of Fauna and Flora (CITES), and the Convention on Migratory Species of 1979. Additionally, we examine the Ramsar Convention of 1971 and the Community-Based Natural Resource Management (CBNRM) approaches, considering their domestication and impact on wildlife tourism's evolution in Southern Africa. While this approach may not encompass all relevant aspects, it is important to note that it provides valuable insights into the critical role that international law could play in conservation and its interplay with the tourism sector in the Southern African context.

Tourism plays a crucial role in driving economic development in countries across Southern Africa (Mochankana et al., 2023; Rogerson & Rogerson, 2019). The Southern African region is particularly attractive to tourists, accounting for approximately 30% of all tourists visiting Africa (Christie et al., 2013). For example, in 2018, Botswana welcomed around 1.8 million international tourists, while Namibia and South Africa received 1.6 million and 14 million international tourists respectively (UNWTO, 2022). However, the COVID-19 pandemic had a significant impact on the tourism industry in the SADC region, causing a decline in tourism figures until 2020. It is anticipated that the sector will recover in the post-COVID-19 era (UNWTO, 2023). Nonetheless, the tourism sector is poised for substantial growth, especially when tourism activities are effectively managed. Tourism in the SADC region has made a significant economic contribution, representing up to 2.8% of the total GDP in 2017. This positive trend of growth continued, with the sector experiencing a 4% increase in 2019 (United Nations World Tourism Organisation [UNWTO], 2020). This growth underscores the economic importance of tourism in the region, highlighting its potential for further development and economic impact.

Most of the region's tourist attractions are nature-based, with a strong focus on wildlife which generates significant economic impacts in the tourism industry (World Travel and Tourism Council, 2004). Southern Africa boasts a diverse range of wildlife species that draw tourists, including springboks, impala, kudu, eland, gemsbok, as well as various predators and birds. Additionally, the region is home to charismatic megafauna species that are particularly popular in nature-based tourism, collectively known as "The Big Five." These iconic species include the elephant, buffalo, rhinoceros, lion, and leopard (Van der Merwe et al., 2021).

Wildlife tourism in Southern Africa is extensive due to the region's diverse ecosystems, which boast a wide array of large mammals, predators, and ungulates, making it a unique and sought-after tourist destination (Goodwin & Leader-Williams, 2000; Grünewald et al., 2016). These wildlife populations have faced historical threats from habitat loss, poaching, and extreme climate events, resulting in a decline from 1970 to 1990. These changes in habitat and demographics primarily coincided with the emergence of colonialism, which commercialised wildlife for both trade and tourism (Newmark, 2008). Local populations had previously maintained a harmonious balance as primarily pastoralists (Holechek & Valdez, 2018). However, establishing national parks and conservation areas spurred the growth of a thriving tourist industry and contributed to improvements in animal populations (Saayman & Saayman, 2010). Additionally, concerns for biodiversity conservation (Child, 2004), protection of specific endangered species and their habitats (Ntiamoa-Baidu, 1995), reduction of human-wildlife conflicts (Marker & Boast, 2015), promotion of sustainable consumption (Van Hoven, 2015), and the encouragement of sustainable tourism have led to the emergence of various governance regimes (Ngeta & Child, 2014; Duffy, 2006), including regional and international laws aimed at promoting the sustainable use of wildlife.

For instance, in 1992, the World Summit on Sustainable Development adopted the Rio Conventions, with the primary goal of promoting the protection and conservation of biodiversity (Vinuales, 2015; Raustiala & Victor, 1996). This significant event coincided with the Southern Africa Development Community (SADC) treaty signed on 17 August 1992, emphasising cooperation among SADC member states in various areas, including conservation efforts (Lubbe, 2008). Several other pieces of legislation have since emerged in the last decades. This chapter explores the pertinent international legislation that advocates for the sustainable utilisation of wildlife and assesses its influence on the wildlife tourism sector.

International wildlife and tourism-related laws are complementary to the above goals and ensure none of the pieces of legislation are implemented at the expense of the other. This chapter analyses wildlife tourism and the sustainable development nexus in Southern Africa and further examines the relation in practice and guidelines through the lenses of international laws. Clarity on the above is presented through practical examples from critical international laws and some relevant laws to which countries in southern Africa are signatories. In the subsequent section, we provide a brief overview on wildlife tourism in Southern Africa. We then identify and discuss key regional and international laws that Southern African states have adopted and have a significant impact on the management and sustainable utilisation of wildlife and nature-based resources in the region. Next, we delve into the specific impact of these legislations on wildlife tourism in Southern Africa, examining how these laws shape and influence the wildlife tourism sector. Finally, the chapter concludes by summarising the key findings and implications of these legislations for wildlife tourism in the Southern African context.

14.2 An Overview of the Laws, Treaties and Conventions Governing Wildlife Management in Southern Africa

International law is defined as a collection of rules that sovereign nation-states mutually recognise as binding in their interactions, and it constitutes a "framework comprising norms, standards, principles, institutions, and procedures within a 'society' formed not by individual human beings but by these sovereign entities" (Bedjaoui, 1991, p. 2). International laws are "a body of norms, in written form or otherwise, intended to discipline the relationship of states among themselves. Thus, in principle it regulates the conduct of states and not that of individuals" (Bedjaoui, 1991, p. 2). Consequently, international law operates as a "voluntary" legal system in which the governed nations govern themselves to safeguard their respective sovereignties (Tanner, 2004, p. 26). In this case international law is seen as a law of coordination between the sovereign jurisdictions than a law of subordination which regulates its subjects through, where necessary coercion through state apparatus (Bedjaoui, 1991). Despite its nature the reality as observed by Louis Henkin, "almost all nations observe all principles of international law and almost all of the obligations almost all the time" (Henkin, 1968; Koh, 1997, p. 2599). This statement also suggests that there may be times when states do not obey international law which creates challenges for interstate coordination. Southern African countries have actively signed various regional and global treaties to foster wildlife management and promote tourism for example.

Article 21(e) of the SADC treaty is central to these agreements, which was established in August 1992 as a regional legislative instrument. This article calls upon member states to collaborate on diverse issues, including managing natural resources and environmental preservation. The SADC treaty serves as the overarching framework within which SADC member states have successfully developed and overseen transfrontier conservation parks. Currently, there exist 18 transfrontier parks in the Southern African region, with eight of them formalized, four in the emerging stage, and six in the conceptual phase. Collectively, these parks cover over half of the designated conservation area, spanning over a million square miles (Peace Parks Foundation, 2022). The influence of the agreements that have established these Transfrontier Conservation Areas (TFCAs) on wildlife preservation and the expansion of the tourism industry has been profound. These impacts will be explored in detail in the subsequent section.

The Protocol on Wildlife Conservation and Law Enforcement, an integral component of the SADC Treaty, effected in 2003 and has since been ratified by several member states, including Malawi, Mauritius, Mozambique, South Africa, Namibia, Zambia, Lesotho, Tanzania, Botswana, with Angola and Zimbabwe having signed the Protocol. The primary objective of this protocol is to establish a common framework, harmonizing approaches to wildlife conservation and sustainable resource utilization within the context of each party's national legislation. Additionally, it aims to bolster the effective enforcement of laws governing wildlife resources (SADC, 2003). Furthermore, several other regional treaties have been enacted, such as the African Convention on the Conservation of Nature and Natural Resources (African Union, 2003) and the Lusaka Agreement on Cooperative Enforcement Operations aimed at combating illegal trade in wild fauna and flora, established in 1994. These regional instruments collectively constitute an essential component of the legal framework designed to safeguard wildlife and promote sustainable tourism. Southern African countries have now devised the SADC Law Enforcement and

Anti-Poaching Strategy (SADC LEAP) which is embedded in the Protocol on Wildlife Conservation and Law Enforcement (PWCLE) mainly to combat poaching.

The International Trade in Endangered Species of Wildlife Flora and Fauna (CITES) plays a pivotal role in safeguarding endangered species by implementing strict controls and regulations on trade through export permit systems. As a prominent guide in international law, CITES provides a structured framework for the regulated transboundary trade of wildlife derivatives (Abensperg-Traun, 2009; Heppes & McFadden, 1987). All Southern African states are members and signatories to CITES with Angola having joined in 2022. In response to international legal frameworks, the tourism industry has been adapting its operations to mitigate adverse environmental impacts and promote sustainable development. For instance, at the 19th Conference of the Parties to CITES held in Panama in 2022, Southern Africa countries such as South Africa, Namibia, and Zimbabwe submitted a proposal seeking permission to trade their existing ivory stock to finance conservation efforts (Smit, 2019). However, this proposal did not receive support from the CITES Secretariat. This action by the Southern African countries underscores the potential conflict between international laws, regional and local conservation efforts.

A crucial piece of international legislation facilitating the cross-border management of migratory species is the Convention on the Conservation of Migratory Species of Wild Animals (CMS), established in Bonn in 1979. This convention has the overarching goal of conserving terrestrial, marine, and avian migratory species throughout their entire range, necessitating cooperation among countries that host these species as they cross international boundaries (Lyster, 1989). Under this convention, participating states are obligated to take measures for the conservation and restoration of habitats, as well as to eliminate or minimize obstacles to the migration of these species (Baldwin, 2011). There are also provisions to address factors that pose threats to these migratory animals (Caddell, 2005). The significance of this international law in the context of tourism cannot be underestimated, particularly in Southern Africa. The region is home to a diverse array of migratory species, including over 100 migratory bird species (BirdLife South Africa, 2019), bat migrations in Zambia, and other notable animal migrations such as the great Zebra migration, the great Wildebeest migration in Kenya and Tanzania, and the elephant migration from Botswana to the north. Migration is a critical ecological process that faces threats from human activities and habitat fragmentation (Purdon et al., 2018). As tourists are drawn not only to observe individual animals but also to the spectacle of these migrations, the impact of this international law on tourism in the region is substantial.

The Rio de Janeiro in 1992 Conventions encompass the United Nations Framework Convention on Climate Change (UNFCCC), the Convention on Biological Diversity (CBD), and the United National Convention to Combat Desertification (UNCCD). The CBD's primary objective, in particular, is to preserve biological diversity, promote the sustainable utilization of its components, and ensure the fair and equitable sharing of benefits derived from commercial use (Chandra & Idrisova, 2011). These conventions have collectively aimed to guide signatory countries, including those in Southern Africa, in addressing the disparities in global environmental priorities that have historically favoured Northern regions, often at the expense of Southern regions such as the SADC area (Swiderska, 2002).

In Southern Africa, two notable approaches to climate change adaptation have emerged: ecosystem-based adaptation and community-based adaptation. These strategies promote non-consumptive community-based tourism and other ecosystem services as income-generating activities aimed at enhancing the resilience of local communities to cope with the impacts of climate change (Reid, 2016). It's essential for Southern African countries to consider the possibility of a decline in international tourists, particularly those interested in nature-based and wildlife-based tourism when devising tourism plans. The hospitality sector, a significant component of the tourism industry, has historically contributed to environmental degradation and pollution, primarily through increased waste management burdens and excessive consumption of energy and water (Mbasera et al., 2016). The implementation of international and national climate mitigation policies, such as the Paris Agreement, which aims to reduce greenhouse gas emissions, may also have implications for tourist flows. Southern African countries like Namibia, Botswana, and South Africa, being long-haul destinations, could potentially be affected by these policies (Saarinen et al., 2022). This highlights the need for a comprehensive approach to sustainable tourism development that considers both climate change adaptation and mitigation measures.

However, when it comes to issues related to the environment, social development, human rights, and equitable benefits from wildlife tourism, there can be some ambiguity, particularly regarding free trade through the World Trade Organization (WTO) as a regulatory body (Schroeder et al., 2008). This can create grey areas in the realm of international conservation laws, including commitments to the three Rio Conventions above and CITES. Programmatic issues stemming from these conventions may not have been fully addressed, especially in Southern Africa, where wildlife tourism can directly impact the environment. This is especially relevant given the current and emerging trends in the exploitation of natural resources in developing countries, such as rhino poaching and controversial trophy hunting practices, which have generated divided sentiments within Southern Africa (Lindsey et al., 2007). Despite these challenges and potential legal gaps, wildlife tourism still holds significant opportunities to thrive, strengthen its role as a key economic driver, and enhance its impact on livelihoods.

Furthermore, the Southern African countries are active members of the International Union for the Conservation of Nature (IUCN), an organization dedicated to the conservation of endangered species. The IUCN has established global criteria for assessing the risk of extinction, which are universally adopted and applied (Miller et al., 2007). This is highly relevant to the development of tourism in the SADC region, where countries place a strong emphasis on wildlife tourism. Additionally, numerous tourism and travel companies and operators in Southern Africa have aligned themselves with the World Travel and Tourism Council (WTTC) to enhance tourism development within their respective countries and regions (Higginbottom, 2004). Given that tourism in Southern Africa is intricately linked to nature and often situated within delicate ecosystems, it becomes imperative for tourism-related legislation to address sustainability comprehensively across the region, encompassing both protected areas and areas outside of formal conservation zones. This holistic approach is essential for ensuring the long-term viability and responsible management of tourism activities in the region.

14.3 International Laws and Wildlife Tourism in Southern Africa

Assessing the influence of international laws on wildlife tourism in Southern Africa through four criteria provides a comprehensive framework to evaluate their impact. These criteria are based on our own reflections and research into tourism developments in Africa. Much of the literature on the impact of International Law on Wildlife focuses on assessing provisions for community participation (see for example Dhliwayo et al., 2009), using extent of ratification to effectiveness (Cirelli & Morgera, 2010) and habitat protection and population growth. We build on this research to produce a criterion that looks at ratification; wildlife population and habitat protection; tourism enterprise growth and tourist arrivals. These criteria can offer valuable insights into the effectiveness and significance of these laws in shaping the ratification of these laws. Here's a brief overview of each criterion:

We provide a schematic representation of the inter-relationships between international law and wildlife tourism below in Fig. 14.1.

Domestication and Ratification This criterion assesses how international laws have been incorporated into domestic legal frameworks and the establishment of coordination structures to ensure compliance. It measures the extent to which Southern African countries have adopted and implemented these laws at the national level.

(i) **Wildlife Population Growth and Habitat Protection**: This criterion evaluates the impact of international laws on the conservation of wildlife populations and the protection of their natural habitats. It considers whether these laws have contributed to the preservation and growth of wildlife populations and the safeguarding of critical habitats.
(ii) **Tourism Enterprise Growth**: This criterion focuses on the growth and development of tourism enterprises related to wildlife tourism. It assesses whether international laws have fostered an environment conducive to the expansion of tourism businesses that rely on wildlife attractions.
(iii) **Tourist Arrivals in Southern Africa**: This criterion measures the number of tourists visiting Southern Africa for wildlife tourism experiences. It explores whether international laws have influenced the attractiveness of the region as a wildlife tourism destination, leading to increased visitation.

Fig. 14.1 Inter-relationships between international law and wildlife tourism

By applying these criteria, it becomes possible to gauge the multifaceted influence of international laws on wildlife tourism in Southern Africa, providing a comprehensive understanding of their role in shaping the industry and contributing to conservation efforts.

14.4 Domestication and Ratification

The Table 14.1 below summaries dates on which various pieces of international and regional laws were ratified and came into force in specific countries in the Southern African region.[1]

The table above shows there is widespread ratification of regional and international laws in Southern Africa. This is important to create governance structures for coordination to promote and protect wildlife. Protection of wildlife and their habitats is a key imperative for the growth of tourism.

[1] Based on data collected from Ecolex (www.ecolex.org) as of 5 September 2023 and google searches.

Table 14.1 Southern African Countries and Ratification of International laws

	SADC-Treaty	PWCLE	SADC LEAP	CBD	CITES	Ram	CMS	WHC	UNWTO
Angola	1992	1999	2013	1998	2013	2021	2006	1992	1989
Botswana	1992	1999	2013	1996	1978	1997		1999	1995
Dem. Rep of Congo	1998	1999	2013	1994	1976	1996	1990	1974	1979
Lesotho	1992	1999	2013	1995	2003	2004		2004	1981
Madagascar	2005	2001	2013	1996	1975	1999	2007	1983	1975
Malawi	1992	1999	2013	1994	1982	1997	2019	1982	1975
Mauritius	1998	1999	2013	1992	1975	2001	2004	1995	1975
Mozambique	1992	1999	2013	1995	1981	2004	2009	1983	1975
Namibia	1992	1999	2013	1997	1991	1995		2000	1997
Seychelles	1998	1999	2013	1992	1977	2005	2005	1980	1991
South Africa	1994	1999	2013	1996	1975	1975	1991	1997	1994
Eswatini	1992	1999	2013	1994	1997	2013	2013	2005	1999
Tanzania	1992	1999	2013	1996	1980	2000	1999	1977	1975
Zambia	1992	1999	2013	1993	1981	1991		1984	1975
Zimbabwe	1992	1999	2013	1995	1981	2013	2012	1982	1981

Source: Author computation based on data collected from Ecolex (www.ecolex.org) as of 5 September 2023 and google searches.

14.5 Wildlife Population Growth and Habitat Protection

The ratification of regional and international laws has indeed played a significant role in wildlife population and habitat protection, particularly in Southern Africa. However, the broader global context reveals alarming trends in species population declines and habitat loss. Globally, there has been an average decline of 69% in species populations since 1970–2018, largely driven by factors such as land use changes and climate change (WWF, 2022). In Africa, the Living Planet Index (LPI) shows a slightly lower decline of 66% compared to the global average (WWF, 2022). These statistics underscore the urgent need for coordinated efforts to address these threats, which have led to the potential extinction of at least 1 million plant and animal species and the loss of 10 million hectares of forests annually (WWF, 2022). Below we review population trends for three of the big five (Lion, Elephant and Rhino) which are major source of attraction for tourists.

Southern Africa is home to some of the world's largest remaining populations of elephants and rhinos. Recent aerial surveys in the Kavango Zambezi Transfrontier Park (KAZA-TFCA), covering five countries, have shown stable or increasing elephant populations, reflecting successful wildlife management practices and coexistence efforts (Bussiere & Potgieter, 2023). However, this success is not without challenges, as animal migration routes are increasingly threatened by land use changes, emphasizing the need for continued conservation efforts. Moreover, the data collected from these surveys can also be a source of tension among Southern African states, particularly regarding proposals to sell elephant ivory (CITES, 2022; Pinnock, 2021). These proposals have faced scrutiny and denial from the CITES secretariat, highlighting the complexities and debates surrounding wildlife conservation and international trade (CITES, 2022). The conservation and protection of elephants and rhinos in Southern Africa have been a focal point for both regional and international efforts. Within the framework of CITES and local and regional domestic policies, a range of strategies and action plans have emerged to safeguard these iconic species. Such include:

1. **Southern Africa Regional Elephant Conservation and Management Strategy, 2005**: This strategy aims to coordinate elephant conservation and management efforts across Southern African countries, promoting cooperation and sustainability.
2. **Botswana: National Policy and Strategy for the Conservation and Management of Elephants in Botswana, 2003**: Botswana, with its significant elephant population, has implemented a comprehensive policy and strategy to ensure the conservation and responsible management of elephants within its borders.
3. **Mozambique: Strategy and Action Plan for the Conservation and Management of Elephants in Mozambique 2010–2015**: Mozambique's action plan outlines measures to protect and manage elephant populations within the country.

These strategies and action plans exemplify the commitment of Southern African nations to the conservation of these species. Additionally, they are aligned with international conventions and agreements, such as CITES, which provide a global framework for addressing the threats faced by elephants and rhinos. Botswana's thriving elephant populations have not only contributed to conservation but have also become a significant draw for tourists, particularly in regions like Chobe, where herds of elephants congregate. This has led to the organization of tourism activities by global lead companies and tour operators, attracting visitors from South Africa and Victoria Falls for elephant and other wildlife viewing experiences. While elephants have experienced relatively stable or increasing populations in Southern Africa, the rhino, on the other hand, remains critically endangered (WWF, 2022). Southern Africa serves as a crucial refuge for a substantial rhino population, with a conservatively estimated count of at least 23,290 rhinos actively under conservation efforts (International Rhino Foundation, 2023, p. 6).

In addition to the rhino, the lion is another species facing vulnerability and the risk of extinction in many parts of Africa. Between 1993 and 2014, lion populations across the African continent saw a concerning decline of 43% (AWF, 2020). Lions are particularly threatened in North Africa and severely depleted in Central and

West Africa, with the highest lion populations found in Southern Africa, largely due to dedicated conservation efforts. Remarkably, Southern Africa is the sole region where fenced lion populations have seen a noteworthy increase of 12%, primarily attributed to conservation endeavours undertaken in countries such as South Africa, Botswana, Zimbabwe, and Namibia (AWF, 2020). This underscores the critical role of conservation efforts in preserving these iconic species and highlights the positive impact achievable through dedicated initiatives.

14.6 Tourism Enterprise growth

Hall and Page (1999) assert that tourism has a strong dependency on the environment, with tourists typically gravitating toward destinations boasting superior environmental quality. Moreover, numerous scholars contend that one of the primary draws for tourists to Southern Africa lies in the presence of the 'Big Five' and the presence of protected World Heritage sites (RETOSA, 2008). Rogerson (2009) further highlights how the strength of tourism products rooted in ecotourism has been a catalyst for the tourism industry's growth in Southern Africa. Namibia serves as a compelling example where CBNRM has spurred the establishment of joint ventures between communities and private sector entities. These ventures grant communities rights over wildlife utilization within a conservation framework. Between 1990 and 2022, there were more than 65 joint ventures and 87 conservancies Namibia blending tourism with wildlife conservation (NACSO, 2022). These developments are underpinned by conservation policies.

South Africa has also witnessed the growth of ecotourism enterprises; however, concerns have arisen over their tendency to cater primarily to high-end markets while potentially excluding black operators. The rise of tourism enterprises relying on nature-based experiences has paralleled the increase in tourist arrivals in Southern Africa. Notably, there has been a surge in tourism within protected areas (KC, 2022). For instance, Victoria Falls City has experienced an expansion of tourist-related businesses, capitalizing on its status as a protected World Heritage Site, safeguarded by the UNESCO World Heritage Convention (Zimbabwe Tourism Authority, 2022). Similar growth patterns can be observed in Tanzania, which boasts seven UNESCO World Heritage sites and has witnessed a proliferation of tourist enterprises (Chirikure et al., 2021; Kyara et al., 2021). These tourism enterprises also extend to regional protected area networks, particularly the eighteen (18) Trans frontier Conservation Areas (TFCAs). These TFCAs serve as magnets for tourist visits and foster the development of emerging ecotourism ventures. It is evident that conservation efforts have played a pivotal role in the emergence and growth of tourism in the Southern African region, underscoring the profound connection between environmental preservation and the tourism industry.

14.7 Tourist Arrivals in Southern Africa

Tourist arrivals in Southern Africa have experienced significant growth over the years (see Table 14.2). In 1990, the region welcomed 3 million tourists, a number that surged to 13.2 million by 2005 (Rogerson & Viser, 2006). Notably, Southern Africa's countries are all members of UNWTO, an organization that establishes industry standards and marketing campaigns for tourism businesses. Moreover, more than a third of tourists visiting the region are drawn to nature-based activities, establishing a clear connection between conservation laws and tourist arrivals. For example, in Botswana more than 40% of the land is designated for Wildlife which explains why over 90% of its visitors come for wildlife purposes (Magole & Gojamang, 2015) In Namibia a survey by the Ministry of Tourism showed that over 65% of visitors stated that wildlife was a driver for their choice to visit the country (Ministry of Environment and Tourism, 2013). While in Zimbabwe a visitor survey for 2015–16 showed that about 53.4 of activities engaged by tourists were nature based (Zimbabwe Tourism Authority, 2016) As Bonner (1993, p. 219) has observed "people don't come to Africa to see the cities and the factories and the farms, they want to see the buffaloes and the elephants." In essence, the success of conservation measures is reflected in the increasing number of visitors to the region.

While 2020 witnessed a sharp decline in these numbers due to the COVID-19 outbreak, the 2022 figures show a promising upward trend, indicating that Southern Africa's tourism efforts are making a robust comeback.

Table 14.2 Tourism Arrivals in Southern Africa

Country	1995[a]	2005	2019	2020	2022
South Africa[b,c]	4,676,000	7,368,742	10,228,593	2,802,320[d]	5,698,062
Botswana[e]	636,000	1,684,000	1,555,399	358,225	
Zimbabwe[f]	1,297,000	1,558,501	2,294,259	639,356	1,043,781
Namibia[g]	300,000	777,890	1,681,336	192,026[h]	461,027
Tanzania[i,j]	295,000	613,000	1,527,230	620,867[k]	1,454,920

[a]Ashley, C. and J. Barnes (1996). *Wildlife Use for Economic Gain.—The Potential for Wildlife to Contribute to Development in Namibia*, Research Discussion Paper No. 12, Directorate of Environmental Affairs, Ministry of Environment and Tourism, Government of Namibia, Windhoek.
[b]tourism.gov.za/AboutNDT/Publications/State of Tourism Report (STR) 2019-20.pdf
[c]Statistical release (statssa.gov.za)
[d]The South African tourism sector struggled in 2020 | Statistics South Africa (statssa.gov.za)
[e]Tourism Statistics Annual Report 2019.pdf (statsbots.org.bw)
[f]Tourism Statistics Dashboard (zimbabwetourism.net)
[g]TOURIST-STATISTICAL-REPORT-2019.pdf (visitnamibia.com.na)
[h]Tourist Statistical Report 2021_final.pdf (gov.na)
[i]Ministry of Natural Resources and Tourism (maliasili.go.tz)
[j]International tourism, number of arrivals - Tanzania | Data (worldbank.org)
[k]PRESS_RELEASE_TOURISTS_ARRIVALS_JAN_TO_APRIL_2022.pdf (nbs.go.tz)

14.8 Conclusion

This chapter has delved into the intricate relationship between international laws and wildlife tourism, with a particular focus on their impact in Southern Africa. While our examination is not exhaustive, we have identified the most crucial regional and international legislation that holds relevance for wildlife conservation in this region. Additionally, we propose a framework consisting of four key areas for assessing the legislation's influence on wildlife tourism. Firstly, domestication and ratification, an essential measure lies in the domestication and ratification of these international laws by Southern African countries. This concerted effort has resulted in state coordination and the establishment of governmental structures, contributing to enhanced ecological connectivity. Secondly, through wildlife population growth and habitat protection. The impact on wildlife populations and habitat protection serves as another significant criterion. The ratification of these laws has facilitated the creation of transboundary conservation areas within the Southern African Development Community (SADC), promoting wildlife growth and safeguarding natural habitats. Notably, over half of the SADC region's land is now under conservation, largely due to the establishment of peace parks. Thirdly, through tourism enterprise growth and tourist visits: The legislation's effect on the growth of tourism enterprises and tourist visits is a vital indicator of its success.

Notably, initiatives like the Kavango-Zambezi Transfrontier Conservation Area (KAZA) exemplify the tangible outcomes of these laws, as KAZA, the world's largest transfrontier conservation area, encompasses 36 national parks and 3 UNESCO World Heritage sites, all aimed at promoting tourism. Lastly, we assessed the application of International Agreements: we observed the practical application of international agreements such as CITES in regulating wildlife trade, the World Heritage Convention (WHC) in the establishment of heritage sites, Ramsar in the protection of wetlands, and the utilization of SADC protocols for animal protection. Several challenges exist in the application of these laws that relate to state capacity and compliance.

Finally, this chapter underscores the widespread ratification of identified laws by Southern African countries and their consequential positive effects. This includes the creation of transboundary conservation areas, which have significantly advanced conservation efforts and provided fertile ground for tourism growth. The successful operation of international agreements further reinforces the substantial impact of these legal frameworks on wildlife conservation and tourism in Southern Africa.

References

Abensperg-Traun, M. (2009). CITES, sustainable use of wild species and incentive-driven conservation in developing countries, with an emphasis on southern Africa. *Biological Conservation, 142*(5), 948–963.

African Union. (2003). *African convention on the conservation of nature and natural resources*. Retrieved from https://au.int/sites/default/files/treaties/41550-treaty-Charter_ConservationNature_NaturalResources.pdf

AWF. (2020). *Recovering Africa's lost lion populations*. AWF. Retrieved August 30, 2023, from https://www.awf.org/blog/recovering-africas-lost-lion-populations

Baldwin, E. (2011). Twenty-five years under the convention on migratory species: Migration conservation lessons from Europe. *Environmental Law, 41*, 535.

Bedjaoui, M. (1991). *International law: Achievements and prospects*. UNESCO.

BirdLife South Africa. (2019). *Bird life South Africa annual report 2019*. Bird Life South Africa. Retrieved August 29, 2023, from https://www.birdlife.org.za/wp-content/uploads/2020/06/BirdLife_South_Africa_Annual_Report_2019.pdf

Bonner, R. (1993). *At the hand of man: Perils and hope for Africa's wildlife*. New Vintage Books.

Bush, G., Hanley, N., & Colombo, S. (2008). *Measuring the demand for nature-based tourism in Africa: A choice experiment using the "cut-off" approach* (Stirling Economics Discussion Paper 2008–06). Retrieved from https://core.ac.uk/download/pdf/9049784.pdf

Bussiere, E., & Potgieter, D. (2023). *KAZA elephant survey 2022 Volume II: Stratum reports*. KAZA TFCA Secretariat.

Caddell, R. (2005). International law and the protection of migratory wildlife: An appraisal of twenty-five years of the Bonn Convention. *Colo. J. Int'l Envtl. L. & Pol'y, 16*, 113.

Chandra, A., & Idrisova, A. (2011). Convention on biological diversity: A review of national challenges and opportunities for implementation. *Biodiversity and Conservation, 29*, 3295–3316.

Child, G. (2004). Growth of modern nature conservation in Southern Africa. In B. Child (Ed.), *Parks in Transition* (p. 21). Routledge.

Chirikure, S., Ndoro, W., & Bugarin, F. (2021). Usable pasts forum: UNESCO and heritage tourism in Africa. *African Archaeological Review*, 513–533. https://doi.org/10.1007/s10437-021-09454-6

Christie, I., Fernandes, E., Messerli, H., & Twining-Ward, L. (2013). *Tourism in Africa: Harnessing tourism for growth and improved livelihoods*. World Bank. Retrieved from http://www.worldbank.org/afr/tourism

Cirelli, M., & Morgera, E. (2010). *Wildlife law in the southern African development community* (FAO Legal Paper Online 84). www.fao.org/legal/prs-ol

CITES. (2022). *Current rules on commercial international trade in elephant ivory under CITES and proposals to CITES CoP17*. Convention on International Trade in Endangered Species of Wild Fauna and Flora. Retrieved from https://cites.org/eng/news/Current_rules_commercial_international_trade_elephant_ivory_under_CITES_Proposals_CITES_CoP17_200716

Dhliwayo, M., Breen, C., & Nyambe, N. (2009). Legal, policy, and institutional provisions for community participation and empowerment in transfrontier conservation in Southern Africa. *Journal of International Wildlife Law & Policy, 12*(1–2), 60–107. https://doi.org/10.1080/13880290902938153

Duffy, R. (2006). The potential and pitfalls of global environmental governance: The politics of transfrontier conservation areas in Southern Africa. *Political Geography, 25*(1), 89–112. https://doi.org/10.1016/j.polgeo.2005.08.001

Goodwin, H. & Leader-Williams, N. (2000). Tourism and protected areas – distorting conservation towards charismatic megafauna? In A. Entwhistle, & N. Dunstone (Eds.), *Priorities for the conservation of mammalian diversity: Has the panda had its day?* (pp. 257–275). Cambridge.

Grünewald, C., Schleuning, M., & Böhning-Gaese, K. (2016). Biodiversity, scenery and infrastructure: Factors driving wildlife tourism in an African Savannah National Park. *Biological Conservation, 201*, 60–68. https://doi.org/10.1016/j.biocon.2016.05.036

Hall, C., & Page, S. (1999). *The geography and tourism and recreation; environment, place and space*. Routledge.

Henkin, L. (1968). *How nations behave: Law and foreign policy*. Columbia University Press.

Heppes, J. B., & McFadden, E. J. (1987). The convention on international trade in endangered species of wild fauna and flora: Improving the prospects for preserving our biological heritage. *BU Int'l LJ, 5*, 229.

Higginbottom, K. (2004). Wildlife tourism: An introduction. *Wildlife tourism: Impacts, management and planning*, 1–14.

Holechek, J., & Valdez, R. (2018). Wildlife conservation on the rangelands of eastern and southern Africa: Past, present, and future. *Rangeland Ecology & Management, 71*(2), 245–258. https://doi.org/10.1016/j.rama.2017.10.005

International Rhino Foundation. (2023). *State of the rhino report 2023*. Retrieved from https://rhinos.org/wp-content/uploads/2023/09/State-of-the-Rhino-2023_Final_Updated-9212023-.pdf

Kalvelage, L., Revilla Diez, J., & Bollig, M. (2022). How much remains? Local value capture from tourism in Zambezi, Namibia. *Tourism Geographies, 24*(4–5), 759–780. https://doi.org/10.1080/14616688.2020.1786154

KC, B. (2022). Complexity in balancing conservation and tourism in protected areas: Contemporary issues and beyond. *Tourism and Hospitality Research, 22*(2), 241–246. https://doi.org/10.1177/14673584211015807

Koh, H. (1997). Why do nations observe international law. *Yale Law Journal, 106*(8), 2599–2660.

Kyara, V., Rahman, M., & Khanam, R. (2021). Tourism expansion and economic growth in Tanzania: A causality analysis. *Heliyon, 7*(5). https://doi.org/10.1016/j.heliyon.2021.e06966

Lindsey, P., Frank, L., & Alexander, R. (2007). Trophy hunting and conservation in Africa: Problems and one potential solution. *Conservation Biology: The Journal of the Society for Conservation Biology, 21*(3), 880–883. Retrieved from https://doi.org/10.1111/j.1523-1739.2006.00594.x.

Lubbe, W. (2008). Straddling borders and legal regimes: The case for cooperative transfrontier biodiversity conservation in the southern African development community. *Yearbook of International Environmental Law, 18*(1), 126.

Lyster, S. (1989). The convention on the conservation of migratory species of wild animals (The Bonn convention). *Natural Resources Research, 29*, 979.

Magole, L., & Gojamang, O. (2015). The dynamics of tourist visitation to national parks and game reserves in Botswana. *Botswana Notes and Recors, 37*, 80–96.

Marker, L., & Boast, L. (2015). Human–wildlife conflict 10 years later: Lessons learned and their application to cheetah conservation. *Human Dimensions of Wildlife, 20*(4), 302–309. https://doi.org/10.1080/10871209.2015.1004144

Mbasera, M., Du Plessis, E., Saayman, M., & Kruger, M. (2016). Green management in hotels: A supply-side analysis. *Journal of Environmental Management & Tourism, 7*(2), 205–215.

Ministry of Environment and Tourism. (2013). *Report on the Namibian tourist exit survey*. Ministry of Tourism and Environment.

Miller, R. M., Rodríguez, J. P., ANISKOWICZ-FOWLER, T. H. E. R. E. S. A., Bambaradeniya, C., Boles, R., Eaton, M. A., ... & Pollock, C. (2007). National threatened species listing based on IUCN criteria and regional guidelines: Current status and future perspectives. *Conservation biology, 21*(3), 684–696.

Mochankana, L., Garekae, H., Bapadile, J., & Mbaiwa, J. (2023). Tourism commodification of traditional basket weaving in the Okavango Delta, Botswana. *African Geographical Review*, 1-16.

NACSO. (2022). *Database for joint ventures in Namibia*. Namibian Association of CBNRM support organisations.

Newmark, W. (2008). Isolation of African protected areas. *Frontiers in Ecology and the Environment, 6*(6), 321–328. https://doi.org/10.1890/070003

Ngeta, K., & Child, B. (2014). Wildlife governance in Africa. In G. Barnes & B. Child (Eds.), *Adaptive cross-scalar governance of wildlife*. Routledge.

Ntiamoa-Baidu, Y. (1995). Conservation education in threatened species management in Africa. *Bird Conservation International, 5*(4), 455–462. https://doi.org/10.1017/S0959270900001179
Peace Parks Foundation. (2022). *2021 annual review*. Peace Parks Foundation. Retrieved August 28, 2023, from https://www.peaceparks.org/wp-content/uploads/2022/08/Peace-Parks-Foundation-Annual-Review-2021.pdf
Pinnock, D. (2021, November 17). South Africa: Environmentalists and hunters slam government over proposals for the trophy hunting of elephants, leopards and rhinos. *Journal of African Elephants*. Retrieved from https://www.africanelephantjournal.com/south-africa-environmentalists-and-hunters-slam-government-over-proposals-for-the-trophy-hunting-of-elephants-leopards-and-rhinos/
Purdon, A., Mole, M., & Chase, M. (2018). Partial migration in savanna elephant populations distributed across southern Africa. *Scientific Reports, 8*. Retrieved from. https://doi.org/10.1038/s41598-018-29724-9
Raustiala, K., & Victor, D. (1996). Biodiversity since rio: The future of the convention on biological diversity. *Environment: Science and Policy for Sustainable Development, 38*(4), 16–45. https://doi.org/10.1080/00139157.1996.9929252
Reid, H. (2016). Ecosystem-and community-based adaptation: Learning from community-based natural resource management. *Climate and Development, 8*(1), 4–9.
RETOSA. (2008). *The ultimate travel guide to southern Africa*. Retrieved from www.retosa.co.za
Rogerson, C. (2009). Tourism development in southern Africa: Patterns, issues and constraints. In J. Saarinen (Ed.), *Sustainable tourism in southern Africa: Local communities and natural resources in transition*. Channelview.
Rogerson, C. M., & Visser, G. (2006). International tourist flows and urban tourism in South Africa. *Urban Forum, 17*, 199–213. https://doi.org/10.1007/s12132-006-0005-8
Rogerson, C. M., & Rogerson, J. M. (2019). Tourism, local economic development and inclusion: Evidence from overstrand local municipality, South Africa. *Geo Journal of Tourism and Geosites, 25*(2), 293–308.
Saarinen, J., Fitchett, J., & Hoogendoorn, G. (2022). *Climate change and tourism in southern Africa*. Routledge.
Saayman, M., & Saayman, A. (2010). Regional development and national parks in South Africa: Lessons learned. *Tourism Economics, 16*(4), 1037–1064. https://doi.org/10.5367/te.2010.0007
SADC. (2003). *Protocol on wildlife conservation and law enforcement*. SADC. Retrieved from www.sadc.int/sites/default/files/2021-08/Wildlife_Conservation.pdf
Schroeder, R., St. Martin, K., Wilson, B. & Sen, D. (2008). Third world environmental justice, society & natural resources, *21*(7), 547–555. https://doi.org/10.1080/08941920802100721
Smit, E. (2019). *Namibia submits CITES proposals*. Retrieved from Save the Elephants: https://www.savetheelephants.org/news/namibia-submits-cites-proposals/
Swiderska, K. (2002). *Implementing the rio conventions: implications for the south*. International Institute for Environment and Development. Retrieved from http://www.jstor.com/stable/resrep16617
Tanner, R. J. (2004). Transfrontier conservation areas of Southern Africa and International Law in the context of Community involvement.
UNWTO. (2020). *World tourism barometer no 18*. UNWTO. Retrieved from https://www.unwto.org/world-tourism-barometer-n18-january-2020
UNWTO. (2022, October 7). *African tourism united to transform sector for growth and opportunity*. Retrieved from www.unwto.org: https://www.unwto.org/news/african-tourism-united-to-transform-sector-for-growth-and-opportunity
UNWTO. (2023). *Tourism recovery accelerates to reach 65% of pre-pandemic levels*. UNWTO. Retrieved from https://www.unwto.org/news/tourism-recovery-accelerates-to-reach-65-of-pre-pandemic-levels
Van der Merwe, P., Saayman, A., & Jacobs, C. (2021). Assessing the economic impact of COVID-19 on the private wildlife industry of South Africa. *Global Ecology and Conservation, 28*, e01633. https://doi.org/10.1016/j.gecco.2021.e01633

Van Hoven, W. (2015). Private game reserves in southern Africa. In R. Van der Duim, M. Lamers, & J. van Wijk (Eds.), *Institutional arrangements for conservation, development and tourism in eastern and southern Africa* (pp. 101–118). Springer. https://doi.org/10.1007/978-94-017-9529-6_6

Vinuales, J. (2015). *The rio declaration on environment and development*. Oxford University Press.

World Travel and Tourism Council. (2004). *World travel and tourism: Forging ahead*. World Travel and Tourism Council.

WWF. (2022). *The living planet report: Building a nature-positive society*. Retrieved from. https://wwflpr.awsassets.panda.org/downloads/lpr_2022_full_report.pdf

Zimbabwe Tourism Authority. (2016). *Zimbabwe visitor exit survey*. Zimbabwe Tourism Authority.

Zimbabwe Tourism Authority. (2022). *Zimbabwe destination development program: A market analysis report: Markets and segments of focus and optimal marking budget allocation*. IFC and ZTA.

Chapter 15
Transboundary Conservation Areas and Tourism

Moren T. Stone

15.1 Introduction

To advance the cause of conservation in the Southern and parts of East Africa, 18 terrestrial transboundary natural resources management (TBNRM) have been established, covering at least 750,000 km^2, an area comparable in size to the surface areas of Germany, Italy and Portugal combined (Cumming, 2011; Retief et al., 2023; Thomson et al., 2013). The foundation for transboundary conservation is broad-ranging and includes the primary purpose of "the conservation of biological diversity, the sustainable use of its components and the fair and equitable sharing of the benefits arising out of the utilization of genetic resources" (Braack et al., 2006, p. 1). In the SADC context, TBNRM are defined as, "… areas or components of large ecological regions that straddle the boundaries of two or more countries, encompassing one or more protected areas, as well as multiple resource use areas" (Ron, 2007, p. 4). Transboundary conservation includes both large-scale ecosystem-wide approaches, and much smaller-scale focused interventions that recognise the work in protected areas and in areas within the wider production landscape (Braack et al., 2006). The formation of TFCAs involves merging formally isolated protected areas (mostly national parks), freehold, state and wildlife management areas and communal land across international political boundaries to be managed as a bioregion (Sandwith et al., 2001; Ramutsindela, 2009).

The term Transboundary Conservation is an inclusive concept. Literature indicates that there are several forms which include transboundary natural resource management area (TBNRM); transboundary conservation and development area (TBCA); transfrontier conservation area (TFCA); transboundary protected area

M. T. Stone (✉)
Department of Environmental Science, University of Botswana, Gaborone, Botswana
e-mail: stonemt@ub.ac.bw

© The Author(s), under exclusive license to Springer Nature Switzerland AG 2024

L. S. Stone, M. T. Stone (eds.), *Wildlife Tourism Dynamics in Southern Africa*, Geographies of Tourism and Global Change,
https://doi.org/10.1007/978-3-031-57252-4_15

(TBPA); transfrontier park (TFP) and parks for peace (Cumming, 2011; Thomson et al., 2013; Braack et al., 2006; Griffin et al., 1999; SADC protocol on wildlife conservation and law enforcement, 1999; Sandwith et al., 2001). Table 15.1 summarizes different forms of transboundary conservation regimes.

TFCAs are widely distributed in southern and eastern Africa. The geographical distribution of the TFCAs is influenced by the common knowledge that eastern and southern Africa have the widest diversity and number of large ungulates and associated predators on earth (Thomson et al., 2013). These wildlife populations (in terms of numbers and diversity) are not only a priceless global asset, they are also vital to international tourism (Thomson et al., 2013), hence, the development of TFCAs is not isolated from wildlife-based tourism development. The development of TFCAs is premised on the notion that natural resources transcend borders, therefore, their

Table 15.1 Types of transboundary conservation regimes

Type of transboundary conservation regime	Description
Transboundary natural resource management area (TBNRM area)	These are areas in which co-operation to manage natural resources occurs across boundaries (Griffin et al., 1999). It should be noted that this concept does not necessarily involve any formal protected areas and the emphasis falls mainly on the processes by which natural resource management can be sustained in the region straddling an international boundary (Braack et al., 2006)
Transboundary conservation and development area (TBCA)	These are areas, which may or may not include protected areas, which straddle one or more borders between states (Braack et al., 2006)
Transfrontier conservation area (TFCA)	These are essentially the same concept as TBCAs, reflecting the southern African regional usage of the term "transfrontier" (Braack et al., 2006). The Southern African Development Community (SADC) defined these as "an area or component of a large ecological region that straddles the boundaries of two or more countries, encompassing one or more protected areas, as well as multiple resource use areas" (SADC protocol on wildlife conservation and law enforcement, 1999, p. 13)
Transboundary protected area (TBPA)	An area of land and/or sea that straddles one or more boundaries between states, sub-national units such as provinces and regions, autonomous areas and/or areas beyond the limits of national sovereignty or jurisdiction, whose constituent parts are especially dedicated to the protection and maintenance of biological diversity, and of natural and associated cultural resources, and managed through legal or other effective means (Sandwith et al., 2001)
Transfrontier park (TFP)	Essentially the equivalent of a TBPA, implying that all or most of the areas comprising the jointly-managed Transfrontier Park are high-status formal protected areas, usually of National Park status, with perhaps one or more smaller areas serving as linking corridors (Braack et al., 2006)
Parks for peace	Parks for Peace are transboundary protected areas that are formally dedicated to the protection and maintenance of biological diversity, and of natural and associated cultural resources, and to the promotion of peace and cooperation (Sandwith et al., 2001)

management needs cross border joint management (Mogende, 2016). TFCAs are seen as the latest evolution of a more holistic approach to transnational environmental management which integrates social, economic and ecosystem sustainability (Spenceley, 2008). To mention but a few, the enactment of TFCAs development in Southern Africa is attributed to the following: the management of protected areas and the overpopulation of elephants (Ramutsindela, 2004); to promote conservation, cooperation and developmental goals in the region (Zunckel, 2014); expectations of the tourism sector to develop as a reason for TFCA expansion and to invigorate some of the otherwise isolated, desolate or barren areas and make them highly regarded tourist destinations, instead of a region for wildlife habitat (Duffy, 2000); availability of funds as International Union for Conservation of Nature (IUCN) and Conservation International (CI) channel funds and provide technical assistance in TFCA projects (Thomson et al., 2013).

This chapter examines the broader concept of TFCAs by exploring what TFCAs are, policy framework that facilitated their development the role of different stakeholders in their implementation, the benefits they yield to stakeholders, the potential of tourism development as a transboundary resources and implementation challenges. By so doing, the chapter reviews, integrates and synthesises the growing body of literature and empirical evidence on the development and performance of transfrontier conservation areas (TFCAs).

15.2 Methods and Materials

This chapter used available secondary data sources. Secondary data were mainly from the Google Scholar internet search engine. Peer reviewed journal articles and SADC reports and policies were used to inform this chapter. This chapter aims to review, integrate and synthesises the growing body of literature and empirical evidence on the development and performance of transfrontier conservation areas (TFCAs). Materials gathered focuses on the benefits, challenges and tourism potential of TFCAs as transboundary resources.

15.2.1 The Genesis of Transfrontier Conservation Areas (TFCAs) in Southern Africa

The development and evolution of transboundary parks is traced back to the beginning of the twentieth century, with the first peace park created between Canada and the USA in 1926 known as Glacier-Waterton International Peace Park (Mayoral-Phillips, 2002; Quinn et al., 2012). By 1969, approximately 121 TFCAs were established around the world (Mogende, 2016). In Southern Africa, the first TFCA to be established was Kgalagadi Transfrontier Park (KTP), created between South Africa and Botswana (Hall-Martin & Modise, 2002; Ramutsindela, 2004). To date, there are 18 TFCAs identified in the SADC region these differ in terms of their level of development (Mogende, 2016).

These established TFCAs are Ai/Ais-Richtersveld TP (Namibia and South Africa); Kgalagadi Transfrontier Park (Botswana and South Africa); Kavango Zambezi TFCA (Angola, Botswana, Namibia, Zambia and Zimbabwe); Great Limpopo TFCA (Mozambique, South Africa and Zimbabwe); Malawi-Zambia TFCA (Malawi and Zambia); Lubombo TFCRA (Mozambique, South Africa and Eswatini); Maloti-Drakensberg TFCDA (Lesotho and South Africa) and Iona-Skeleton Coast TFCA (Angola and Namibia). Emerging TFCAs include the Greater Mapungubwe TFCA (Botswana, South Africa and Zimbabwe); Chimanimani TFCA (Mozambique and Zimbabwe); Mayombe Forest TPA (Angola and DRC) and Niassa-Seulos TFCA (Mozambique and Tanzania). Conceptual TFCAs include Liuwa Plains-Mussuma TFCA (Angola and Zambia); Lower Zambezi-Mana Pools TFCA (Zambia and Zimbabwe); Zimoza TFCA (Mozambique, Zambia and Zimbabwe); Kagera TFCA (Rwanda, Tanzania and Uganda); Mnazi Bay-Quirimbas TFCMA (Mozambique and Tanzania) and Western Indian Ocean TMP (Comoros, France, Madagascar, Mauritius, Mozambique Seychelles and Tanzania). These transfrontier conservation initiatives are created as 'peace-parks,' envisaged to usher in a new conservation paradigm and policies premised on supporting biodiversity, sustainable economic development and regional peace and stability (Andersson et al., 2013; Bhatasara et al., 2013).

One of the arguments in favour of transfrontier conservation and sustainable tourism development is that they have a great capacity for providing employment opportunities for local communities, especially poor people in the borderlands of low and middle-income countries (Peace Parks Foundation, 2006; UNWTO, 2014). The potential of transfrontier conservation to boost tourism, protect biodiversity and uplift communities led to them booming in southern Africa (Chiutsi & Saarinen, 2019). The potential of transfrontier parks to support sustainable tourism development is widely supported by the World-Wide Fund for Nature (WWF), which underscores that transfrontier conservation areas can strategically develop sustainable tourism to sustain the costs of conservation management, while also providing employment and entrepreneurial opportunities for people in developing countries (Visser, 2011). The creation of TFCAs manifests a new kind of conservation thinking that is based on contemporary modes of neoliberal governance and political economy, often serving both non-local and local needs in development (Büscher, 2013). Proponents of TFCAs in Southern Africa do not want them to be seen as supporting the continuation of the fortress model of conservation which thrived earlier on the impoverishment and marginalisation of local communities (Murphree, 2009).

15.2.2 Policy Frameworks that Support and Facilitate the Development of Transfrontier Conservation Areas (TFCAs) in Southern Africa

The development of TFCAs in Southern Africa is premised to develop SADC into a functional and integrated network of TFCAs where shared natural resources are sustainably co-managed and conserved to foster socioeconomic development and regional integration for the benefit of people living within and around TFCAs in the SADC region and the world (Secretariat SADC, 2017). To attain this mission, SADC devised policy instruments to facilitate the realisation of TFCAs mandate. The development of several policies and legal frameworks to provide an enabling environment for the establishment and development of TFCAs in the SADC region include the SADC Protocol on Wildlife Conservation and Law Enforcement (1999), the SADC Protocol on Forestry (2002), the SADC Protocol on Shared Water Courses (2002) and the SADC Regional Biodiversity Strategy (2006) (Secretariat SADC, 2017). The SADC Regional Indicative Strategic Development Plan (RISDP) has identified sustainable development, conservation of wildlife and transboundary natural resources as a priority for SADC (Ramsamy, 2004). The SADC Secretariat is therefore mandated by its SADC Member States to facilitate and support SADC's efforts in the establishment and development of TFCAs.

To implement the Protocol on Wildlife Conservation and Law Enforcement in view of TFCAs, SADC established the "SADC Programme for Transfrontier Conservation Areas" in 2013 (Secretariat SADC, 2017, p. 2). The SADC Programme for TFCAs highlights seven action areas for reaching its overarching vision: policy harmonisation and advocacy, sustainable financing, Capacity building, data and knowledge management, local livelihoods, climate change vulnerability, and TFCAs as marketable tourism products. It is these seven action areas that guide SADC in its execution and implementation of TFCAs.

15.2.3 Contributions of Transboundary Natural Resources Management (TBNRM) to Conservation, Peace, Security, and Tourism Development

Since their establishments, TFCAs and peace parks have contributed to peace and cooperation between countries and nations in many ways. The following are some of their contributions: promoting discussions among neighbouring countries that have environmental, social, and economic interests in the zones over political boundaries, providing control and security over resources in the border land to benefit the real owners, generating common opportunities for sustainable development and ecotourism, and creating a sustainable net of relationships between protected areas' managers of involved countries, governments and local actors such as international NGOs (Hammill & Besancon, 2007). Due to the difficulties faced by southern African countries in managing resources, there has been a new development of militarising natural resources due to the escalations of wildlife poaching; diverting resources that could be developing the region to guard against wildlife resources causes a concern. TBNRM is envisaged to provide a non-military model for addressing conflicts and promoting stability in the region (Katerere et al., 2001). Moyo and Tevera (2000) argue that transboundary natural resource management is an option with potential to broaden benefits from natural resources management and facilitate a culture of non-military approaches to natural-resource-based conflicts.

Human-induced factors such as natural resources degradation, pollution, and resources overexploitation contribute to human insecurity in the region and furthermore inter-state conflicts and competition over control of and access to natural resources such as water are likely to escalate as the region's population increases (Katerere et al., 2001). The introduction of the transboundary resources management on shared resources between states is seen as providing a potential remedy in the co-management of shared resources.

Since transboundary protected areas (TBPAs) straddle the borders of at least two countries, they also became known as "Peace Parks" to encourage cooperation among states characterised by historical and/or current conflict (Trogisch & Fletcher, 2022). Therefore, 'peace parks', offer the potential to harness 'peace tourism' as a market-based, business-oriented solution to these challenges (D'Amore, 2009).

TBNRM brought cooperation between states that share the same resources. Cooperation at a regional level is important for the effective implementation of global and regional international commitments (Selier et al., 2016). To exemplify the essence of cooperation, countries sharing the same resources implement uniform penalties that assist in preventing the bypassing of international treaties like CITES rules, preventing situations that could result in invariant practices whereby certain countries have choices that can easily conflict with international treaties. This is further envisaged in the SADC resources shared protocol that provides a collective conservation framework for the protection and sustainable use of wildlife populations that extend to and fulfil lifecycles between two or more countries (Selier et al., 2016). The SADC Protocol represents one of the most advanced efforts towards regional harmonisation of wildlife legislation that is being experimented on around the world (Di Minin et al., 2013). Diversified natural resources present a significant basis for socioeconomic activities in the region, in particular, tourism based on wildlife and other resources which is considered to be an industry with high growth potential, especially in areas that have marginal value for agriculture (Katerere et al., 2001).

15.2.4 Transboundary Natural Resources Management (TBNRM) and the Role of Different Stakeholders

Transnational conservation management programmes are complex initiatives with a diverse set of stakeholders, which facilitates and/or hinders their implementation in practice (Duffy, 2006). Hamm (2017) reiterates that modern natural resource management places increasing emphasis on advancing voluntary sources of cooperation, compliance and trust, which often are difficult to get from all stakeholders due to the uncertainty of human behaviour. Whilst community participation is widely recognised in relation to the transfrontier management of natural resources, many local inhabitants neither comprehend nor trust the participatory process (Briedenhann & Wickens, 2004). The indulgence of many stakeholders ranging from communities, NGOs, donors, governments, and their agencies compound the implementation of transboundary conservation areas as stakeholders have different interests and expectations.

According to Briedenhann and Wickens (2004), for conservation and tourism to sustain itself in transboundary conservation areas, all stakeholders must be willing partners. The core elements and ideas of inclusivity and sustainability often get lost in the process if the understanding level of all stakeholders is not reached (Utting, 2015). Conditions that may promote conflict in communities include major differences in power and status, which could be reflected in different access to resources, sometimes leading to exploitation (getting more than others in a common activity), accumulation (avoiding the sharing of surpluses) and the possible splitting or breakdown of communities (Chiutsi & Saarinen, 2019; IUCN, 2004). Undesirably, right from the onset, there have been concerns about the implementation of transfrontier conservation areas in southern Africa, this has largely been characterised by a cursory reference to communities with planning essentially remaining technocratic and top-down (Muboko & Murindagomo, 2014). There is need for sustained political support for TBNRM and a policy framework that supports transboundary cooperation (Chiutsi & Saarinen, 2019).

15.3 Challenges Faced by SADC TFCAs

Several challenges and issues have emerged as Southern Africa endeavours to implement its TFCAs. Among the challenges are stakeholders' engagement, balancing livestock production and wildlife conservation in and around TFCAs, conservation and peace tourism contradictory outcomes. These challenges are discussed in the section that follows.

15.3.1 Stakeholders' Engagement and Participation

The implementation of the TFCAs reveals that the provision for stakeholder engagement and participation is weak. Some indications are that very limited provision is made for participation and/or stakeholder engagement within the institutional structures as described in the TFCAs agreements (Retief et al., 2023). The weaknesses of highly centralized decision-making structure of TFCAs have been highlighted with a call for the voices of local stakeholders, as the de facto resources managers to be more prominent (Cumming et al., 2013). There is also very limited information sharing among TFCA stakeholders and conflict is only dealt with at a very late stage, mostly driven by emergencies or crises such as outbreak of fires or incidences of poaching rare species (Malan, 2015). Retief et al. (2023) argue that a need has been identified in practice to allow individuals affected by the operational rules of the TFCA to participate in modifying these rules, but the opaqueness of the current decision-making system described in the agreements does not allow for such participation.

Moreover, there is no clear methodology on which broader stakeholders, more-especially local communities, are to be meaningfully involved. In view of the debate on stakeholders' engagement and participation, Katerere et al. (2001) caution that while trying to take advantage of the opportunities presented by TBNRM, care must be taken not to inadvertently marginalise communities in pursuit of the mega-scale of TBNRM, as the real danger posed by the TBNRM approach is to further isolate the poor by pooling common transboundary resources while de-facto privatising associated benefits.

15.3.2 Balancing Livestock Production and Wildlife Conservation in and Around Transfrontier Conservation Areas (TFCAs)

The development of TFCAs takes place in areas that host livestock and crop production at subsistence and commercial levels. The balancing of livestock production and wildlife conservation in and around Southern Africa's TFCAs has to some extent proven to be problematic. Biodiversity conservation, which the TFCAs movement is an integral part of, and more effective livestock production/trade are pivotal to future rural development in southern Africa. Hence, it is crucial to mitigate the obstacles that have impede the progress towards the coexistence of these two sectors (Thomson et al., 2013). Livestock husbandry has traditionally been considered the most important element of rural development in southern Africa and consistent with this view, indications are that on-going efforts, dating from colonial times, aimed at the commercialisation of livestock production and securing access to high-value European markets, particularly for beef (Thomson et al., 2013) have been at the forefront. Therefore, many more people in the SADC region, perhaps exceeding 100 million, are partially dependent on livestock (Thornton et al., 2002). Consequently, the emergent constraint to the successful integration of wildlife conservation and livestock production systems in southern Africa within the TFCAs landscape concerns the abundance and range of infectious agents that co-evolved with wildlife (Bengis et al., 2004).

While innovative ways of managing some of these diseases have been developed through vaccine development in the mid-twentieth century, effective control of most of them has historically focused on the separation of wildlife and livestock to prevent the transmission of infections (Thomson et al., 2013) through veterinary cordon fences (VCFs) (see Chap. 3 on the impacts of VCFs on wildlife resources and tourism). The need to separate livestock from wildlife to create zones free from diseases that constrained livestock production and market access such as foot and mouth disease (FMD) resulted in extensive fencing systems, particularly in Botswana, Zimbabwe, Namibia and South Africa (Perkins, 2020). The VCFs pose a threat to the development and success of TFCAs as they act as barriers to the movement of wildlife, closing wildlife movement corridors. As a result, southern Africa faces an on-going loss of biodiversity as well as limited access to high-value markets for livestock products with little prospects of ever being able to compete effectively in those markets (Thomson et al., 2013). Conversely, the impact of the prevalence VCFs and diseases affect the development of wildlife-based tourism. Thomson et al. (2013) argue that livestock owners in these locations need to be able to generate a reasonable return on their most valuable personal resource in order to reduce unacceptable levels of poverty that cannot be addressed by tourism alone.

15.3.3 Conservation and Peace Tourism Contradictory Outcomes

In bringing together current debates over 'peace tourism' with critical conservation studies, Trogisch and Fletcher (2022) offer a pioneering analysis on the role of tourism in peace-building efforts within TBPAs in conflict-affected regions of Greater Virunga Transboundary Collaboration (GVTC), a TBPA composed of Virunga National Park in the DRC, Volcanoes National Park in Rwanda and Mgahinga Gorilla National Park in Uganda. Due to its history of intra- and interstate warfare, including the Rwandan genocide and its aftermath characterized by persistent violent conflict in the DRC, the Dutch government financed the intergovernmental GVTC to institutionalise transboundary conservation within Virunga Conservation Area (VCA) on the model of a 'Peace Park' (Prunier, 2011). The VCA is mainly centred on the conservation of Gorrila to promote gorilla tourism as a peace-building strategy founded on socio-economic and political cooperation among the three countries. The assessment of the performance of VCA in view of conservation and peace tourism is that there is a prevalence of various kinds of violence, from reported killings of mountain gorillas, tourists and park rangers (Jenkins, 2008; Maekawa et al., 2013; Marijnen & Verweijen, 2016). Other forms of violence range from social exclusion to imprisonment and the murder of civilians living adjacent to the TBPA (Trogisch & Fletcher, 2022).

Trogisch and Fletcher (2022) further indicate that the transboundary collaboration efforts centred on gorilla tourism development have paradoxically spurred increased economic competition among the three states that has in turn precipitated political distancing as well as militarisation within the three National Parks constituting the VCA. The consequential segregation among the three states weakens collaboration efforts and trust among national conservation authorities and tourism actors and in so doing contradicts the promoted 'peace tourism' narrative in the TBPA treaty.

In the context of Botswana, Mayoral-Phillips (2002) argues that during the planning and implementation phases of the Kgalagadi TFCA, no reference was given to the Bushmen community by the Botswana government, and no economic management plan that included tourism development made reference to community involvement. To some extent, TBRM marginalizes communities further; limits transboundary rural tourism opportunities to uplift them; and compromises the much needed product diversification and future complementary strategies for rural development (Manyane, 2017). For Lane (2005, p. 13) TFCA sustainability 'cannot be successfully based on a narrow pro-nature conservation ethic, it should help protect the culture and character of communities, habitats, the rural economy, the tourism industry, and even develop an understanding among decision-makers/stakeholders so they may work towards a balanced and diversified rural economy.' On a different note, Ntuli et al., 2019 posit that transboundary wildlife resources, such as elephants, require huge tracts of land for forage and migration in response to seasonal variations in rainfall, however, the establishment of TFCAs has increased conflicts between people and wildlife, as communities' ancestral land, on which they reside, has become corridors facilitating the movement of wildlife between different parks.

15.3.4 Implications of Transboundary Natural Resources Management (TBNRM) on Tourism Development

The challenges faced or created by transboundary conservation area have implications on tourism development. Conservation benefits such as habitat expansion and connectivity seem clear and straight forward, providing an opportunity for tourism development modelled on the expanse provided by the TBNRM areas (Metcalfe, 2003). Therefore, the development of transboundary conservation areas as tourism destinations taps into the world's fastest growing global industry (Ashley, 1995). Potential benefits are envisioned to include an increase in employment opportunities; stimulation of rural economic development through outsourcing of services to local communities; collection of firewood, medicinal plants and cutting of thatch grass; and the use of agricultural land for conservation will be more beneficial from a financial and employment perspective (Metcalfe, 2003). Due to the implementation of TBNRM, cross-border investments in tourism have become an attractive window for foreign investors (Katerere et al., 2001). While this is an opportunity for tourism development within the TBNRM landscape, this trend may eventually increase the foreign control of land and tourism resources in the region with limited local participation and benefits (Ntuli et al., 2019).

Local rural communities lack business acumen and finances to start tourism businesses promoted within the TBNRM designated areas. Katerere et al. (2001) argue that to facilitate investment, many private sector interventions are disguised as "partnerships" with communities and many of these agreements are inappropriately labelled "joint ventures" and "partnerships" when they are nothing more than lease agreements to gain access to valuable tourism resources. Communities become desperate and enter into less than perfect "partnership" arrangements with the private sector in order to generate some revenue (Chiutsi & Saarinen, 2019). Mudenda (2000) concluded that the anticipation by communities on tourism beneficiation has led to a proliferation of joint venture partnerships, but many partnership arrangements that are hailed as success stories are not what the communities want, but more of a private sector agenda. Nevertheless, there are indications that transboundary tourism cooperation has strengthened international relations, facilitated the development of tolerant relations between people, search for understanding between nations, and have encouraged the overcoming of xenophobia among other challenges (Kornus et al., 2017).

Ntuli et al. (2019) identify several benefits realised by different stakeholders, including: communities benefitting from income from trophy hunting and photographic tourism (i.e., consumptive and non-consumptive tourism), governments getting revenue generated from issuing tourism operational licenses, tourism business taxes, and the tourism business industry opening tourism businesses that generate income and directly and indirectly hiring employees. In the process the tourism value chain is created regionally and reinforces regional tourism development. In respect to tourist flows and related investments, transboubdary parks, especially a chain of nature and game reserves, have become one of the major tourist attractions in southern Africa and are a key economic consideration for local development and livelihood sustainability (Chiutsi & Saarinen, 2017).

15.4 Lessons Learnt and Future Research

Drawing from the results of this chapter we can conclude that transboundary conservation areas, transfrontier conservation areas and peace parks have all been packaged within global environmental rhetoric and as such a few benefits have accrued towards community development and rural livelihood empowerment (Mayoral-Phillips, 2002). More research is needed on the local communities' perceptions of TBNRM, important determinants of the success of conservation efforts in Southern

Africa. This is in line with Ntuli et al. (2019) argument that these perceptions affect people's attitudes and behaviour towards conservation. The involvement of local communities in transboundary conservation should be viewed as an integral part of regional development initiatives. With this background, and with the composition of communities in mind, IUCN (2004) cautions that it is imperative that regional and international approaches to conservation and community engagement in transfrontier conservation areas take cognisance of the growing complexity of communities and the relationships between them to harmonise social needs and the developmental aspirations of local communities.

Future research should be geared towards the growing complexities of communities and assessments of whether their needs can be met by the mandate of TFCAs. TBNRM development must have legitimacy with a clear methodology that creates space for the broader involvement of key stakeholders, particularly local communities, and must offer a real opportunity to redistribute property rights and tangible benefits. Meaningful involvement and partnerships should not be envisioned by other stakeholders as being imposed on certain segments of participants, but it should be inclusive and equitable in both participation and beneficiation as a way forward for conservation and as a solution to poverty and economic development in the region.

TFCAs and development policies can only be effective if they incorporate ecological and social mechanisms that can positively impact the socio-ecological systems of transboundary conservation areas.

References

Andersson, J. A., de Garine-Wichatitsky, M., Cumming, D. H., Dzingirai, V., & Giller, K. E. (2013). *Transfrontier conservation areas: People living on the edge*. Earthscan/Routledge.

Ashley, C. (1995). *Tourism, communities and potential impacts on local incomes and conservation* (Research Discussion Paper No. 10). Ministry of Environment and Tourism, Namibia.

Bhatasara, S., Nyamwanza, A. M., & Kujinga, K. (2013). Transfrontier parks and development in southern Africa: The case of the Great Limpopo Transfrontier Park. *Development Southern Africa, 30*, 629–639.

Braack, L., Sandwith, T., Peddle, D., & Petermann, T. (2006). *Security considerations in the planning and management of transboundary conservation areas*. IUCN.

Briedenhann, J., & Wickens, E. (2004). Tourism routes as a tool for the economic development of rural areas: Vibrant hope or impossible dream? *Tourism Management, 25*(1), 71–79.

Büscher, B. (2013). *Transforming the frontier: Peace parks and the politics of neoliberal conservation in southern Africa*. Duke University Press.

Chiutsi, S., & Saarinen, J. (2017). Local participation in transfrontier tourism: Case of the Sengwe community in Great Limpopo transfrontier conservation area, Zimbabwe. *Development Southern Africa, 34*(3), 260–275.

Chiutsi, S., & Saarinen, J. (2019). The limits of inclusivity and sustainability in transfrontier peace parks: Case of Sengwe community in Great Limpopo transfrontier conservation area, Zimbabwe. *Critical African Studies, 11*(3), 348–360.

D'Amore, L. (2009). Peace through tourism: The birthing of a new socio-economic order. *Journal of Business Ethics, 89*(S4), 559–568.

Di Minin, E., Hunter, L. T., Balme, G. A., Smith, R. J., Goodman, P. S., & Slotow, R. (2013). Creating larger and better connected protected areas enhances the persistence of big game species in the Maputaland-Pondoland-Albany biodiversity hotspot. *PLoS One, 8*(8), e71788.

Duffy, R. (2006). Global environmental governance and the politics of ecotourism in Madagascar. *Journal of Ecotourism, 5*(1&2), 128–144.

Griffin, J., Cumming, D., Metcalf, S., t'Sas-Rolfes, M., Singh, J., Chonguica, E., Rowen, M., & Oglethorpe, J. (1999). *Study on the development of transboundary natural resource management areas in southern Africa*. Biodiversity Support Programme. Available at www.bsponline.org

Hall-Martin, A., & Modise, S. (2002). *Status report: Existing and potential transfrontier conservation areas in the SADC region*. Peace Park Foundation.

Hamm, J. A. (2017). Trust, trustworthiness, and motivation in the natural resource management context. *Society & Natural Resources, 30*(8), 919–933.

Hammill, A., & Besancon, C. (2007). Measuring peace park performance: Definitions and experiences. In *Peace parks: Conservation and conflict resolution* (p. 23). MIT Press.

IUCN. (2004). *Indigenous and local communities and protected areas, towards equity and enhanced conservation*. The World Conservation Union.

Jenkins, M. (2008). *Who murdered the Virunga Gorillas?* http://ngm.nationalgeographic.com/2008/07/virunga/jenkins-text

Katerere, Y., Hill, R., & Moyo, S. (2001). *A critique of transboundary natural resource management in southern Africa* (Paper no. 1, IUCN-ROSA series on transboundary natural resource management). IUCN.

Kornus, A. O., Siutkin, S. I., Kornus, O. H., & Danylchenko, O. S. (2017). Event tourism for transboundary cooperation. Doctoral dissertation, Сумський державний університет.

Lane, B. (2005). Sustainable rural tourism strategies: A tool for development and conservation. *Interamerican Journal of Environment and Tourism, 1*, 12–18.

Maekawa, M., Lanjouw, A., Rutagarama, E., & Sharp, D. (2013). Mountain gorilla tourism generating wealth and peace in post-conflict Rwanda. *Natural Resources Forum, 37*(2), 127–137.

Manyane, R. M. (2017). Rethinking trans-boundary tourism resources at the Botswana-North West Province border. *South African Geographical Journal, 99*(2), 134–151. https://doi.org/10.1080/03736245.2016.1208579

Marijnen, E., & Verweijen, J. (2016). Selling green militarization: The discursive (re)production of militarized conservation in the Virunga National Park, Democratic Republic of the Congo. *Geoforum, 75*, 274–285.

Mayoral-Phillips, A. J. (2002). Transboundary areas in southern Africa: Meeting the needs of conservation or development. In *Ninth conference of the international association for the study of common property, Victoria Falls, Zimbabwe*. 1721, p. 2002.

Metcalfe, S. (2003, September). Impacts of transboundary protected areas on local communities in three southern African initiatives. In *Workshop on transboundary protected areas in the governance stream of the 5th world parks congress, Durban, South Africa*, pp. 12–13.

Mogende, E. (2016). *The politics of Kavango-Zambezi (KAZA) TFCA in Botswana*. Master's thesis. University of Cape Town.

Moyo, S., & Tevera, D. (2000). Regional environmental security in southern Africa. In D. Tevera & S. Moyo (Eds.), *Environmental security in southern Africa* (pp. 3–24). Southern African Regional Institute for Policy Studies.

Muboko, N., & Murindagomo, F. (2014). Wildlife control, access and utilization: Lessons from legislation, policy evolution and implementation in Zimbabwe. *Journal for Nature Conservation, 22*, 206–211.

Mudenda, G. (2000). Cross-border white farmer migration in southern Africa: The Zambian experience. In D. Tevera & S. Moyo (Eds.), *Environmental security in southern Africa*. SAPES Trust.

Murphree, M. (2009). The strategic pillars of communal natural resource management: Benefit, empowerment and conservation. *Biodiversity and Conservation, 18*, 2551–2562.

Ntuli, H., Jagers, S. C., Linell, A., Sjöstedt, M., & Muchapondwa, E. (2019). Factors influencing local communities' perceptions towards conservation of transboundary wildlife resources: The case of the Great Limpopo trans-frontier conservation area. *Biodiversity and Conservation, 28*, 2977–3003.

Peace Parks Foundation. (2006). *Peace parks foundation ten years review 1997–2006*. Peace Parks Foundation.

Perkins, J. S. (2020). Take me to the river along the African drought corridor: Adapting to climate change. *Botswana Journal of Agriculture and Applied Sciences, 14*(1), 60–71.

Prunier, G. (2011). *Africa's world war: Congo, the Rwandan genocide, and the making of a continental catastrophe*. Oxford University Press.

Quinn, M. S., Broberg, L., & Freimund, W. (Eds.). (2012). *Parks, peace, and partnership: Global initiatives in transboundary conservation*. University of Calgary Press.

Ramsamy, P. (2004). The Regional Indicative Strategic Development Plan (RISDP) as a strategic framework for deeper economic integration and social development in the SADC Region.

Ramutsindela, M. (2004). *Parks and people in postcolonial societies: Experiences in southern Africa*. Kluwer.

Ramutsindela, M. (2009). Transfrontier conservation and local communities. In J. Saarinen, F. Becker, H. Manwa, & D. Wilson (Eds.), *Sustainable tourism in southern Africa, local communities and natural resources in transition*. MPG Books Group.

Retief, F. P., Alberts, R. C., Lubbe, W. D., Roos, C., & Cilliers, D. P. (2023). A critical evaluation of international agreements towards a revised categorization for transfrontier conservation areas (TFCAs). *Environmental Management, 72*, 1–12.

Ron. (2007). *Southern Africa development community (SADC) proposed framework for transfrontier conservation areas (TFCAs)*. SADC Secretariat.

Sandwith, T. S., Shine, C., Hamilton, L. S., & Sheppard, D. A. (2001). *Transboundary protected areas for peace and co-operation*. IUCN. Available at www.iucn.org/themes/wcpa/pubs/guidelines.htm

Secretariat SADC. (2017). *Southern African Development Community transfrontier conservation areas monitoring and evaluation framework (SADC TFCAs M&E Framework)*. Final Draft.

Selier, S. J., Slotow, R., Blackmore, A., & Trouwborst, A. (2016). The legal challenges of transboundary wildlife management at the population level: The case of a trilateral elephant population in southern Africa. *Journal of International Wildlife Law & Policy, 19*(2), 101–135.

Southern African Development Community (SADC). (1999). *Protocol on wildlife conservation and law enforcement*. Southern African Development Community.

Spenceley, A. (Ed.). (2008). *Responsible tourism: Critical issues for conservation and development*. Earthscan.

Thomson, G. R., Penrith, M. L., Atkinson, M. W., Atkinson, S. J., Cassidy, D., & Osofsky, S. A. (2013). Balancing livestock production and wildlife conservation in and around southern Africa's transfrontier conservation areas. *Transboundary and Emerging Diseases, 60*(6), 492–506.

Trogisch, L., & Fletcher, R. (2022). Fortress tourism: Exploring dynamics of tourism, security and peace around the Virunga transboundary conservation area. *Journal of Sustainable Tourism, 30*(2–3), 352–371. https://doi.org/10.1080/09669582.2020.1857767

United Nations Word Tourism Organization (UNWTO). (2014). *World tourism highlights*. UNWTO.

Utting, P. (2015). Forewords. In P. Utting (Ed.), *Revisiting sustainable development* (pp. 1–15). UNRISD.

Visser, G. (2011). Sustainable tourism in southern Africa: Local communities and natural resources in transition. *South African Geographical Journal, 93*(1), 119–120.

Zunckel, K. (2014). *Southern African Development Community transfrontier conservation guidelines: The establishment and development of TFCA initiatives between SADC Member States*. South African Development Community (SADC).

Chapter 16
Protected Areas and Conservation Conflicts: Impacts on Sustainability

Kgosietsile Velempini and Hesekia Garekae

16.1 Introduction

Protected Areas (PAs) evolved in a bid to manage and conserve biodiversity, more especially wildlife (Dudley, 2008; Possingham et al., 2006). There is abundant literature on protected areas, with noted success and failures across the world and over time (Brockington & Igoe, 2006; Maisels et al., 2007; Schmidt-Soltau, 2004). Notwithstanding this, the establishment of PAs was marked with the dispossession and displacement of local communities who inhabited the ancestral areas (Agrawal & Redford, 2009; Claridge & Kobei, 2023). Relocations and evictions of local communities within areas designated for the establishment of PAs have been noted across the world, e.g., India (Lasgorceix & Kothari, 2009), Tanzania (The Oakland Institute, 2018), and South Africa (Thondhlana et al., 2011). These displacements are accompanied by socio-economic costs which impoverished most communities (Agrawal & Redford, 2009). Communities' rights and access to resources are often limited and/or curtailed. This affects communities' main sources of livelihood.

In Botswana, some communities were relocated to pave way for the establishment of PAs though this was largely undocumented. In northern Botswana, local people who resided in their ancestral lands were relocated when Moremi Game Reserve was established (Mbaiwa et al., 2008). Similarly, the promulgation of the first Transboundary Protected Area (TBPA) in Southern Africa, the Kgalagadi Transfrontier Park (KTP) culminated in the displacement and relocation of local

K. Velempini (✉)
Faculty of Education, Department of Languages and Social Sciences Education, Environmental Education Unit, University of Botswana, Gaborone, Botswana

H. Garekae
Department of Environmental Science, University of Botswana, Gaborone, Botswana
e-mail: garekaeh@ub.ac.bw

© The Author(s), under exclusive license to Springer Nature Switzerland AG 2024

L. S. Stone, M. T. Stone (eds.), *Wildlife Tourism Dynamics in Southern Africa*, Geographies of Tourism and Global Change,
https://doi.org/10.1007/978-3-031-57252-4_16

communities (‡*Khomani* San and Mier communities) (SADC Secretariat, 2018). In 1999, the Government of Botswana and South Africa signed a historic bilateral agreement to manage adjacent parks, Mabuasehube National Park and Kalahari Gemsbok National Park respectively, as a single ecological unit commonly known as a peace park (Botswana Tourism Organization, 2021).

This study focuses on the period from the late 1930s when the Botswana Gemsbok National Park was proclaimed in 1938 by the Bechuanaland government. This study investigates the lived experiences of local people focusing on *elders* from the BORAVAST villages about their relocation from ancestral lands (currently the KTP). In this study, *Elders* are defined according to the document received from the local village chief showing Ages (73–93 years old) of "people that may be contacted on the history of the coloured community that once lived in the Kgalagadi Transfrontier Park". BORAVAST villages consist of Bokspits, Rappelspan, Vaalhoek, and Struizendam (including Ganahoek and Welkom farms) in Kgalagadi south district of Botswana. "The district has an average annual rainfall of 150mm in the south to 250mm in the north. There is no surface water except in seasonal shallow pans and the Molopo and Nossop fossil valleys." (United Nations Department of Economic and Social Affairs, 2008, p.14). This study contributes to knowledge on the opportunities and challenges experienced by local communities living adjacent to their transformed ancestral lands in Africa and globally. Furthermore, lessons learnt can guide local community trusts, wildlife authorities, tourism experts, educators, and other practitioners on important interrelated issues on protected areas, conservation, relocation of local and indigenous people, and sustainable development.

Relocations for conservation purposes and to boost the travel and tourism industry continues to haunt many local land claimants (Thondhlana et al., 2012) in the BORAVAST villages of Botswana, who are relatively powerless and lack adequate resources and platforms to raise their voices (van Rooyen, 1998). The travel and tourism industry is one of the highest contributors to Botswana's Gross Domestic Product [GDP]. The KTP, which is mainly visited by self-driven campers is close to the BORAVAST villages and is one of the major tourist destinations in the southern part of Botswana (Botswana Tourism Organization, 2021). Research shows that local communities adjacent to KTP trace their historical roots from diverse *Bantu* groups (e.g., Coloureds, the Mier, and the San), who used to live freely and survive in the park, their ancestral lands (Van Wyk & Le Richie, 1984). Writing about the local people in the BORAVAST villages, Van Wyk and Le Richie (1984) opined, "on the Botswana side of the Nossob there were a number of Coloured hamlets stretching from the confluence to almost as far as *Rooiputs* in the present park" (p. 21). Moreover, Campbell (2004) indicated that, "in 1983, 280 people under Titus Matthys were resettled between *Tweerivieren* on the Nossob River and Rappelspan on the Molopo River" (p. 57). This shows that local people used to live inside the KTP prior to the relocation.

This study assesses the relocation of communities from KTP and how it impacted their livelihoods. The study addressed the following research questions; (1) what are the lived experiences of local people in the BORAVAST villages? and (2) what impacts did the relocation have on their livelihoods?

16.2 Research Study Site

This study took place in the Kgalagadi south, Botswana, where the KTP spans the Kalahari Desert region of Botswana and South Africa. The region has a population of 35,347 people. Respondents in the study were from the villages of Bokspits (population = 710), Rappelspan (population = 352), Vaalhoek (population = 558), Struizendam (population = 746) (Statistics Botswana, 2022). Figure 16.1 below shows the location of KTP in Botswana and South Africa.

Sites such as *Leeuwdrill*, *Rooiputs*, *Kij Kij* and *Kameelsleep* were identified as ancestral lands by land claimants who were interviewed in this study (see Fig. 16.2).

16.3 Research Methods

This study employed a qualitative research design approach to generate reliable, trustworthy and credible data focusing on the lived experiences of local people. Respondents were drawn from KTP land claimants. A document about the history

Fig. 16.1 The location of KTP in Botswana and South Africa. (Source: Adapted from Encyclopaedia Britannica, 2004)

Fig. 16.2 Sites identified as ancestral lands by land claimants. (Source: Adapted from Kalahari Safaris, 2023)

of the coloured community who once inhabited the KTP was sourced from the chief of Struizendam village. The document profiled the details of 10 land claimants, who were all (except one) residents of BORAVAST. Nine (9) participants were purposively sampled for semi-structured interviews (see Table 16.1). Respondents selected for the study included KTP land claimants who had resettled in BORAVAST. On the 31st of August 2017, the selected participants were issued a permit by the Department of Wildlife and National Parks (DWNP) in Botswana to enter the park, visit their ancestral lands and take photographs and videos. Ancestral places visited included Rooiputs, Leeuwdrill, Kameelsleep, Kij Kij, and Stalaasie Kolk.

Semi-structured interviews were conducted with key informants (the elderly) from the villages of BORAVAST. Some interviews were conducted in 2018 with follow-ups through telephone and intermittent visits to BORAVAST in 2020 and 2022. Interviews were conducted in the Afrikaans language (the indigenous language). Responses were then translated to English.

With consent from respondents, a TECHNO POP 4 digital phone was used to record interviews. Interviews took place at residential homes of participants who were accompanied by family members who were at home during the interviews. Some family members shared their stories and were also included in the study (see Table 16.1).

Photo-voice is a qualitative research technique that this study also utilised. The origins of photovoice as a research method can be largely traced to health sciences research (Wang et al., 1998; Wu et al., 1995). The method has gained wider traction and has been employed in other professions such as sociology, education, and communications (Allen, 2012). Photovoice is constantly used in research with

Table 16.1 Respondents' demographics

Interviewee code	Gender	Village or farm	Age (at the time of interview)	Experienced living inside the current KTP
Participant A	Male	Bokspits	91	Yes
Participant B (*assisted in interpretation*)	Male	Gannahoek farms		No
Participant C (*family member to participant A*)	Male	Bokspits	Not recorded	No
Participant D	Male	Vaalhoek	93	Yes
Participant E	Male	Rappelspan	Not recorded	Yes
Participant F (*family member to participant E*)	Female	Rappelspan	Not recorded	No
Participant G (family member to participant E)	Male	Rappelspan	Not recorded	No
Participant H	Male	Welkom farms	Not recorded	Yes
Participant I	Male	Struizendam	82	Yes

Table 16.2 Descriptions of photos used

Picture number	Description
1.	Some of the representatives of Kgalagadi south land claimants at two Rivers, KTP border gate
2.	One of the historical water wells inside KTP. The well was a primary source of water for household consumption and drinking livestock
3.	Another abandoned water well in KTP. In the picture, is one of the local chiefs in BORAVAST
4.	Local people are seen throwing stones into one of the abandoned water wells to assess water availability
5.	A renovated historical water well, with solar panels and reservoir installed by the Department of Wildlife and National Parks
6.	Some of the representatives of the land claimants resting on a modified structure inside KTP. In the picture, a foundation/slab of a remnant building denotes that people lived in the area before it was declared a protected area
7.	Some of the representatives standing by a remnant site, with red sand bricks. This could have been a residential area
8.	One of the elderly representatives who indicated that he lived in the area as a young person prior to establishment of the park
9.	Some of the land claimants leaning on a structure written '*Kammelsleep*', which was described as an ancestral site. In the picture, *Oryx gazelle* (Gemsbok) could be seen in the background

indigenous communities. It is a process of viewing and learning from photos and videos that are taken at a spatial place where the research is being conducted (Wang & Burris, 1997). The images are taken by participants themselves, hence promoting the co-production of knowledge. The essence of photovoice research goes beyond the appealing nature of images but to the connotations participants attach to each image. This method promotes inclusivity and enables the participation of marginalised communities who may not be reached through other conventional research methods such as interviews through electronic platforms (email, telephones). Therefore, this method can give a 'voice' to the 'voiceless' in society who are mostly excluded in decision-making processes (Evans-Agnew & Rosemberg, 2016). Land claimants captured photos (n = 27) and video clips (n = 14) during their visit to the park (See Table 16.2). However, for this paper, only nine (9) photos were selected for analysis based on communities' experiences with relocation from KTP and how it impacted their livelihoods. All photos and print screen photos from videos are used with permission from the land claimants.

The photos described in Table 16.2 above are shown below (see Fig. 16.3).

Secondary data were also used and included official documents about the formation of the KTP and relocation of people. Furthermore, they also investigated the connection between the formation of the park in the 1930s, conservation, the relocation of local people, natural resource use conflicts and their impacts on the livelihoods of local people.

Fig. 16.3 Photos captured land claimants. (Source: Kgalagadi south land claimants, 2017)

Data analysis included the transcription of recorded interviews. Data were analysed through thematic analysis which involved deriving themes from the data. Photos and video clips were also scrutinised to learn about the historical sites and infrastructure before the KTP.

16.4 Findings

Five (5) emerging themes are presented in this study about the relocation of local people from the contemporary KTP for nature conservation and tourism development purposes. These are discussed below.

16.4.1 Lived Experiences in the Ancestral Lands

Respondents showed that they still hold onto past experiences when they lived in their ancestral lands; they still have memories about how relocations were carried out. For instance, when asked to share how they lived in what is now the park and what they remember, participant A responded as follows: "*I will explain what I know because I was there. It is not something that I was told. There was no park by then. We did not stay in a park*". He continued to state that they left the ancestral area when he was eleven (11) years old. According to the respondent,

> It was the on the 11th of May 1938 when we moved from there. We came to settle in Bokspits because by that time the pastures and water was sufficient in some areas of Bokspits.

On the other hand, participant E indicated, "*I was born in Two-Rivers in Kay-Kay. We lived by eating the meat of wild animals*". Regarding his experiences at the ancestral lands, Participant H stated:

> I was not born in the ancestral lands, but my father lived there. We even have a document which allowed him to search for water in the ancestral lands. At that time, we did not drill boreholes, but we dug wells. The first one was Rooiwell on the South African side and another one was dug at Wolfkraans, just opposite Struizendam on the South African side. The other well was dug at Welkom and my father was digging for other people. There was also the one in Kamfersboom. There was less water in Kamfersboom.

The above sentiment from Participant H was corroborated by some of the pictures (Picture 2, 3, and 4 in Fig. 16.3) which depict historical water wells which were used by inhabitants prior to the establishment of the park. Some of the wells have now been refurbished and are not drinking holes for wildlife. According to participant D, local people lived in the ancestral lands. He indicated that "*some people used to say that they stayed in Kameelsleep, and then moved to Kay-Kay. I grew up in Kay-Kay. That is where I was at the age of 14 years when we left*". For his part, participant B said:

> People were staying in the ancestral lands from the early 1920's and then the South Africans established on their side the Gemsbok National Park. Since people were staying in the Bechuanaland side, the Warden complained that local people are poaching wild animals on the South African side. It might have been true or it might just have been a complaint. Then the South African government and the British government reached an agreement in 1938 that the Botswana side, the area north of the Nossob River be declared part of the National

> Park. This is also why people were instructed to leave the area because it was now declared a national park.

The above suggest that some local people in the Kgalagadi south lived in the contemporary KTP before it was designated a protected area. The 1938 declaration of the national park and instructions for people to leave the area affected their livelihoods and denied them access and rights to resources, consequently leading to social conflict and grievances as presented below.

16.4.2 Displacement and Relocation from the Protected Area

The displacement and relocation of local people from wildlife areas is a recurrent and common issue in wildlife tourism dynamics in Africa. In the southern African region, this transition is accompanied by conservation, tourism and socio-economic goals and objectives. Regarding consultation, participant A stated that, *there was no consultation. We were not consulted or asked if we agreed or not. Some of the elderly men that were there were just instructed to leave as it was already agreed by the two governments*. This suggests that the consensus to relocate local people was made by the South African government and the British Protectorate without the knowledge, involvement and engagement of local people. Participant A continued to explain as follows:

> The South African government and the British government reached an agreement in 1938 that the Bechuanaland side, the area north of Nossob riverbed, should be included or declared part of the National Park… This is why people were instructed to leave the area. We left the area when I was 11 years old.

To corroborate this, participant D also said,

> The South African Government (Boers) together with the British government agreed that the area be declared a national park. Some people, especially the younger generation, did not agree to move out. A District Commissioner came from the British Government and told them to move out; otherwise, they would be removed forcefully.

Similarly, participant I indicated, "*They told us that they want the area to be declared a park, but it would only be for five years, thereafter people would be allowed to go back*". He explained that however, the local people did not insist on going back. Participant B, reacted as follows about the relocation, "*Since the area was declared a park, there was no way that people would go back again because it was a park then. The story they told them that after five years they can go back was not true*". On the same question about the relocation, Participant E stated, "*The British government is the one that moved us out*". However, Participant F (a female family member) highlighted that, "*they had the feeling they must move due to pressure. Something was not good for them… he was only 8 years when the parents moved from the park, he was just a young person*". Equally, participant H said that the South African authorities forced them out of the ancestral lands sometime in May 1938. He stated, "*We were harassed by the South African law enforcement authorities, even chickens were tied, and food was lost on the way and my uncle's cow gave birth on the way*".

When asked about his impression on the displacement and relocation of people, participant A reacted as follows,

> No, we cannot be happy because it was done in an unfair way. My parents suffered to make developments and after doing that…working so hard… they were told to leave and come to a place with no developments…we cannot be happy. If we can go back there, I can still be able to show you the places, like Rooiputs, Leeuwdril and others…The people who stayed there developed themselves and had to leave properties behind.

Likewise, participant B said,

> … people had property on the Botswana side. They were informed that it was only for five years, after five years they would go back, so people lived with the mentality that they left their properties which they were not compensated for. All those people staying there passed away. They have been complaining but no one has listened to their complaints.

The experiences above suggest an unfair process of relocating local people for purposes of wildlife conservation and tourism development.

16.4.3 Sustainable Livelihood Developments Abandoned Inside the Protected Area

The findings in this study suggest that the loss of assets during the displacement and relocation from the protected area had a major impact on the local community. Assets constitute human livelihood development (Natarajan et al., 2022; Scoones, 1998). In many countries, the transformation of public lands into protected areas led to heightened poverty, social conflict, and environmental deterioration in the long run. According to participant A, some of the livelihood sources that disappeared during relocation include livestock such as goats, sheep, cattle, donkeys and horses. In relation to livelihoods, he also stated that: *what I am going to say is what I exactly know. For example, those mud bricks that were used at the boreholes, I was part of making the bricks, by mixing the clay using our feet. The young boys used to mix the mud to make bricks* (see picture 7 on Fig. 16.3). On the other hand, participant C stated that he used to work inside the contemporary KTP, but on the side of South Africa. He indicated, "*These people lost good water. The water in KTP tastes good. Since they came from that side, they are suffering to get good water. The place that they are currently practising livestock farming at has salty water*".

Participant B said,

> So, that's why we asked for permission, went there, went off the road to those places. We saw the pit wells, and we went out of dry rivers to see where people used to live. We could see with our eyes that there were developments because some of the foundations were made of clay and some foundations were made with cement (see picture 6, Fig. 16.3).

In relation to property abandoned inside the protected area (the ancestral lands), participant D responded as follows,

> At Kay-Kay, they were three (3) pit-wells and one (1) borehole that are still visible today. The houses were not properly constructed because in those times people used grass, except for the headman whose house was built with clay bricks. There is a possibility that the foundation of clay bricks may still be there. But they had to leave the houses just like that. They left and drove away with some of the wild animals they had (see picture 2, Fig. 16.3).

The analysis of photos and videos show developments such as fully fledged pit wells constructed of rocks, cement, and with dry logs on the surface (to signify a water point and to fend off animals that might fall inside, see picture 2, 3, and 4 on Fig. 16.3). Red-brownish bricks on the land signifying fallen structures (see picture 7, Fig. 16.3), existing water sources where gemsboks drank from (see picture 9 Fig. 16.3) and possible water sources close to an indigenous Camel thorn tree (see picture 5 Fig. 16.3). The water source has been modified and now includes a solar panel. Moreover, the water source is close to a wind wheel for possible pumping of water underground (see picture 2, 4, 6, and 8 Fig. 16.3).

Photos 3 and 7 show some BORAVAST elders viewing the ancestral water sources and lands that they used to occupy around the 1930s. According to Van Wyk and Le Richie (1984, p. 23), "there were only four watering points in the Nossob, two in the southern part dug by the Coloureds…a borehole by Kij Kij… and a well operated by a hand-pump at Unie-End". Participants in this study reported about ten (10) water points.

On being asked how local people sustained themselves in their ancestral lands, participant I responded as follows, "*We hunted and were allowed to hunt…they hunted and many people had lots of animals. But the problem is that there was no market, the only market was in South Africa*". Regarding buying and hunting participant D said,

> The nearby shops were on the South African side called Kopernoerd and Andresville. We walked to these places to buy. We made a living in the park. We hunted and were allowed to hunt. Even the ostrich skins were brought from the park and sold at Kopernoerd and Andresville. Animal skins were used to make robes for milking cows and handling horses. These were prepared from the park and brought to the shops. After preparing the thongs and skins, parents came to shops to sell to shop owners and got money to buy sugar, milk, and tea.

On why he thinks his story about relocation is important to children and the general public, participant I indicated that,

> This history is important because our parents made some developments like pit wells, and then also constructed houses that side, but it was mostly pit wells. Water was important by then and even today. They left their property that side.

Participant I further indicated, "*Something important to me is that we left property like water points. We moved this side where there was no water*".

16.4.4 Heritage Education as a Contribution to Tourism and Sustainable Development

Respondents in this study believe that their local environmental histories and past cultural experiences in the protected area can contribute immensely to the contemporary education curriculum, diversify the tourism industry and consequently enhance local sustainable development. Responding to a question on why it is important to know his story of living in the protected area and how relocation was done, participant A responded,

> I think it is important that history should be carried from generation to generation. History is not only for today, but it can also take millions of years. Therefore, I think it is important that this knowledge that I have, and the way we lived that side in the current KTP, how we were instructed to leave without compensation should be known. All this knowledge can be documented and carried to the next generation.

Regarding the importance of his story, participant B also said,

> … to keep this history of the coloured people in the national archives. Currently, there is nothing in the archives about this knowledge. This history is important because people had been staying there and had property and they were moved from there. That is why I said in my initial contribution that this information should be documented and kept in government archives so that at the end of the day people can learn that there are people who lived in that area.

On the other hand, participant G explained that:

> It is very important. Even at my age I have a lot to know about our family and background. These are the last elders who can bring this story to us. In fact, I am very fond of history. Last time, I asked them why they relocated. They told me they were moved by British colonisers and the compensation they received was just 10 pounds harmers. They were given those items to go dig other boreholes as they moved out.

Similarly, participant H explained that it is essential for people to learn and know about this story because a lot was done in the park and the local people were not compensated when they relocated. The compensation was a pick, a spade, and a robe for rolling around the gathered items.

16.4.5 Compensation for Conservation and Tourism Development

There are increasing concerns regarding compensation given to communities that are relocated from their ancestral lands for conservation and tourism development purposes. On the issue of compensation, participant A indicated:

> I think it was not good that we were not compensated. The other thing is that local people had developed their lives. One such development was water, if you are a farmer, you need water and pastures. People did not struggle to get water and there was sufficient grazing. Life was easy, even though we know by then people did not have enough money. But for

> livestock rearing, there was no problem because there was good water and pastures. Life was better that side.

He continued to state that he became old in the current Bokspits village, and his animals dwindled because there was no grazing. He lamented,

> Now as an old man, I have nothing anymore, I only get the Old Age Pension,[1] about BWP 430.00.[2] I have to survive on that old age pension alone. If I had stayed in my ancestral lands where my parents had left animals, made developments, I could have been a rich man.

In relation to compensation, participant D stated:

> We were not compensated. The only compensation was a plough, two harmers, and two (2) picks. We want compensation because we left our pit-wells and other property. We left donkey carts. We had to leave some of our utensils in the park because we used donkeys to leave the park and you can't take everything. I feel we were not compensated for what we had accumulated in the present day park.

On being asked about the existence of tourism during their time in the ancestral lands, participant D indicated that, "*There were no visitors, no tourists by then, just sometimes people walking around and seen in donkey carts and those in oxen pulled vehicles*".

The experiences from land claimants raised important issues including being relocated permanently after being told that they would return after 5 years and livelihood assets that were abandoned in ancestral lands. Land claimants emphasized that their experiences can contribute to heritage education and benefit learners and future generations who will get knowledge that local people lived in the protected area where the present day Kgalagadi Transfrontier Park is located. Furthermore, in the twenty-first century, heritage education can also enhance the ecotourism sector. Land claimants also expressed dissatisfaction because there was no compensation for them relocating from their ancestral lands.

16.5 Discussions, Conclusion, Recommendations

The findings demonstrate that there were people who inhabited the area prior to the establishment of Botswana's Gemsbok National Park, which later merged with South Africa's Kalahari Gemsbok National Park to form the current Kgalagadi Transfrontier Park. Back then, the inhabitants engaged in land-based livelihood activities such as pastoral farming and hunting and gathering. Pastoral farming was conducive due to the abundance and diversity of pastoral resources in the area. Moreover, the participants claimed that the area was endowed with good ground

[1] Old age pension is for Botswana citizens only who are 65 years old and above…Currently (2023), the pension allowance is BWP530.00 per month. https://www.gov.bw/allowances/old-age-pension-allowance

[2] BWP 430.00 is the amount that was reported by the Participant A in 2017. BWP 430 is equivalent to 32.64 USD. https://bwp.currencyrate.today/convert/amount-430-to-usd.html

water compared to the salty one they have access to now. This made pastoral farming thrive in their ancestral lands compared to where the community was relocated to. Similarly, the diversity of plant and wildlife resources was integral to human livelihoods. Wildlife was a source of protein, which was key in diversifying the diets of locals. The contribution of wildlife to rural livelihoods is well established in Botswana (Mbaiwa, 2005) and elsewhere (Rao et al., 2011).

Although protected areas are welcome developments for conservation purposes, they are synonymous with the relocation and displacements of local communities (Lasgorceix & Kothari, 2009; Agrawal & Redford, 2009). This is also the case in this study. In 1938, the locals were relocated from the area they inhabited to pave way for the establishment of the Gemsbok National Park. Notwithstanding this, there was no community consultation and or participation in the decision-making processes for the establishment of a protected area in the land they inhabited. Moreover, the locals opined that they were deceived into thinking that they would return to their ancestral area after a short period of time post the establishment of the PA. The lack of consultation from the Botswana Government could be a result of pressure exerted by the South African government, who had long declared part of the landscape a PA and felt that local communities in Botswana were intruding and engaging in illicit activities. Consequently, this sparked social conflict and grievances from land claimants in the BORAVAST villages of Botswana.

The relocation disrupted the locals' livelihoods and assets. The locals left some of their assets in the park, only taking a handful with them under the auspices that they will be returning to the area. However, that was never the case. Moreover, the locals lost productive land which was favourable for pastoral farming. Regulations for managing the park were promulgated but prohibited free access to the PA unless sanctioned by the officials through the issuance of entry permits. Therefore, access to the park for land-based livelihood activities was curtailed, i.e., foraging livestock, and hunting and gathering. Beyond access, the locals were also detached from their ancestral land. This affected their spiritual well-being, as limited access hindered them from connecting with the spirits of their forefathers. Besides the resultant losses from relocation, there was no meaningful compensation accorded to locals to remediate the losses they incurred. In the Peruvian Amazon, Peru, the establishment of Allpahuayo-Mishana National Reserve displaced and dispossessed locals of land they inhabited (Cardozo, 2011). This development disrupted the Peruvian community's lifestyle and the livelihoods and local people got banned from accessing the land to collect resources which they subsisted on prior to establishment of the PAs. Consequently, this influenced local communities to harbour negative attitudes towards the management and conservation of resources in the reserve.

This paper contributes to the established body of knowledge on protected areas and conflicts, seldom understood in Botswana. Most of the literature in Botswana places particular emphasis on the socio-economic benefits derived from protected areas (Stone, 2013; Lekgau & Tichaawa, 2020). This underscores the importance of this study, which presents insights on the lived experiences and rich environmental histories of the Kgalagadi South land claimants, who were displaced from their ancestral land to pave way for the establishment of a protected area.

These experiences can transform the contemporary education curriculum, diverse tourism industry and enhance local livelihoods. Indeed, participants contend that the historical sites in KTP could be explored for heritage and cultural tourism. Cultural and educational tours could be conducted to enlighten tourists and learners about historical sites in the park, and the associated traditional and cultural lifestyles. The community could be the one entrusted with conducting the tourism activities in the park. Hence, this approach could redress some of the losses incurred by the community during relocation. Heritage and cultural tourism has the potential to diversify Botswana's tourism offering, which has largely been dominated by wildlife-based tourism. Future studies should elucidate on how local communities adapt to resettled areas.

Acknowledgements The authors would like to thank local community members in the BORAVAST especially the participants for their time and sharing their experiences. Some of the participants in this study passed away after sharing this important knowledge; may their souls continue to rest in peace.

References

Agrawal, A., & Redford, K. (2009). Conservation and displacement: An overview. *Conservation and Society, 7*(1), 1–10.

Allen, Q. (2012). Photographs and stories: Ethics, benefits and dilemmas of using participant photography with Black middle-class male youth. *Qualitative Research, 12*(4), 443–458.

Botswana Tourism Organization. (2021). *Kgalagadi Transfrontier Park*. https://www.botswanatourism.co.bw/explore/kgalagadi-transfrontier-park

Brockington, D., & Igoe, J. (2006). Eviction for conservation: A global overview. *Conservation and Society, 4*(3), 424–470.

Campbell, A. (2004). Establishment of Botswana's National Park and game reserve system. *Botswana Notes and Records, 36*, 55–66.

Cardozo, M. (2011). Economic displacement and local attitude towards protected area establishment in the Peruvian Amazon. *Geoforum, 42*(5), 603–614.

Claridge, L., & Kobei, D. (2023). Protected areas, Indigenous rights and land restitution: The Ogiek judgment of the African Court of Human and Peoples' Rights and Community Land Protection in Kenya. *Oryx, 57*(3), 313–324.

Dudley, N. (Ed.). (2008). *Guidelines for applying protected area management categories*. IUCN.

Encyclopaedia Britannica. (2004). Kgalagadi Transfrontier Park. kids.britannica.com/students/article/Kgalagadi-Transfrontier-Park/607320

Evans-Agnew, R. A., & Rosemberg, M.-A. S. (2016). Questioning photovoice research: Whose voice? *Qualitative Health Research, 26*(8), 1019–1030.

Kalahari Safaris. (2023). https://www.kalaharitour.co.za/chalet-tour-3-6-days-kgalagadi-transfrontier-park/.

Lasgorceix, A., & Kothari, A. (2009). Displacement and relocation of protected areas: A synthesis and analysis of case studies. *Economic and Political Weekly, 44*(49), 37–49.

Lekgau, R. J., & Tichaawa, T. M. (2020). Community perceptions on the socio-economic impacts of wildlife tourism from the Kgalagadi Transfrontier Park in Botswana: The case of Tsabong. *African Journal of Hospitality, Tourism and Leisure, 9*(6), 1044–1059. https://doi.org/10.46222/ajhtl.19770720-67

Maisels, F., Sunderland, T., Curran, B., von Loebenstein, K., Oates, J., Usongo, L., Dunn, A., Asaha, S., Balinga, M., Defo, L., & Telfer, P. (2007). Central Africa's protected areas and the purported displacement of people: A first critical review of existing data. In K. H. Redford & E. Fearn (Eds.), *Protected areas and human displacement: A conservation perspective* (pp. 75–89).

Mbaiwa, J. E. (2005). Wildlife resource utilisation at Moremi game reserve and Khwai community area in the Okavango Delta, Botswana. *Journal of Environmental Management, 77*(2), 144–156.

Mbaiwa, J. E., Ngwenya, B. N., & Kgathi, D. L. (2008). Contending with unequal and privileged access to natural resources and land in the Okavango Delta, Botswana. *Singapore Journal of Tropical Geography, 29*(2), 155–172.

Natarajan, N., Newsham, A., Rigg, J., & Suhardinman, D. (2022). A sustainable livelihoods framework for the 21st century. *World Development, 155*, 1–15.

Possingham, H., Wilson, K., Andelman, S., & Vynne, C. (2006). Protected areas: Goals, limitations, and design. In M. J. Groom, G. K. Meffe, & C. R. Carroll (Eds.), *Principles of conservation biology* (3rd ed., pp. 507–549). Sinauer Associates.

Rao, M., Zaw, T., Htun, S., & Myint, T. (2011). Hunting for a living: Wildlife trade, rural livelihoods and declining wildlife in the Hkakaborazi National Park, North Myanmar. *Environmental Management, 48*, 158–167. https://doi.org/10.1007/s00267-011-9662-z

SADC Secretariat. (2018). *Transfrontier conservation areas*. Southern African Development Community. https://www.sadc.int/sites/default/files/2022-07/SADC_TFCA_Brochure.pdf

Schmidt-Soltau, K. (2004). Is forced displacement acceptable in conservation projects? *id21 Insights*. http://www.Id21.org/insights. Accessed 01 August 2023.

Scoones, I. (1998). *Sustainable rural livelihoods. A framework for analysis* (IDS Working Paper 72). https://opendocs.ids.ac.uk/opendocs/bitstream/handle/20.500.12413/3390/Wp72.pdf?sequence=1

Statistics Botswana. (2022). *Population & Housing Census 2022. Population of cities, towns, villages & associated localities*. Statistics Botswana.

Stone, M. T. (2013). *Protected areas, tourism and rural community livelihoods in Botswana*. Unpublished Ph.D. dissertation. Arizona State University.

The Oakland Institute. (2018). *Losing the Serengeti: The Maasai land that was to run forever*. www.oaklandinstitute.org

Thondhlana, G., Shackleton, S., & Muchapondwa, E. (2011). Kgalagadi Transfrontier Park and its land claimants: A pre and post-land claim conservation and development history. *Environmental Research Letters, 6*, 1–2.

Thondhlana, G., Vedeld, P., & Shackleton, S. (2012). Natural resource use, income and dependence among San and Mier communities bordering Kgalagadi Transfrontier Park, southern Kalahari, South Africa. *International Journal of Sustainable Development & World Ecology, 19*(5), 460–470.

United Nations Department of Economic and Social Affairs. (2008). The Real Jewels of the Kalahari. Dryland Ecosystem Goods and Services in Kgalagadi South District, Botswana. Background Paper No. 7. https://www.un.org/esa/sustdev/csd/csd16/documents/bp7_2008.pdf

van Rooyen, A. F. (1998). Combating desertification in the southern Kalahari: Connecting science with community action in South Africa. *Journal of Arid Environments, 39*(2), 285–297. https://doi.org/10.1006/jare.1998.0407

Van Wyk, P., & Le Richie, E. A. N. (1984). The Kalahari Gemsbok National Park: 1931–1981. *Supplement to Koedoe, 1984*, 21–31.

Wang, C., & Burris, M. A. (1997). Photovoice: Concept, methodology, and use for participatory needs assessment. *Health Education & Behavior, 24*(3), 369–387.

Wang, C. C., Yi, W. K., Tao, Z. W., & Carovano, K. (1998). Photovoice as a participatory health promotion strategy. *Health Promotion International, 13*(1), 75–86.

Wu, K., Burris, M., Li, V., Wang, Y., Zhan, W., Xian, Y., Yang, K., & Wang, C. (Eds.). (1995). *Visual voices: 100 photographs of village China by the women of Yunnan Province*. Yunnan People's Publishing House.

Chapter 17
Tourism Joint Venture Partnerships and Conservation in Southern Africa

Gaofenngwe Dibotlhale, Hesekia Garekae, and Moren T. Stone

17.1 Introduction

Despite the effects of COVID-19, tourism remains one of the world's largest industries. There is growing awareness in the international development community of the potential of tourism to stimulate regional development, reduce poverty, empower marginalized communities, and help protect the environment (Mogomotsi et al., 2023; Torres et al., 2011). Local communities tend to possess relatively limited knowledge (including technology), so they are likely to have a greater need to enhance their knowledge base with more advanced knowledge possibly transferred from established entities that have tourism business experience (Yun et al., 2018). There have been calls from scholars and practitioners to involve local communities to take advantage of tourism business opportunities, however, the reality is that in the Global South there is a paucity of successful models for engaging local communities to optimally participate in tourism business development. To bridge this gap, a call for joint venture partnerships (JVPs) between private tourism investors and local communities has been proposed (Ashley & Jones, 2001; Stone, 2015a, b; Stronza, 2008). Tourism JVPs between communities and private investors are an emerging trend in Southern Africa and they take different forms in each country (Ashley & Jones, 2001).

One strategy that indigenous leaders throughout the world are using to expand their knowledge, skills and management capacity in tourism is to forge JVPs with other actors (Forstner, 2004). These partnerships are purported to enable residents

G. Dibotlhale (✉) · H. Garekae · M. T. Stone
Department of Environmental Science, University of Botswana, Gaborone, Botswana
e-mail: garekaeh@ub.ac.bw; stonemt@ub.ac.bw

© The Author(s), under exclusive license to Springer Nature Switzerland AG 2024

L. S. Stone, M. T. Stone (eds.), *Wildlife Tourism Dynamics in Southern Africa*, Geographies of Tourism and Global Change,
https://doi.org/10.1007/978-3-031-57252-4_17

with no experience in tourism to participate and benefit from tourism (Stronza, 2008). In such partnerships, local communities link their knowledge, land, labour and social capital with the investment capital, business acumen and managerial experience of outside tourism operators or non-governmental organisations (NGOs) (Stronza, 2008). In Southern Africa, tourism JVPs between the community and private companies are usually established through the initiative of Community Based Natural Resources management (CBNRM) (Ashley & Jones, 2001; Lepetu et al., 2008). CBNRM was developed around the 1980s as a more inclusive approach to conservation that involves local people in the management of natural resources (Wali et al., 2017; Zielinski et al., 2020).

Ashley et al. (2001) posit that a 'joint venture' in the context of southern Africa is not used in the strict legal sense as used in commerce, when two or more companies form a joint venture, but is defined as a contractual partnership between a community or local institution and a private investor who work together in establishing and operating a single tourism or hunting enterprise. The parties may not both/all own the company assets in legal terms; both/all may have rights and responsibilities to contribute to the benefit from the enterprise (Ashley & Jones, 2001).

CBNRM was funded by the United States Agency for International Development (USAID) in many countries in Southern Africa such as Botswana, Namibia, Zimbabwe, Zambia, Mozambique and South Africa (Cobbinah, 2015; Cobbinah et al., 2015). CBNRM is perceived as a win-win approach premised on achieving both the conservation of natural resources and improving the livelihoods of communities (Jewitt et al., 2014; Sebele, 2010; Sène-Harper & Séye, 2019). Empirical evidence indicates that if local communities participate in the management of natural resources in their areas, they get direct benefits and develop positive attitudes towards conservation (Ashley & Jones, 2001; Musavengane & Kloppers, 2020). However, scholars such as Adeyanju et al. (2022) and Büscher et al. (2012) have questioned whether CBNRM can achieve biodiversity conservation and improve community livelihoods. In Southern Africa, tourism depends on wildlife; trophy hunting generates money for local communities involved in CBNRM. If more wildlife is killed through trophy hunting to increase community benefits, conservation will be compromised. On the other hand, if CBNRM activities focus more on the conservation of biodiversity then community benefits might decrease.

As a result of the critics of CBNRM, JVPs between local communities and private companies were suggested because it was realized that although local communities have the rights to manage resources in their areas, they are not well equipped with management expertise, skills, knowledge and capital to manage them (Ashley & Jones, 2001). Communities that take part in CBNRM activities are encouraged to get into JVPs with private companies to improve the efficiency of community-based tourism businesses. This chapter assesses the relationships and performance of joint venture partnerships in Southern Africa.

17.2 Methods

This study relied on the use of secondary data sources. Journal articles and reports on CBNRM and JVPs in Southern Africa provided data used in the chapter. Information on different types of JVPs and examples of JVPs in Southern Africa were derived from these sources.

17.3 Overview of JVPs in Southern Africa

The Department of Wildlife and National Parks (1999, p. 1) defines a joint venture as "a business activity undertaken between two or more partners for their mutual benefit. Partners in a community joint venture will be rural people, who have user rights to the natural resources occurring in an area and established private sector companies that recognize an area's potential for business development".

JVPs or contractual agreements are seen as a remedy because it is difficult for local communities to develop natural resources on their own for tourism activities because they lack expertise in tourism business management (Ashley & Jones, 2001). Ideally, private companies have better knowledge of tourism business management and finances, so joint ventures give local communities opportunities to tap into the skills and knowledge from private companies. Local communities in countries such as Zimbabwe, Namibia and Botswana that have privileges to use wildlife resources were encouraged to formally register community-based organizations (CBOs) and get into joint venture partnerships with private companies to manage wildlife together. Trophy hunting is involved in joint ventures that promote conservation and wildlife management (Minja et al., 2022).

While JVPs in Southern Africa may take different forms there are two common examples of JVPs between local communities and private companies. The first type occurs when a community owns a wildlife hunting quota in their area, and they sell the quota to a trophy hunter (Mbaiwa, 2015). Communities that live adjacent to wildlife management areas and protected areas are normally given a wildlife quota for trophy hunting; they get into agreements to sub lease the quota to a private tourism operator (Mbaiwa, 2015). In return the private tourism operator pays for the wildlife quota (Ashley et al., 2001). The other common type is when the private company agrees to establish a tourism infrastructure such as a safari lodge in the community's land, and the private company pays lease fees to the community (Ashley et al., 2001). The partnerships normally bind private companies to provide job opportunities, skills transfer and capacity building for local communities (Gujadhur, 2001).

17.4 Established JVPs for Wildlife Conservation in Southern Africa

17.4.1 Botswana

There are five different types of JVPs in Botswana. These include jointly owned and run lodges, jointly owned and run hunting operations, jointly owned and run photographic operations, CBO owned tourism operations with marketing and management provided by the private sector, and jointly owned and run operations which process and markets veld products (Van der Jagt et al., 2000). An example of jointly owned and run lodges is the Ngoma lodge situated in the Chobe District area and the Santawani Lodge in the Okavango area (Lepetu et al., 2008). With jointly owned and run hunting operations, examples include the Sankuyo Tshwaragano Management Trust (STMT) in the Okavango area. The communities in Sankuyo have been allocated controlled hunting areas (CHAs) in NG33 and NG34 for photographic and hunting purposes and they lease their hunting rights to private tourism operators (Mokgalo, 2020).

17.4.2 The Case of Ngoma Lodge, Botswana

In Botswana the Chobe Enclave Conservation Trust (CECT) was the first trust to enter into a joint venture agreement with a private safari company, it is also the first community-based organization to earn large amounts of money from natural resource utilisation (Jones, 2002). The trust got into a JVP with a private company called Ngoma Management Company to establish a conservation lodge called the Ngoma Safari Lodge (Black & Cobbinah, 2017). The agreement started in 2010 with the construction, development and management of the lodge, which is situated in Chobe Forest Reserve, about 55 km from Kasane. CECT was able to mobilise significant donor support from USAID and African Wildlife Foundation contributed \$1.77 million (about 50% of the total capital value of the lodge) and their own capital towards constructing the lodge (Chevallier & Harvey, 2016; Stone, 2015b). Ngoma lodge provides accommodation and nature-based experiences in the form of game drives (Mbaiwa & Tshamekang, 2012). The local community has a 20 year head lease and receive land rental fees of BWP 60,000.00 with an annual increment of BWP 15,000.00 and variable community fees of 6% (Year 1–5), 8% (Year 6–10), 9% (Year 11–15) and 10% (Year 16–20) of all total lodge revenue (Mbaiwa & Tshamekang, 2012). According to the agreement, the lodge has to employ at least 16 people from Chobe Enclave villages; in 2016 the lodge employed 36 people (Chevallier & Harvey, 2016). In 1997, the CECT earned BWP 46,400 and by 2011, this figure had increased to BWP 578,685,[1] more than 10 times the original annual income (Stone, 2015b). The CECT has been managing the annual wildlife quotas received from the Department of Wildlife and National Parks since 1994 with a 50-year head lease from the Land Board to conduct wildlife and tourism activities (Chevallier & Harvey, 2016).

In Botswana, the stumbling block for implementing JVPs is the lack of implementation guidelines. The Botswana government in realising this challenge, developed a series of formal guidelines for the establishment and implementation of joint ventures (Ashley & Jones, 2001). These guidelines serve a good purpose of guiding government agencies that assist communities in their negotiations with private companies and evaluating tender bids. This has developed the community political capital as they now have a voice during the tendering process. This has enabled them to decide on which companies to partner with.

17.4.3 Damaraland Camp in Namibia

Through the CBNRM initiative, some communities are given rights over some wildlife species; they are then encouraged to form and legally register CBOs called conservancies under the Nature Conservation Act of 1996 (Jones, 1999; Turpie & Letley, 2021). A conservancy must have a registered membership, legal constitution, representative management committee and defined wildlife management boundaries (Ashley & Jones, 2001). After communities receive concession rights over wildlife, they get into JVPs with private companies for photographic tourism and trophy hunting. The first joint venture to be formed in 1996 was between Wilderness Safari and communities in the Kunene region through Tora conservancy, for the management of Damaraland camp (Ashley & Jones, 2001). It is an up-market camp consisting of a 16-bed tented lodge which provides jobs for the local people. The camp is situated in the Huab river valley where there are numerous wildlife species. The partnership required Wilderness Safari to train local people for management positions, to pay bed levy and flat fees to the community (Salole, 2007). In return, the community provides the Wilderness Safari with exclusive rights to use the lodge area.

To pursue the JVP, the parties signed a contract that included a commitment by Wilderness Safaris to recruitment staff from the local community; train community members up to management level; pay 10% of net accommodation fees to the community in the form of a levy per bed per night; pay 300 British Pound rent for the lodge site per month; an option for the community to purchase the lodge assets and continue alone after 10 years; and a provision for the lodge to outsource goods and services from the community (Rylance & Spenceley, 2016). Through the JVP arrangement, Wilderness Safaris funded the total cost of constructing the lodge, but

[1] 1 pula = 0.073 USD (11 September 2023).

ownership was transferred to the community at the rate of 20% per year, beginning on the tenth year of the partnership; once the community owned 100% of the camp and its assets, it sold 60% of it back to Wilderness Safaris (Rylance & Spenceley, 2016). The community retained the remaining 40% equity stake, and the lodge was subsequently upgraded, funded by both Wilderness Safaris and the Torra Conservancy (Rylance & Spenceley, 2016). According to Snyman (2012) the partnership has contributed to poverty reduction and economic development in the Kunene region.

17.4.4 The Mahenye Lodge in Zimbabwe

The Department of National Parks and Wildlife Management (DNPWLM) in Zimbabwe introduced the CBNRM programme in 1982 through the Communal Area Management Programme for Indigenous Resources [CAMPFIRE] (Rihoy & Maguranyanga, 2007; Shereni & Saarinen, 2021). CAMPFIRE was established to mediate human wildlife conflicts which rose because of the displacement of local communities from their ancestral land to establish protected areas; local people were not even included in the management of wildlife. CAMPFIRE promotes JVPs between local communities and private companies, especially those in the safari hunting sector.

A joint venture agreement between the local Shangaan people and African Sun Limited (formerly Zimsun Limited) was formed to manage the Mahenye lodge located in Ndowoyo Communal Land (Mudzengi et al., 2021). The Zimsun leased land from the Mahenye community to develop two lodges with a contractual agreement to pay annual lease fees to the community (Goodwin, 2002). The company agreed to contribute a percentage of its gross receipts to the project, increasing from 8% to 12% over the life of the contract. After the lodges opened for business in the mid-1990s, revenue from lodge receipts and hunting safaris produced annual cash disbursements to approximately 800 households, representing an increment of 5–10% above incomes from subsistence agriculture and animal husbandry (Mudzengi et al., 2022). Unlike most CAMPFIRE communities that relied exclusively on revenue from sport hunting, the Mahenye project through the JVP, branched out into game-viewing tourism (Mudzengi et al., 2021). The main tourism activities include game drives, bird watching, fishing and boat cruises. The partnership has employed local community members who are employed as tour guides and to work in the lodges. The partnership has contributed to infrastructural improvements such as roads, water connections, electricity connections at the local clinic, telephone connections at the school and the community center (Chiutsi et al., 2011; Murphree, 2001). The lodges have also provided training opportunities for the local people. The Mahenye CAMPFIRE committee also initiated environmental education programmes to raise environmental awareness (Mudzengi et al., 2021).

17.4.5 The Makuleke in South Africa

Under South Africa's Strategic Development Initiative, a wide range of joint ventures have been developed on communal land, inside protected areas, and on land that residents have reclaimed from within protected areas through the legal land claim process (Ashley & Jones, 2001). National Park authorities have found ways to increase community benefits through conservation and the development of joint venture partnerships (Chiutsi et al., 2011). The Makuleke community through the Makuleke Communal Property Association (CPA) got into a partnership with SanParks to co-manage the region that constitutes the Luvuvhu, Limpopo and Pafuri rivers. Due to the region's proximity to Kruger National Park, there is a variety of wildlife in the region (Turner, 2004). Tourism activities in the Makuleke region include safari hunting, lodging, cultural and village tourism. The Makuleke community signed an agreement with Matswani Safaris to establish lodges in the area. In return, local people have been hired to work in the lodge and Matswani safaris offered vocational training to the Makuleke community (Turner, 2004).

The community is also provided with game meat during trophy hunting activities (Chiutsi et al., 2011). Employment through the JVP arrangement was found to be one of the main benefits for the people of Makuleke. For instance, those employed as tour guides earned a salary of approximately R7500, housekeepers got R5500, cooks earned R5000 and those employed by the project as administrators earned approximately R8000, accountants were paid R6000, the project secretary/receptionist earned R5000, and project drivers earned R4000 per month (Matiku et al., 2020). Matiku et al. (2020) conclude that the employment benefits provided through the JVPs have improved the livelihoods of many households in the villages.

17.5 Impacts of JVPs on Community-Based Tourism Ventures

17.5.1 Socio-economic Contribution of JVPs to Local Communities

JVPs between the community and private companies can bring several benefits and opportunities for both parties. For local communities, joint ventures can create opportunities for economic development because they can tap into resources such as management skills, specialized skills, advanced technology and financial resources to enhance their natural resource management practices (Ashley & Jones, 2001). This will increase financial benefits that communities get from natural resource management and improve their livelihoods.

In Botswana, JVPs have benefitted local communities through the creation of employment opportunities, income generation, and heightened knowledge of conservation and the protection of local heritage sites and resources (Black & Cobbinah, 2017; Elijah et al., 2022; Mbaiwa & Tshamekang, 2012). Elijah et al. (2022) indicates that about 88% of the Goo-Moremi community has benefited from a JVP with a private investor, through job opportunities and income generation of P100, 000 (US$10000.03) given to the community annually through the village trust. Communities that own hunting rights through JVPs can sell their wildlife quotas to private tourism operators and the money they get is used for community development and wildlife conservation. According to Johnson (2009) between 2006 and 2009 trophy hunting in Botswana generated USD 413,014 through CBOs. In 2018, CECT communities made BWP 4571897.92 (USD 335,356.44) from selling wildlife quota through JVPs (Stone & Stone, 2020).

Tourism activities such as game drives generate income for the partnerships and local communities get to benefit from them. For instance, in Zimbabwe, Gonarezhou National Park generates money for the local communities through wildlife JVPs (Musakwa et al., 2020). From 1995 to 2002 revenue from hunting in CAMPFIRE projects were maintained at USD1.5-2.0 million annually (Arntzen et al., 2003). According to Namibian Association of CBNRM Support Organisations (NACSO) (2010), in 2009 JVPs in Namibia managed to contribute NAD 19.9 million (57% of all income) to conservancies. Local people also get employment opportunities created by JVPs; they work in conservation lodges such as Ngoma lodge in Botswana, Chilo and Mahenye in Zimbabwe and Damaraland camp in Namibia (Arntzen et al., 2003).

According to Mbaiwa (2015), revenue from joint ventures is used to provide social services such as provision of scholarships, support for local sports activities, building of water standpipes, provision of communication tools such as radios, funeral assistance and building houses for orphans and the elderly. CBOs also provide monthly allowance to the needy and disabled persons. According to Mbaiwa and Stronza (2010), CBNRM activities in the Okavango area (Khwai, Sankoyo and Mababe) funded local people with scholarships amounting to between BWP 7000 to BWP 35,000 per village. In Namibia, Torra Conservancy distributed cash dividends to individual conservancy members; they have managed to contribute to the building of a crèche and provided grants to a local school (Arntzen et al., 2003).

17.5.2 JVPs Contribution to Capacity Building, Skills Transfer and Environmental Management

Private companies provide training and capacity building as part of partnership agreements (Mbaiwa & Tshamekang, 2012). Training programmes and knowledge sharing are organized to enhance the conservation skills of community members. By investing in capacity building, JVPs contribute to the long-term sustainability of the tourism industry. Community members tap into the opportunity and use natural resources for sustainable conservation practices and improve resource monitoring. The communities are also empowered to have a sense of ownership of their resources. According to Snyman (2012), JVPs in Namibia have contributed to improved planning skills and knowledge on the resource rights of communities. JVPs in Torra conservancy have empowered the local communities to make decisions regarding the management of their natural resources (Snyman, 2012).

JVPs encourage integrated resource management because there are different stakeholders involved. Mbaiwa and Stronza (2010) and Arntzen et al. (2007) have noted that local communities have positive attitudes towards wildlife and conservation in CBNRM areas in Botswana because of the financial benefits they get from wildlife quotas. CBOs have hired community escort guides (CEGs) to ensure compliance with trophy hunting and to control poaching (Mbaiwa, 2015). CEGs make patrols in community concession areas, they accompany hunters in their safari hunting activities, and they enforce the rules of conservation in their CHA; this contributes to wildlife conservation. Another joint venture activity that promotes conservation is the wildlife quota system that is used for controlling wildlife species. Only wildlife species that are in abundance are included in the annual wildlife quota, whereas species that are threatened are not. In Zimbabwe, JVPs have organized conservation training workshops and conservation awareness in schools (Chiutsi & Mudzengi, 2012).

17.5.3 Challenges Faced by JVPs Arrangements

Several studies have discussed challenges faced by joint venture partnerships between communities and private companies (Dolezal & Novelli, 2022; Gujadhur, 2000; Mbaiwa & Tshamekang, 2012; Rozemeijer & van der Jagt, 2000). Although JVPs give communities an opportunity to participate in natural resources management and tourism businesses, there is minimal transfer of skills from the private sector to communities (Gujadhur, 2000; Rozemeijer & van der Jagt, 2000). Local communities still lack skills, information and knowledge on the management of natural resources and running tourism businesses (Dolezal & Novelli, 2022). Schumann (2008) argues that the absence of long-term or strategic planning has been a serious stumbling block for the success of joint ventures between communities and the private sector.

There are claims that JVPs formed under community-based tourism are not as effective as anticipated (Dolezal & Novelli, 2022; Mbaiwa & Tshamekang, 2012; Rozemeijer and van der Jagt (2000). It was anticipated that JVCs would result in the transfer of skills to communities to enable them to run tourism businesses independently (DWNP, 1999). However, Gujadhur (2001), Mbaiwa and Tshamekang (2012) and Rozemeijer and van der Jagt (2000) point out that there are no true collaborations and transfer of skills. According to Gujadhur (2001) local communities have little to do with the planning, management and monitoring of tourism businesses.

Furthermore, there are many stakeholders involved in JVPs, so there is lack of trust (Cassidy & Jansen, 2000; Rozemeijer & van der Jagt, 2000) and it becomes hard to satisfy the interest of all stakeholders involved.

Local communities are not actively involved in decision making processes because they normally lack funds, skills and knowledge, so their voice is not heard. Private companies take an upper hand in decision making even when natural resources involved belong to local communities. Mbaiwa and Tshamekang (2012) noted that in Botswana CBOs such as CECT who have partnerships with private companies derive less benefits than private sector companies.

17.5.4 JVPs, Community-Based Conservation and Sustainability

The agenda 2030 and Sustainable Development Goals (SDGs) advocate for the inclusion of rural communities in sustainable development efforts (UNWTO & UNDP, 2017). SDGs also emphasize the importance of public-private partnerships in empowering marginalized communities and giving them a platform to take part in the management of their natural resources (Dolezal & Novelli, 2022). Community based tourism (CBT) allows local communities to establish tourism businesses using natural resources in their areas, and to get into JVPs with private tourism operators to run tourism businesses together. CBT implies that local people are responsible for the conservation of natural resources in their areas (Karacaoglu & Birdir, 2017), and are also responsible for limiting the number of visitors to avoid negative impacts on the community, local cultures and the environment (Prasiasa et al., 2020). Hence CBT is regarded as a sustainable form of tourism because it contributes to environmental conservation while at the same time improving community livelihoods.

Musakwa et al. (2020) agree that partnerships yield better conservation and community benefits, so there is need for transparency, trust and accountability to increase the sustainability of JVPs. JVPs also have the potential to provide economic benefits and decision-making powers to host communities (Gutierrez, 2022; Lapeyre, 2010) through capacity building, stakeholder inclusion and community participation (Dangi & Jamal, 2016; Stone, 2015a, b). By supporting local businesses and disadvantaged members of the community, JVPs in CBT can alleviate poverty, hence supporting the SDG number one (1) of no poverty.

However, for long term sustainability (habitat restoration, wildlife conservation and community development) to be achieved in JVPs, there should be clear benefit sharing mechanisms to ensure that community interests are protected, effective communication and trust building between the contractual parties (Rozemeijer & van der Jagt, 2000). If there is no proper planning and management of tourism activities in JVPs, it can destroy the environment and negatively affect communities. Although Graci (2013) highlights that collaborations and joint ventures can help tourism to move towards sustainability, the efforts of sustainable tourism are hindered by lack of proper collaboration between key stakeholders.

The interest in community involvement in the conservation of natural resources has increased across the world (Selfa & Endter-Wada, 2008). There is also an increase in protected areas and tourism activities which may increase improvements in community livelihoods and the conservation of wildlife resources if indeed JVPs are to serve their mandate. To make such partnerships possible, continued strengthened capacity building for local communities is needed to ensure they ultimately assume full control of tourism operations.

17.6 Conclusions

This chapter assessed JVPs in Southern Africa. JVCs have been considered a better initiative to use in wildlife conservation and wildlife improvement. They have yielded better economic benefits for local communities in Southern Africa, however, they have faced many challenges. Results show that although local communities are involved, they do not actively participate in tourism development and the management of wildlife resources. For JVCs to work effectively, there is a need for trust, communication and transparency between partners. If managed well, JVCs can result in capacity building and skills transfer to 0ol.local communities and the creation of opportunities for private sector investors to access new business opportunities using assets owned by communities. Tourism companies need to demonstrate their commitment to local development and justify the existence of what has been seen as an elitist industry by finding a winning formula in the implementation of JVPs. Joint ventures that have been developed in the southern Africa region reflect the governments' willingness to seek new ways to develop sustainable local economies, with a focus on tourism and conservation. JVPs therefore, symbolize equity and inclusiveness in the tourism industry; a industry long dominated by capitalist corporations at the disadvantage of local communities who are the custodians of wildlife resources. A critical tenet of JVPs is that community members should be actively involved in the enterprise, not only as staff, but as owners, planners and administrators (Stronza, 2008).

It is evident that communities do not play any meaningful role in the operations of JVPs. There is limited capacity building and skills transfer from the private sector to the community. This is contrary to the ideals of JVP, which are to instill the community with business acumen. Among other things, JVPs are premised on capacitating the local community with requisite skills and knowledge to run business enterprises. It is anticipated that the community will gradually graduate from the partnership after acquiring the necessary skills and knowledge to run business enterprises independently. However, this is seemingly far-fetched since communities have been relegated to collecting paltry rental fees and other meagre benefits accruing from their partnership with private investors. Therefore, this calls for active community involvement in the operations and decision-making process of JVPs.

For example, communities should be involved in the planning and development of business portfolios and reinvestment of the proceeds. To achieve this, there is a need to formulate holistic regulations to govern the implementation of JVPs. Moreover, there is a need for the development of action plans to guide knowledge and skills transfer between private investors and local communities. The study calls for future research to develop models/frameworks geared towards improving the performance of JVPs.

References

Adeyanju, S., O'connor, A., Addoah, T., Bayala, E., Djoudi, H., Moombe, K., et al. (2022). Learning from community-based natural resource management (CBNRM) in Ghana and Zambia: Lessons for integrated landscape approaches. *International Forestry Review, 23*(3), 273–297.

Arntzen, J. W., Molokomme, D. L., Terry, E. M., Moleele, N., Tshosa, O., & Mazambani, D. (2003). Main findings of the review of CBNRM in Botswana. *Occasional paper*, (14).

Arntzen, J., Buzwani, B., Setlhogile, T., Kgathi, D. L., & Motsolapheko, M. K. (2007). Community-based resource management, rural livelihoods and environmental sustainability. Centre for Applied Research.

Ashley, C., & Jones, B. (2001). Joint ventures between communities and tourism investors: Experience in southern Africa. *International Journal of Tourism Research, 3*(5), 407–423.

Ashley, C., Goodwin, H., & Roe, D., (2001). Pro-poor tourism: Expanding opportunities for the poor. *PPT Policy Briefing No 1*, 4pp.

Black, R., & Cobbinah, P. (2017). On the rim of inspiration: Performance of AWF tourism enterprises in Botswana and Rwanda. *Journal of Sustainable Tourism, 25*(11), 1600–1616. https://doi.org/10.1080/09669582.2017.1296454

Büscher, B., Sullivan, S., Neves, K., Igoe, J., & Brockington, D. (2012). Towards a synthesized critique of neoliberal biodiversity conservation. *Capitalism Nature Socialism, 23*(2), 4–30.

Cassidy, L., & Jansen, R. (2000). *National conference on community based natural resource management in Botswana: Report of conference proceedings*. IUCN.

Chevallier, R., & Harvey, R. (2016). *Is community-based natural resource management in Botswana viable?* (Policy Insights 31) (pp. 1–12). South African Institute of International Afairs (SAIIA).

Chiutsi, S., & Mudzengi, B. K. (2012). Community tourism entrepreneurship for sustainable tourism management in Southern Africa: Lessons from Zimbabwe. *International Journal of Academic Research in Business and Social Sciences, 2*(8), 127.

Chiutsi, S., Mukoroverwa, M., Karigambe, P., & Mudzengi, B. (2011). The theory and practice of ecotourism in southern Africa. *Journal of Hospitality Management and Tourism, 2*(2), 14–21.

Cobbinah, P. (2015). Contextualising the meaning of ecotourism. *Tourism Management Perspectives, 16*, 179–189.

Cobbinah, P. B., Black, R., & Thwaites, R. (2015). Biodiversity conservation and livelihoods in rural Ghana: Impacts and coping strategies. *Environmental Development, 15*, 79–93.

Dangi, T., & Jamal, T. (2016). An integrated approach to "sustainable community-based tourism". *Sustainability, 8*(5), 1–32.

Dolezal, C., & Novelli, M. (2022). Power in community-based tourism: Empowerment and partnership in Bali. *Journal of Sustainable Tourism, 30*(10), 2352–2370.

DWNP. (1999, April). *Joint ventures, A guide to developing natural resource based business ventures* (Revised edition). The Department of Wildlife and National Parks.

Elijah, B., Moswete, N. N., & Mpotokwane, M. A. (2022). Assessment of costs and benefits of joint venture partnerships in community-based tourism between the private sector and Goo-Moremi residents, Botswana. In J. Saarinen, B. Lubbe, & N. N. Moswete (Eds.), *Southern African perspectives on sustainable tourism management. Geographies of tourism and global change*. Springer. https://doi.org/10.1007/978-3-030-99435-8_7

Forstner, K. (2004). Community ventures and access to markets: The role of intermediaries in marketing rural tourism products. *Development Policy Review, 22*(5), 497–514.

Goodwin, H. (2002). Local community involvement in tourism around national parks: Opportunities and constraints. *Current Issues in Tourism, 5*(3-4), 338–360.

Graci, S. (2013). Collaboration and partnership development for sustainable tourism. *Tourism Geographies, 15*(1), 25–42. https://doi.org/10.1080/14616688.2012.675513

Gujadhur, T. (2000). *Organisations and their approaches in community based natural resources management in Botswana, Namibia, Zambia and Zimbabwe* (No. 1). Iucn.

Gujadhur, T. (2001). *Joint venture options for communities and safari operators in Botswana* (CBNRM Support Programme web edition). IUCN.

Gutierrez, E. L. M. (2022). Participation in tourism cases on community-based tourism (CBT) in The Philippines. *Ritsumeikan Journal of Asia Pacific Studies, 37*(1), 23–36.

Jewitt, S., Nasir, D., Page, S., Rieley, J., & Khanal, K. (2014). Indonesia's contested domains. Deforestation, rehabilitation and conservation-with-development in Central Kalimantan's tropical peatlands. *International Forestry Review, 16*(4), 405–420.

Johnson, S. (2009). State of CBNRM report 2009. In *Botswana National CBNRM Forum*. Botswana.

Jones, B. T. (1999). Policy lessons from the evolution of a community-based approach to wildlife management, Kunene region, Namibia. *Journal of International Development: The Journal of the Development Studies Association, 11*(2), 295–304.

Jones, B. T. (2002). *Chobe Enclave, Botswana: Lessons learnt from a CBNRM project 1993–2002* (CBNRM Support Programme Occasional Paper 7) (p. 11). IUCN.

Karacaoglu, S., & Birdir, K. (2017). Success factors of community based tourism (CBT) perceived by local peoples: The case of % 100 Misia project. *International Rural Tourism and Development Journal, 1*(2), 53–61.

Lapeyre, R. (2010). Community-based tourism as a sustainable solution to maximise impacts locally? The Tsiseb Conservancy case, Namibia. *Development Southern Africa, 27*(5), 757–772. https://doi.org/10.1080/0376835X.2010.52283

Lepetu, J., Makopondo, R., & Darkoh, M. (2008). Community-based natural resource management and tourism partnership in Botswana: Which way forward? *Botswana Notes and Records, 39*, 113–124.

Matiku, S., Zuwarimwe, J., & Tshipala, N. (2020). Community-driven tourism projects' economic contribution to community livelihoods—A Case of Makuleke Contractual Park Community Tourism Project. *Sustainability, 12*(19), 8230.

Mbaiwa, J. (2015). Ecotourism in Botswana: 30 years later. *Journal of Ecotourism, 14*(2-3), 204–222.

Mbaiwa, J. E., & Stronza, A. L. (2010). The effects of tourism development on rural livelihoods in the Okavango Delta, Botswana. *Journal of Sustainable Tourism, 18*(5), 635–656.

Mbaiwa, J., & Tshamekang, T. (2012). Developing a viable community-based tourism project in Botswana: The case of the Chobe Enclave Conservation Trust. *World Sustainable Development Outlook, 2012*, 519–536.

Minja, G., Suh, J., & Tan, Y. (2022). Joint venture wildlife management of the Loliondo game controlled area in Tanzania: A multi-criteria analysis. *South African Geographical Journal*, 1–19. https://doi.org/10.1080/03736245.2022.2129767

Mogomotsi, P. K., Mogomotsi, G., Stone, L. S., & Stone, M. T. (2023). An analysis of residents' willingness to pay for domestic nature-based tourism in Botswana during and post the COVID-19 pandemic. *African Journal of Hospitality, Tourism and Leisure, 12*(1), 257–272. https://doi.org/10.46222/ajhtl.19770720.366

Mokgalo, L. (2020). *A sustainable strategy to mitigate the ban of hunting on local communities in Botswana*. Doctoral dissertation. North-West University (South Africa).

Mudzengi, B., Gandiwa, E., Muboko, N., & Mutanga, C. N. (2021). Towards sustainable community conservation in tropical savanna ecosystems: A management framework for ecotourism ventures in a changing environment. *Environment, Development and Sustainability, 23*, 3028–3047.

Mudzengi, B. K., Gandiwa, E., Muboko, N., & Mutanga, C. N. (2022). Innovative community ecotourism coping and recovery strategies to COVID-19 pandemic shocks: The case of Mahenye. *Development Southern Africa, 39*(1), 68–83.

Murphree, M. (2001). Community, council & client: a case study in ecotourism development from Mahenye, Zimbabwe. In *African wildlife and livelihoods: The promise and performance of community conservation* (pp. 177–194). David Philip.

Musakwa, W., Mpofu, E., & Nyathi, N. (2020). Local community perceptions on landscape change, ecosystem services, climate change, and livelihoods in Gonarezhou National Park, Zimbabwe. *Sustainability, 12*(11), 4610.

Musavengane, R., & Kloppers, R. (2020). Social capital: An investment towards community resilience in the collaborative natural resources management of community-based tourism schemes. *Tourism Management Perspectives, 34*, 100654. https://doi.org/10.1016/j.tmp.2020.100654

NASCO. (2010). *Namibia's communal conservancies: A review of progress. Report*. Namibian Association of Community Based Natural Resource Management Support Organisations (NASCO).

Prasiasa, D., Astawa, I., Widari, S., & Udiyana, G. (2020). Analysis of community-based tourism implementation for sustainability of Tourism Village in Bali Province. *Journal of Tallent Development and Excellence, 12*(1), 152–116.

Rihoy, E., & Maguranyanga, B. (2007). *Devolution and democratisation of natural resource management in southern Africa: A comparative analysis of CBNRM policy processes in Botswana and Zimbabwe*. PLAAS.

Rozemeijer, N., & van der Jagt, C. (2000). *Practical guide for the facilitating CBNRM in Botswana*. IUCN/SNV Support Programme.

Rylance, A., & Spenceley, A. (2016). Applying inclusive business approaches to nature-based tourism in Namibia and South Africa. *Tourism: An International Interdisciplinary Journal, 64*(4), 371–383.

Salole, M. (2007). Merging two disparate worlds in rural Namibia: Joint venture tourism in Torra conservancy. In *Tourism and indigenous peoples: Issues and implications* (pp. 205–219). Routledge.

Schumann, F. (2008). *Private and public sector collaboration in Guam s tourism industry: Is Guam prepared for the future?* Universal-Publishers.

Sebele, L. S. (2010). Community-based tourism ventures, benefits and challenges: Khama rhino sanctuary trust, central district, Botswana. *Tourism Management, 31*(1), 136–146.

Selfa, T., & Endter-Wada, J. (2008). The politics of community-based conservation in natural resource management: A focus for international comparative analysis. *Environment and Planning A, 40*(4), 948–965.

Sène-Harper, A., & Séye, M. (2019). Community-based tourism around national parks in Senegal: The implications of colonial legacies in current management policies. *Tourism Planning & Development, 16*(2), 217–234.

Shereni, N., & Saarinen, J. (2021). Community perceptions on the benefits and challenges of community-based natural resources management in Zimbabwe. *Development Southern Africa, 38*(6), 879–895.

Snyman, S. (2012). Ecotourism joint ventures between the private sector and communities: An updated analysis of the Torra Conservancy and Damaraland Camp partnership, Namibia. *Tourism Management Perspectives, 4*, 127–135.

Stone, M. T. (2015a). Community-based ecotourism: A collaborative partnerships perspective. *Journal of Ecotourism, 14*(2-3), 166–184.

Stone, M. T. (2015b). Community empowerment through community-based tourism: The case of Chobe Enclave Conservation Trust in Botswana. In *Institutional arrangements for conservation, development and tourism in eastern and southern Africa: A dynamic perspective* (pp. 81–100). Springer.

Stone, M. T., & Stone, L. S. (2020). Community-based ecotourism and bush meat consumption dynamics: Implications for conservation and community development. *Journal of Sustainable Tourism*. https://doi.org/10.1080/09669582.2020.1845708

Stronza, A. (2008). Partnerships for tourism development. In *Building community capacity for tourism development* (pp. 101–115). CABI.

Torres, R., Skillicorn, P., & Nelson, V. (2011). Community corporate joint ventures: An alternative model for pro-poor tourism development. *Tourism Planning & Development, 8*(3), 297–316.

Turner, R. (2004). Communities, conservation, and tourism-based development: Can community-based nature. *Development,* 27(4), 629–649.

Turpie, J., & Letley, G. (2021). Would community conservation initiatives benefit from external financial oversight? A framed field experiment in Namibia's communal conservancies. *World Development, 142*, 105442.

UNWTO & UNDP. (2017). *Tourism and the sustainable development goals–journey to 2030. Highlights*. UNWTO & UNDP.

Van der Jagt, C., Gujadhur, T., & Bussel, F. (2000). *Community Benefits through CBNRM in Botswana* (CBNRM Network Occasional Paper No. 2). CBNRM Support Programme SNV/IUCN.

Wali, A., A lvira, D., Tallman, P. S., Ravikumar, A., & Macedo, M. O. (2017). A new approach to conservation: Using community empowerment for sustainable well-being. *Ecology and Society, 22*(4), 6.

Yun, J., Won, D., & Park, K. (2018). Entrepreneurial cyclical dynamics of open innovation. *Journal of Evolutionary Economics, 28*, 1151–1174.

Zielinski, S., Kim, S., Botero, C., & Yanes, A. (2020). Factors that facilitate and inhibit community-based tourism initiatives in developing countries. *Current Issues in Tourism, 23*(6), 723–739. https://doi.org/10.1080/13683500.2018.1543254

Part VI
Conclusion

Chapter 18
Wildlife Tourism Dynamics in Southern Africa: A Synthesis

Moren T. Stone and Lesego S. Stone

18.1 Introduction

Wildlife tourism research epistemology is a topic of ongoing debate and discussion (Benckendorff & Zehrer, 2013). To a certain extent, research on wildlife tourism is fragmented and eclectic (Kanta et al., 2021). In its endeavour to join the debate, this volume reached a dual postulation that wildlife tourism has both positive and negative impacts. The positive impacts include the development of communities and funding for protecting the environment and habitats. In contrast, it also negatively impacts the environment, host communities, and animals in its aim to promote conservation and drive development. There is no doubt that wildlife tourism has grown in recent years, due to the market it keeps attracting, indications are that it will continue to grow as the number of tourists who specifically want to see wildlife is increasing. These tourists display a wide range of heterogeneous attributes such as varied age groups, sex, socioeconomic backgrounds, and motivations (Khanra et al., 2021). This is a promising indication that the wildlife tourism market exists and continues to grow. In its growth trajectories, wildlife tourism has taken diverse forms. The key diverse forms of wildlife tourism are collectively categorized as consumptive and non-consumptive wildlife tourism. Non-consumptive wildlife tourism is generally considered a useful tool for engaging people with nature, to make them understand why nature is important to human beings and to assess how

M. T. Stone (✉)
Department of Environmental Science, University of Botswana, Gaborone, Botswana
e-mail: stonemt@ub.ac.bw

L. S. Stone
Faculty of Business, Department Tourism and Hospitality Management, University of Botswana, Gaborone, Botswana
e-mail: stonel@ub.ac.bw

© The Author(s), under exclusive license to Springer Nature Switzerland AG 2024

L. S. Stone, M. T. Stone (eds.), *Wildlife Tourism Dynamics in Southern Africa*, Geographies of Tourism and Global Change,
https://doi.org/10.1007/978-3-031-57252-4_18

memorable experiences with wildlife and the environment can change the thinking of humankind (Auster et al., 2020; Kanta et al., 2021).

In the same vein, consumptive wildlife tourism is contested as it is varyingly perceived by consumers, as the same tourism products are pursued by people with a wide range of interests, values and meanings, occasionally resulting in controversies over access rights to areas, resource conservation or particular forms of use (Øian et al., 2017). Non-consumptive use is often assumed to be a superior economic and environmental alternative to the consumptive use of wildlife as it involves the direct use or removal of wildlife through activities such as hunting (Meletis & Campbell, 2007). The contestation presents challenges to wildlife tourism as it is practiced on a contested landscape. In line with this, the volume aims to contribute to the wildlife tourism discourse by ensuring readers gain a better understanding of controversies and conflicts to improve sustainability in wildlife tourism dynamisms. Because tourism is increasingly integrated into global and complex market-based systems, the diversity of stakeholders and interests involved are also expansive in nature (Øian et al., 2017), making wildlife tourism a global issue and not one that belongs to the Southern African region only. As a result, it requires global solutions.

While there are contentious debates about which form of tourism (i.e., consumptive and non-consumptive) is more sustainable, some scholars are of the view that these forms of tourism have the same detrimental impacts on wildlife resources. For instance, Tremblay (2001, p. 83) suggests that it seems "non-consumptive activities reflect and convey morally superior values and lead to more intense and desirable experiences". For example, the definition of non-consumptive wildlife tourism adopted by the IUCN refers to 'enjoying' wild plants and animals (Ceballos-Lascuráin, 1996), but Meletis and Campbell (2007) argue that one interpretation of enjoying could include eating traditional dishes made from local ingredients. In the same line of argument, Hinch (1998) argues that tourists undertaking Inuit-hosted eco-tours, for example, might be invited to attend a community feast that includes locally wildlife killed meats. Tourists might also participate directly in hunting and/or fishing (Meletis & Campbell, 2007). Some so-called non-consumptive wildlife tourism destinations include fishing as part of the activities promoted and offered by destinations (Sell, 2004). In line with this debate, Nygren (2003) argues that defining some forms of tourism as non-consumptive perpetuates a western-influenced, pro-preservation and anti-extraction conception of wildlife tourism, and masks the heterogeneous nature of peoples, places, and activities that compose wildlife tourism. Therefore, the consequence of focusing on the 'direct removal of the species' in distinguishing between consumptive and non-consumptive use is that it masks the ways in which non-consumptive wildlife tourists can consume resources and result in broader negative environmental impacts, beyond those that occur on-site in non-consumptive tourism destinations (Meletis & Campbell, 2007).

There are no doubts that this volume has demonstrated and added value to the importance of wildlife tourism dynamics in the portfolio of environmental conservation and the livelihoods of poor rural communities living adjacent to protected areas. Nevertheless, there are critical issues that draw attention in wildlife conservation and tourism development.

18.2 Critical Issues for Southern Africa's Wildlife Tourism

Africa's tourism stories are best explained and understood, not in comparison to those of the Global North as that would be like comparing oranges with apples, but in terms of the continent's unique histories and dynamics that require localized contextualization (Mkono, 2019). Wildlife resources and tourism development entail, among other things, addressing two competing and overlapping goals; conserving biodiversity and improving community livelihoods (Stone & Stone, 2019). The success of wildlife tourism is therefore measured by how it balances resources conservation and community development. Resolving potential conflicts between these two goals is particularly challenging (Karanth & Defries, 2010). Nevertheless, there is no doubt that wildlife is at the heart of southern Africa's tourism product (Mkono, 2019). This wildlife legacy is threatened by changing demographics, increased poaching, habitat fragmentation, and global warming (Holechek & Valdez, 2018), but there are reasons for optimism that rest in adopting wildlife tourism as an intervention measure to mitigate the above highlighted threats. One major problem that engulfs southern Africa is the lack of systematic data on the status quo of wildlife resources and wildlife tourism performance; however, several studies have recently become available on the status and conservation of wildlife, as well as the intricate relationships of people, wildlife, and rangelands in eastern and southern African countries (Holechek & Valdez, 2018). The primary short-term threat for African wildlife is much more a problem of poaching than a lack of habitat. Poaching still occurs even when wildlife tourism development has been adopted as a mitigation measure. This has therefore led to a proliferation of the deployment of military services to provide security for wildlife resources. This development is not immune from criticism. For instance, Lunstrum (2014, p. 827) argues that securitization in the context of wildlife conservation is driven by the Western-directed and colonially derived narratives of sovereignty and land ownership that frame threats to wildlife as threats to "the sovereign nation-state, its territorial integrity, and its borders." This then raises questions of whether interventions adopted, such as CBNRM and community-based tourism enterprises have failed to deliver one of their mandates of protecting wildlife resources.

Jones (2021) argues that the militarization of African wildlife conservation is perpetuating neocolonial subjugation. To provoke further investigation of the ways western capitalist ideologies have driven the militarization of conservation, Lunstrum (2014, p. 817) coined the term 'green militarization' to encompass "the use of military and paramilitary personnel, training, technologies, and partnerships in the pursuit of conservation effort." This is a critical issue that dominates southern African conservation debates. It is not clear yet whether this conservation war will ultimately be won, but the remaining question is at whose benefit and/or cost is the war fought? Nevertheless, in wildlife conservation, legitimate disagreements arise regarding the best way to achieve conservation goals (Corkeron, 2023).

One critical issue that southern Africa wildlife resources and wildlife tourism face is the intensification of climate change. As the global climate crisis intensifies, biodiversity loss is accelerating worldwide. The loss of biodiversity directly affects wildlife-base tourism. The African continent is home to an estimated one-fifth of the world's known species, all of which are projected to decline in number as human activities and climate change continue posing new threats to ecosystems worldwide (Sintayehu, 2018, p. 226). Climate change is a critical yet under-appreciated amplifier of human–wildlife conflict, as it exacerbates resource scarcity, alters human and animal behaviours and distributions, and increases human–wildlife encounters (Abrahms et al., 2023; Mangachena & Pickering, 2023). As the warming of the Earth's climate reaches unprecedented levels, there are increasing calls to incorporate strategies of climate adaptation into current conservation planning and management (LeDee et al., 2021), but most recommendations regarding climate adaptation are difficult to apply at local scales (Prober et al., 2019). Therefore, Sattar et al. (2021) conclude that wildlife currently faces undefeatable problems than in the past due to climate change.

18.3 Southern Africa Wildlife Resources and Tourism: State of Affairs and Lessons

The southern Africa region presents a very interesting case of the state of affairs in managing the endowments of wildlife resources cherished. The main key approach Southern Africa has adopted as far as wildlife resources management is concerned is the win-win approach of conservation and development nexus. The main argument being largely premised on the idea that, if wildlife resources are properly managed, tourism can accommodate local participation and, in the process, yield benefits to compensate costs of living with wildlife resources, this would ultimately promote ownership and enhance conservation practices. While the foregoing remains a highly credible agenda that calls for conservation and development simultaneously, some stern structural, contextual, operational, and socio-cultural challenges make a successful realization of the twin objectives a mammoth obtainable test. For example, to highlight the findings of the chapters contributing to this volume, Zhou asserts that human-wildlife conflicts continue to soar in Zimbabwe, presenting multiple policy challenges on wildlife tourism policy makers even with the adoption of CAMPFIRE to fill the policy vacuum that existed in an endeavor to reconcile the conservation and development agenda.

Southern Africa is facing a dilemma on how to balance available livelihoods options, especially between agriculture and tourism existing on the same landscape. For example, in this volume, Stone highlights that the problem of veterinary fence has have brought a dilemma to both Southern African communities and governments on how to secure farmers' livelihoods through the beef industry while at the same time safeguarding the conservation of wild resources. Fencing aids the fragmentation of landscapes that surround protected areas and results in impassable barriers to the dispersal of highly mobile wildlife species. This in turn affects the performance of wildlife tourism in the region. In the same light, but on a positive note, Tichaawa and Lekgau in their assessment of the performance of Kgalagadi Transfrontier Park found that it constitutes a major role player in the socio-economic development of adjacent communities. Their findings support the positioning of wildlife tourism as an alternative livelihood strategy in such communities. Stone's work on assessing the performance of the TFCAs in Southern Africa indicates that transboundary conservation areas in southern Africa are increasing in number and have contributed to conservation and tourism development. Nevertheless, several challenges have been identified, including issues of stakeholders' engagement, balancing livestock production and wildlife conservation in and around TFCAs, and conservation and peace tourism contradictory outcomes. All these challenges have implications on conservation and tourism development in TFCAs. On assessing consumptive wildlife tourism and its implications for sustainable development, Lyakurwa and Sabuhoro highlight the current controversy regarding utilitarian and ontological perspectives revolving around the trophy hunting industry. Their cautionary stand is that while consumptive tourism contributes to conservation and community development, simplistic media depictions of trophy hunting have the potential to cause negative outcomes for conservation through its impact on public perceptions and political opinions, ultimately impacting negatively on trophy hunting and the state affairs of Southern Africa.

Overall, the many case studies contributing to this volume indicate both positive and negative results on wildlife tourism dynamics in Southern Africa. The contemporary issues, and challenges are lessons that have prospective contributions to the attainment of sustainable development. What is distinctively clear is that the relationships between wildlife resources conservation and tourism development are dynamic, complex and sometimes contradictory, such that there is no single approach, theory, conceptual framework, and methodological foundations to examine the complex relationships and yield conclusive findings. At times, wildlife resources conservation, tourism development and livelihoods seem to be matching and in harmony, however, at other times it appears to be competitive, while at other times, it is in conflicts and wars. The purpose of this book was therefore not to reproduce literature, but to explore, analyze and synthesize by systematically linking wildlife conservation, tourism and community livelihoods so as to establish trade-offs, synergies and way forward to help southern Africa to grow her comparative advantage on natural resources abundance to the benefit of her resources and citizens. The overall results and contributions of this book speak to policy makers, conservation fronts, community development planners and architects who need to recognize how complex it is to implement conservation efforts and development agendas on socio-ecological environments that are not immune from historical settings. The failure to recognize how complex it is to implement conservation and development plans in socio-ecological and complex settings may deem the intended agenda a failure before implementation resumes.

18.4 Policy Matters, Implications and Way Forward

In southern Africa the delegation of power to communities through programs such as CAMPFIRE, conservancies and CBNRM have been in practice but community voices on matters of wildlife tourism policy formulation and its inclusive use remain suppressed (Zhou, 2023). This then questions whether the decentralization of wildlife management from central governments to local communities has not taken place as it was envisaged. Participation in wildlife-based tourism by communities also has not lived to the expectations. One of the challenges of wildlife-based tourism is that resource-poor communities have to engage in 'partnerships' with investors who can finance the construction of tourist lodges and organize hunting expeditions and photo safaris. Seemingly, this runs the risk of 'blatant recycling of colonial fantasies' about wilderness in the design, construction and operation of tourist lodges, where African people are hidden from sight and relegated to the role of servants (Garland, 2008; Nattrass, 2021). Participation in wildlife-based tourism through the rubric of CBNRM, even when perceived to be functioning as democratically and equitably as can be hoped, remains vulnerable to a set of linked concerns pertaining to the power of international tourism operators and hunting outfitters to extract the lion's share of returns (Abrahms et al., 2023). This calls for another chain of policy devices to overcome this resultant outcome that emerges by default and/or through unintended consequences.

As a policy device, several approaches such as partnerships, co-management, and community benefit sharing schemes, community involvement, and devolution have been deployed to manage the relationship between local communities and wildlife tourism enterprises with varying degrees of success (Ashley, 2000). Therefore, sustainable wildlife tourism has been largely framed within a wildlife tourism and conservation paradigm at the expense of local communities' world view (Qin et al., 2020). Local communities' level of involvement and inclusivity in the utilization of wildlife tourism resources remain peripheral issues in the wildlife tourism discourse (Zhou, 2023). According to Cui et al. (2021), Mackenzie et al. (2017) and Zhou (2023) efforts have been invested towards enhancing the co-existence between wildlife tourism enterprises and local communities, however, conflicts continue to surge, creating a need and a call for new approaches. While Mendoza-Ramos and Prideaux (2018) assert that the community empowerment process should include meeting the real and felt needs of the community, community needs are dynamic and not static and changes in needs and wants are inevitable. In the development of strategies that are effective in protecting biodiversity, the role of external actors in strategizing and acting on wildlife conservation in Africa must be reconsidered (Jones, 2021). The external actors' actions seemingly act to internationalize and globalize wildlife conservation in Africa. To truly act in the benefit of both local ecosystems and local livelihoods, the dynamics of conservation must shift to community-oriented approaches (Cassidy & Salerno, 2020). The essence of this volume is in line with Kanta et al. (2021) call that there is a need of a systematic review and diverse research on wildlife tourism to yield scientific evidence which is essential for practitioners and policymakers as they require more information to facilitate their decision-making and actions in the area of tourism development that could improve the competitiveness, innovations of firms and destinations, conservation of the environment, patrimony to next generations, and development in the quality of economic and social life of destination communities in Southern Africa.

References

Abrahms, B., Carter, N. H., Clark-Wolf, T. J., Gaynor, K. M., Johansson, E., McInturff, A., et al. (2023). Climate change as a global amplifier of human–wildlife conflict. *Nature Climate Change, 13*(3), 224–234.

Ashley, C. (2000). *Applying livelihood approaches to natural resources management initiatives: Experiences in Namibia and Kenya* (Working Paper 134). Overseas Development Institute.

Auster, R. E., Barr, S. W., & Brazier, R. E. (2020). Wildlife tourism in reintroduction projects: Exploring social and economic benefits of beaver in local settings. *Journal for Nature Conservation, 58*, 125920.

Benckendorff, P., & Zehrer, A. (2013). A network analysis of tourism research. *Annals of Tourism Research, 43*, 121–149.

Cassidy, L., & Salerno, J. (2020). The need for a more inclusive science of elephant conservation. *Conservation Letters, 13*(5), 1–3. https://doi.org/10.1111/conl.12717

Ceballos-Lascuráin, H. (1996). *Tourism, ecotourism and protected areas*. IUCN.

Corkeron, P. (2023). The militarization of conservation: A different perspective. *Endangered Species Research, 50*, 75–79.

Cui, Q., Ren, Y., & Xu, H. (2021). The escalating effects of wildlife tourism on human-wildlife conflict. *Animals, 11*, 1378. https://doi.org/10.3390/ani11051378. MDPI.

Garland, E. (2008). The elephant in the room: Confronting the colonial character of wildlife conservation in Africa. *African Studies Review, 51*(3), 51–74.

Hinch, T. (1998). Ecotourists and indigenous hosts: Diverging views on their relationship with nature. *Current Issues in Tourism, 1*, 120–124.

Holechek, J., & Valdez, R. (2018). Wildlife conservation on the rangelands of eastern and southern Africa: Past, present, and future. *Rangeland Ecology & Management, 71*(2), 245–258.

Jones, E. (2021). Capital and control: Neocolonialism through the militarization of African wildlife conservation. *Flux: International Relations Review, 11*(2). https://doi.org/10.26443/firr.v11i2.73

Kanta, K. N. M., Kaurav, R. P. S., Allam, U. S., & Srivalli, P. (2021). Wildlife tourism: A synthesis of past, present, and future research agenda. *Enlightening Tourism. A Pathmaking Journal, 11*(2), 390–427.

Karanth, K., & Defries, R. (2010). Nature-based tourism in Indian protected areas: New challenges for park management. *Conservation Letters, 11*, 1–13.

Khanra, S., Dhir, A., Kaur, P., & Mäntymäki, M. (2021). Bibliometric analysis and literature review of ecotourism: Toward sustainable development. *Tourism Management Perspectives, 37*(1), 100777.

LeDee, O. E., Handler, S. D., Hoving, C. L., Swanston, C. W., & Zuckerberg, B. (2021). Preparing wildlife for climate change: How far have we come? *The Journal of Wildlife Management, 85*(1), 7–16.

Lunstrum, E. (2014). Green militarization: Anti-poaching efforts and the spatial contours of Kruger National Park. *Annals of the Association of American Geographers, 104*(4), 816–832. https://doi.org/10.1080/00045608.2014.912545
MacKenzie, C. A., Tumussime, D. M., & Drake, M. (2017). Changing perceptions of protected areas benefits and problems around Kibale National Park, Uganda. *Journal of Environmental Management, 200*, 217–228.
Mangachena, J. R., & Pickering, C. M. (2023). Why are some animals popular with wildlife tourists: Insights from South Africa. *Journal of Ecotourism, 22*(2), 312–328.
Meletis, Z. A., & Campbell, L. M. (2007). Call it consumption! Re-conceptualizing ecotourism as consumption and consumptive. *Geography Compass, 1*(4), 850–870.
Mendoza-Ramos, A., & Prideaux, B. (2018). Assessing, testing, and pushing the wheel of empowerment framework as a measurement tool. *Journal of Sustainable Tourism, 26*, 277–291.
Mkono, M. (2019). The future of tourism in Africa: Optimism in a changing environment. In M. Mkono (Ed.), *Positive tourism in Africa* (pp. 245–251). New York.
Nattrass, N. (2021). Conservation and the commodification of wildlife in the Anthropocene: A southern African history. *South African Historical Journal, 73*(1), 95–116. https://doi.org/10.1080/02582473.2021.1909117
Nygren, A. (2003). Nature as contested terrain: Conflicts over wilderness protection and local livelihoods in Rio San Juan, Nicaragua. In D. G. Anderson & E. Berglund (Eds.), *Ethnographies of conservation: Environmentalism and the distribution of privilege*. Berghan Books.
Øian, H., Aas, Ø., Skår, M., Andersen, O., & Stensland, S. (2017). Rhetoric and hegemony in consumptive wildlife tourism: Polarizing sustainability discourses among angling tourism stakeholders. *Journal of Sustainable Tourism, 25*(11), 1547–1562.
Qin, H., Bass, M., Ulrich-Schad, J. D., Matarrita-Cascante, D., Sanders, C., & Bekee, B. (2020). Community, natural resources and sustainability: Overview of an interdisciplinary and international literature. *Sustainability, 12*, 1061. https://doi.org/10.3390/su12031061
Sattar, Q., Maqbool, M. E., Ehsan, R., Akhtar, S., Sattar, Q., Maqbool, M. E., et al. (2021). Review on climate change and its effect on wildlife and ecosystem. *Open Journal of Environmental Biology, 6*(1), 008–014.
Sell, S. (2004, August 20). 10 great places to tread lightly on Earth. *USA Today*, p. 3D.
Sintayehu, D. W. (2018). Impact of climate change on biodiversity and associated key ecosystem services in Africa: A systematic review. *Ecosystem Health and Sustainability, 4*(9), 225–239.
Stone, M. T., & Stone L. S. (2019). Understanding the relationships between protected areas, tourism and community livelihoods. In Mkono, M (Ed.), *Positive tourism in Africa (pp. 36–52). Routledge.*
Tremblay, P. (2001). Wildlife tourism consumption: Consumptive or non-consumptive. *International Journal of Tourism Research, 3*, 81–86.
Zhou, Z. (2023). Human–wildlife tourism conflict: Towards an inclusive wildlife resources utilisation approach in southern Africa. *GeoJournal, 88*(1), 931–949.

Printed by Printforce, the Netherlands